CRIMINAL JUSTICE, LAW ENFORCEMENT AND CORRECTIONS

DNA FOR DEFENSE ATTORNEYS

Criminal Justice, Law Enforcement and Corrections

Additional books in this series can be found on Nova's website
under the Series tab.

Additional E-books in this series can be found on Nova's website
under the E-book tab.

Law, Crime and Law Enforcement

Additional books in this series can be found on Nova's website
under the Series tab.

Additional E-books in this series can be found on Nova's website
under the E-book tab.

CRIMINAL JUSTICE, LAW ENFORCEMENT AND CORRECTIONS

DNA FOR DEFENSE ATTORNEYS

ROHAN CASSINELLI

EDITOR

nova publishers

New York

For permission to use material from this book please contact us:
Telephone 631-231-7269; Fax 631-231-8175
Web Site: http://www.novapublishers.com

Additional color graphics may be available in the e-book version of this book.

Library of Congress Cataloging-in-Publication Data

ISBN: 978-1-62417-998-3

Published by Nova Science Publishers, Inc. † New York

CONTENTS

PREFACE

Deoxyribonucleic acid, or DNA, is the fundamental building block for an individual's entire genetic makeup. DNA is a powerful tool for law enforcement investigations because each person's DNA is different from that of every other individual (except for identical twins). DNA can be extracted from a number of sources, such as hair, bone, teeth, saliva, and blood. As early as the 1980s, states began enacting laws that required collecting DNA samples from offenders convicted of certain sexual and other violent crimes. The samples were then analyzed and their profiles entered into state databases. Meanwhile, the FBI laboratory convened a working group of federal, state, and local forensic scientists to establish guidelines for the use of forensic DNA analysis in laboratories. This book provides an overview of how DNA is used to investigate crimes and help protect the innocent. It reviews current statutory law on collecting DNA samples, sharing DNA profiles, and providing access to post-conviction DNA testing.

Chapter 1 – The National Institute of Justice is pleased to release DNA for the Defense Bar. This is the fourth publication in a series designed to increase the field's understanding of the science of DNA and its application in the courtroom. The other three publications offer free online training tools for teaching officers of the court about forensic DNA analysis, assisting state and local prosecutors in preparing DNA-related cases, and teaching senior law enforcement officials about policy and practices of effective DNA analysis:

- Principles of Forensic DNA for Officers of the Court: nij.gov/training/dna-officers-court.htm.
- DNA: A Prosecutor's Practice Notebook: nij.gov/training/dna-prosecutors-notebook.htm.
- DNA for Law Enforcement Decision Makers: nij.gov/training/dna-decisionmakers.htm.

DNA for the Defense Bar is specifically designed for criminal defense attorneys. NIJ engaged an impressive multidisciplinary team to produce the most up-to-date information possible in the ever-evolving arena of forensic DNA.

Chapter 2 - Deoxyribonucleic acid, or DNA, is the fundamental building block for an individual's entire genetic makeup. DNA is a powerful tool for law enforcement investigations because each person's DNA is different from that of every other individual (except for identical twins). DNA can be extracted from a number of sources, such as hair, bone, teeth, saliva, and blood. As early as the 1980s, states began enacting laws that required collecting DNA samples from offenders convicted of certain sexual and other violent crimes. The samples were then analyzed and their profiles entered into state databases. Meanwhile, the Federal Bureau of Investigation (FBI) Laboratory convened a working group of federal, state, and local forensic scientists to establish guidelines for the use of forensic DNA analysis in laboratories. The group proposed guidelines that are the basis of current national quality assurance standards, and it urged the creation of a national DNA database. The criminal justice community began to utilize DNA analyses more often in criminal investigations and trials, and in 1994 Congress enacted legislation to authorize the creation of a national DNA database.

Federal law (42 U.S.C §14132(a)) authorizes the FBI to operate and maintain a national DNA database where DNA profiles generated from samples collected from people under applicable legal authority and samples collected at crime scenes can be compared to generate leads in criminal investigations. Statutory provisions also authorize the collection of DNA samples from federal offenders and arrestees, District of Columbia offenders, and military offenders. State laws dictate which convicted offenders, and sometimes people arrested for crimes, will have profiles entered into state DNA databases, while federal law dictates the scope of the national database. Increasing awareness of the power of DNA to solve crimes has resulted in increased demand for DNA analysis, which has resulted in a backlog of casework. Some jurisdictions have started to use their DNA databases for familial searching, which involves using offender profiles to identify relatives who might be perpetrators of crimes. In addition to solving crimes, DNA analysis can help exonerate people incarcerated for crimes they did not commit.

Congress has authorized several grant programs to provide assistance to state and local governments for forensic sciences. Many of the programs focus on providing state and local governments with funding to reduce the backlog of forensic and convicted offender DNA samples waiting to be processed and entered into the national database. Since FY2006, Congress has appropriated approximately $785 million for backlog reduction and laboratory capacity enhancement programs. However, other grant programs provide funding for related purposes, such as offsetting the cost of providing post-conviction DNA testing.

In the 1990s and the early part of the last decade, most of the debate in Congress focused on the scope of DNA databases, reducing the backlog of DNA casework, and providing access to post-conviction DNA testing. Most of the debate about the scope of DNA databases faded away with the enactment of the Violence Against Women and Department of Justice Reauthorization Act of 2005 (P.L. 109-162), which expanded federal collection statutes to include anyone arrested or detained under the authority of the United States. The act also expanded the scope of the national database to include DNA profiles of individuals arrested for state crimes. However, concerns about the backlog of DNA casework and access to post-conviction testing have persisted. In addition, new issues related to the use of DNA in criminal justice have emerged, including whether (1) DNA databases should be used to conduct familial searches, (2) sexual assault evidence collection kits (i.e., “rape kits”) should be standardized, and (3) there should be national accreditation standards for forensic laboratories.

In: DNA for Defense Attorneys
Editor: Rohan Cassinelli
ISBN: 978-1-62417-998-3

Chapter 1

DNA FOR THE DEFENSE BAR*

National Institute of Justice

FOREWORD

The National Institute of Justice is pleased to release DNA for the Defense Bar. This is the fourth publication in a series designed to increase the field's understanding of the science of DNA and its application in the courtroom. The other three publications offer free online training tools for teaching officers of the court about forensic DNA analysis, assisting state and local prosecutors in preparing DNA-related cases, and teaching senior law enforcement officials about policy and practices of effective DNA analysis:

- Principles of Forensic DNA for Officers of the Court: nij.gov/training/dna-officers-court.htm.
- DNA: A Prosecutor's Practice Notebook: nij.gov/training/dna-prosecutors-notebook.htm.
- DNA for Law Enforcement Decision Makers: nij.gov/training/dna-decisionmakers.htm.

DNA for the Defense Bar is specifically designed for criminal defense attorneys. NIJ engaged an impressive multidisciplinary team to produce the most up-to-date information possible in the ever-evolving arena of forensic DNA.

Our sincere gratitude goes to our Technical Working Group members:

Jack Ballantyne
Associate Director of Research
University of Central Florida
Orlando, FL

Catherine Cothran
Palm Beach County Sheriff's Office
Forensic Biology Unit
West Palm Beach, FL

* This is an edited, reformatted and augmented version of a National Institute of Justice Publication NCJ 237975, dated June 2012.

Jules Epstein
Defense Attorney and Law Professor
Widener School of Law
Wilmington, DE

Christine Funk
Assistant State Public Defender
Minnesota Board of Public Defense
Trial Team
Hastings, MN

Chris Plourd
Defense Attorney
San Diego, CA

Vanessa Potkin
Innocence Project
New York City, NY

Ron Reinstein
Arizona Superior Court Judge (retired)
Phoenix, AZ

Edward Ungvarsky
DC Public Defender Service
Washington, DC

I also want to acknowledge the work of the National Clearinghouse for Science, Technology and the Law (NCSTL) at Stetson University College of Law in Tampa, FL — with special thanks to NCSTL's Director of Outreach, Anjali Swienton, who brought her unique background as both a forensic scientist and a lawyer to this comprehensive effort.

Since 2004, NIJ has been responsible for administering public funds to ensure that the nation's criminal justice system maximizes the use of DNA in solving crimes, protecting the innocent, and improving public safety. I hope you will find that DNA for the Defense Bar meets that goal.

John H. Laub, Ph.D.
Director, National Institute of Justice

1. INTRODUCTION

You are defense counsel in a case with DNA evidence or where the absence of DNA evidence may raise factual or legal issues. DNA evidence, when properly collected and analyzed, and when relevant to an alleged crime, can have extraordinary value in the adjudication of a criminal case. This report is designed to help defense attorneys understand:

- The biology of DNA. ~
- Proper collection procedures for DNA evidence.
- Interpretation of DNA analysis and findings.
- When and why an expert is needed.
- Development of case theory in a DNA-based prosecution or in a case where there should be DNA evidence.
- Legal issues for pretrial and trial in cases with DNA evidence.
- Postconviction cases.

Note: DNA testing techniques used in criminal cases have evolved over time. Peers should be consulted to ensure that the most current, relevant and applicable information and case law are being used.

2. DNA Basics: The Science of DNA

Section 1. What Is DNA?

Deoxyribonucleic acid (DNA) is sometimes called a genetic blueprint because it contains all of the instructions that determine an individual's genetic characteristics. A technical explanation of DNA can be found at http://www. genome.gov/ glossary/index.cfm?id=48.

Where Does Nuclear DNA Come From?

Our parents. All human cells with a *nucleus,* except *gamete* cells — egg and sperm cells — have DNA containing the full complement of 46 chromosomes. Each egg and sperm cell carries half of the DNA complement (23 *chromosomes).*

Mixing of genetic markers occurs across the DNA molecule during the formation of sperm cells and egg cells. Because of this mixing process, the DNA in all sperm cells from one man or all egg cells from one woman are not equal halves "split down the middle." Rather, each genetic characteristic has a 50% chance of presenting itself in any given egg or sperm. In humans, very few observable traits are due to inheritance of only one gene. Most observable characteristics are the result of the products of multiple genes interacting. Although actual inheritance of eye color is complex, the following example simplifies the concept of inheritance of eye color for illustrative purposes. Consider a male with brown eyes who inherited a brown eye gene (B) from one parent and a blue eye gene (b) from the other. His "eye genes" will be depicted by geneticists as Bb. (Remember this from high school biology?) Half of his sperm cells will have the B (brown) gene, and half will have the b (blue) gene. Simplistically, the color of his children's eyes will be dictated by two factors: what gene he gives them and what gene their mother gives them. If the mother likewise has brown eyes and carries a Bb profile, their children will statistically be expected to look like this: 25% BB (brown eyes), 25% Bb (brown eyes), 25% bB (brown eyes), and 25% bb (blue eyes). Each of these four profiles reflects the different possible combinations given the genetic characteristics of the original DNA of dad and mom.

What Is DNA Made of?

DNA is found in the cells of all living organisms, except red blood cells. DNA is actually a combination — called a *DNA sequence* — of four *bases:* adenine, cytosine, guanine and thymine, commonly referred to as A, C, G and T (see Figure 1). These *four bases,* in varying combinations, make up yeast, bananas, chickens, rice and people as well as all other living organisms.

The principle of sequence formation is not unlike the principle of the English language. The 26 letters in the alphabet (or the four bases in a DNA sequence) can be combined in various ways to make different words. "The" and "theory" have three letters in common —

both in the specific letters used and the order of the first three letters. Yet the word "the" has no application to the word "theory."

Likewise, in music, there are 12 elements: seven notes (A, B, C, D, E, F and G) and five sharps or flats. Playing these notes in different combinations creates "The Flight of the Bumblebee," Pachelbel's "Canon in D" and the theme song to "Charlie Brown."

With DNA, instead of 26 or 12 elements, there are the four bases mentioned above. Just as the combination of notes dictates what the music sounds like and the combination of letters dictates the word, the combination of As, Cs, Gs and Ts dictates the type of living thing.

The DNA of all human beings is actually nearly identical. Approximately 99.9% of the sequence of As, Cs, Gs and Ts is in the exact same order. This determines common human features such as two eyes, ears on both sides of the head, and long bones in forearms and calves. Although looking at these parts of the DNA molecule might help us determine it is human DNA — rather than, say, banana DNA — it isn't helpful in distinguishing one human from another.

There are, however, places on the human DNA molecule that are different. Of the approximately 3.2 billion *base pairs* in the human genome, a forensic DNA-typing test looks at about 3 thousand base pairs where there are known differences between people.

Source: Christine Funk, Working Group Member.

Figure 1. Nucleotide Base Pairs.

What is a Base Pair?

Picture the DNA molecule as a spiral staircase (see Figure 2). The bases A, C, G and T behave in a predictable pattern of matching and becoming base pairs. A base pair is simply a pair of bases. Bases pair up to form the "steps" of the DNA molecule. The sides of the DNA molecule are made up of sugar and phosphate chains.

Our interest is in the bases themselves. Imagine straightening out the DNA molecule to make a ladder rather than a spiral staircase. Each step of the ladder is a single base pair. As indicated above, there are about 3.2 billion base pairs in the DNA molecules comprising each set of human chromosomes.

Each base pair consists of either an A matched with a T, a T matched with an A, a C matched with a G, or a G matched with a C. That's it. Those are the only four combinations of base pairs that exist. Bases that pair with each other are called *complementary bases.*

These base pairs, about 3.2 billion strong, represent a whole DNA molecule or what is referred to as *nuclear DNA (nDNA).* All cells in the body contain DNA, except for red blood cells, which do not have a nucleus. DNA in blood comes from the *nuclei* of white blood cells.

Back to High School Biology

Picture a chicken egg. An egg is like a cell, except that an egg's outer shell is smoother and more symmetrical than a cell's outer shell or membrane. The yolk of the egg is comparable to the nucleus of a cell.

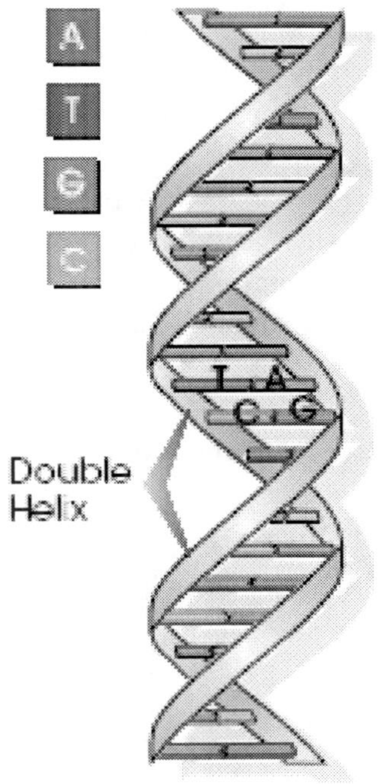

Source: Norah Rudin, Ph.D.

Figure 2. The DNA Double Helix.

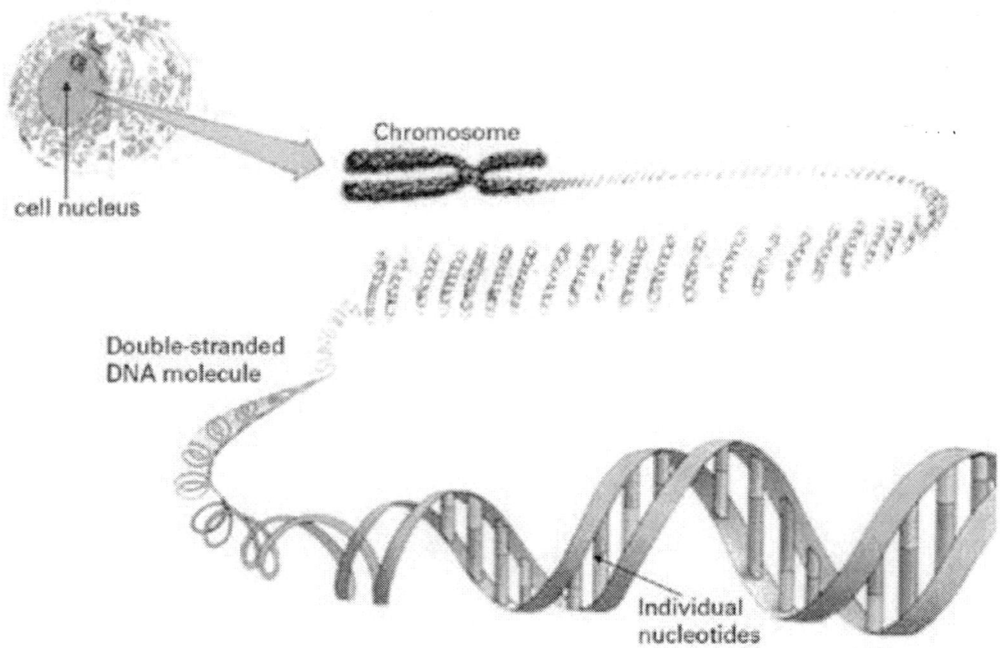

Source: John Butler, National Institute of Standards and Technology.

Figure 3. DNA in the Cell.

The DNA is located inside the nucleus (see Figure 3). The DNA in a single cell is over 6 feet long and is bunched up inside the nucleus of each of our *nucleated cells.* In order for DNA analysts to be able to conduct testing on DNA, they must remove the DNA from the other cellular material that is present, using a process called *DNA extraction* or *DNA isolation.*

Human traits are determined by the particular order of the bases. The first thing the order dictates is that we are human. Second, the order of the base pairs dictates all the physical traits we are born with (such as eye color, face shape, etc.). In addition, there are base pairs that do not "code" for anything and pairs whose coding functions are not yet known.

The DNA looked at in forensic science is not currently known to have any function (such as coding for eye color or the potential predisposition toward a genetically inherited disease) — except for *amelogenin,* which is used in forensic analysis for gender differentiation. The areas at which forensic analysts look are always found in the same spots on the same chromosomes. Each specific location is called a *locus* (pronounced "LOW-cuss"). The forensic science community typically uses a minimum of 13 genetic *loci* (plural for *locus,* pronounced "LOW-sigh"), referred to as the 13 core CODIS *(Combined DNA Index System)* loci. This enables laboratories to search profiles against other profiles already in the CODIS databank (although some laboratories test more than the 13 core CODIS loci). Throughout this training guide, references will be made to the *13 core CODIS loci.*

These core CODIS loci are CSF1PO, D3S1358, D5S818, D7S820, D8S1179, D13S317, D16S539, D18S51, D21S11, FGA, THO1, TPOX, and VWA (TPOX is pronounced "T-Pox." VWA is pronounced V.W.A. Likewise, FGA and CSF are simply pronounced by their individual letters. THO1 is pronounced "Tho One," with a hard "th.").

For the "D" loci, the number following the D (which stands for DNA) indicates the chromosome on which each locus is found. D21S11, for example, is a complete name, which lawyers refer to as D21 for identification. Here "21" refers to the 21st chromosome, S corresponds to the word "single" — meaning there is only one copy of this genetic marker in the human genome, and the number following the S refers to where this locus is found on the 21st chromosome.

Each of the 13 loci was chosen because of its high degree of *polymorphism*, meaning that several different possible genetic types exist for each locus. By examining and identifying these differences, scientists in the laboratory can differentiate between people. To illustrate, the locations on the DNA molecule that dictate for the nose to be in the center of the face are essentially identical among us all. On the other hand, the genes that dictate the shape of one's nose are *polymorphic*. All you have to do is look at 10 people to know that.

At each core CODIS locus, the possible types one can have are labeled by number. At THO1, for example, the types that have been observed are 5, 6, 7, 8, 9, 9.3, 10 and 11. Generally, each person on the planet has two of these: one from mom and one from dad. These types are referred to as alleles (pronounced "uh-LEELS"). If the two alleles in a profile are identical (in other words, the person received a 5 from mom and a 5 from dad), they are homozygous. If the two alleles are different, say, a 5 from mom and an 8 from dad, they are heterozygous at that locus. Rare mutations can and do occur (see, for example, www.cstl.nist.gov/biotech/strbase/).

Short Tandem Repeats (STRs)

The numbers identifying the alleles for the core CODIS loci reflect the number of repeated base pair sequences at that locus. Remember, the locus is polymorphic — it varies from person to person. The way it varies is in its length. A person who has a type 5 has a much shorter length of DNA at that locus than a person who has a type 10.

For example, at THO1, we are not just looking at the 11th chromosome; we are looking for a pattern at a specific location on the 11th chromosome. The pattern looks like this: AATG. We know that everybody has the same AATG sequence on the DNA molecule at THO1. The difference between individuals is how many times the pattern AATG is repeated on both of their 11th chromosomes. Some people have a pattern of the four bases AATG repeated five times, and their DNA type, or allele, for that copy of their 11th chromosome would look like this if it was sequenced by bases: AATGAATGAATG AATGAATG. Other people, however, have other alleles — and the person with five repeats on one of their 11th chromosomes may have a completely different repeat on their other 11th chromosome. For example, the sequence AATG AATGAATGAATGAATGAATGA ATGAA TGAATG shows the pattern of four base pairs repeating nine times. This allele type is a 9. If one additional four-base-pair pattern were repeated, the allele type would be a 10. So what's a 9.3? Although most of the time the DNA we are looking at involves a repeating pattern of four base pairs, sometimes — 9.3 at THO1, for example — this is not the case. We already know that, at this locus, nine repeats of AATG constitute a 9 allele type and 10 repeats make a 10 allele type. A 9.3 reflects nine repeats of AATG and an additional three bases minus one of the As, ATG. If there was an additional A in the same predictable pattern, we'd call it a 10, but because some people have ATG in addition to nine repeats of AATG, an allele type of 9.3 exists.

Not every locus has the repeating pattern of AATG, but every STR locus does have a repeating pattern of base pairs that we look for to identify the allele types for that particular locus.

How Are Loci of Interest Found?

Let's continue to use THO1 as an example. We know it is on the 11th chromosome, and we know where it is on the chromosome. In the lab, DNA test kit *reagents* are combined with a portion of the DNA obtained from a sample. The reagents have several jobs. One is locating the areas of interest (the loci) that we wish to test. *Primers* run along the strands of the DNA molecule, looking for the loci we care about. The primers then identify the DNA strand immediately before and immediately after the region of interest. Think of these areas as the bookends framing the loci. They are called *primer binding sites.*

In addition to identifying the DNA molecule before and after the locus, the *polymerase chain reaction (PCR)* takes place. PCR is a process where potentially billions of copies of the DNA at the locations of interest are made. Some forensic scientists liken this to *molecular Xeroxing*. How is this done? A tube or well — containing the forensic sample (DNA), the primers for the primer binding sites with fluorescent dyes attached, additional bases, and an enzyme to replicate the DNA — is heated. This causes the DNA molecule to split into two strands. Recall our image of the DNA molecule as a straight ladder? The rungs of the ladder are pairs of bases bound together in a very specific way. Bases behave in predictable patterns — A goes with T, T goes with A, C goes with G, and G goes with C. Now imagine that each rung is broken in half, resulting in two half-ladders. If we have half of a ladder, we have a long string of bases without their base pairs. When we cool the test tube down, the bases will seek to pair with their partners. At a high temperature, the DNA strands will stay apart. At a lower temperature, the primers will bind to their corresponding complementary bases on the original DNA strands. At a slightly higher temperature, copies of the DNA are made by the enzyme that replicates the DNA, adding complementary bases to each of the single DNA strands.

Again, let's use THO1 as an example. On one side of the ladder is an AATG pattern. A T base will pair up with the first A base when the temperature is right. Again, a T base will bind to its complementary base (the second A), an A with the T, and finally, a C across from the G. So, where we once had a single strand of DNA, following a single round of heating and cooling with PCR, we have two copies of the strand.

As the PCR process continues, we heat up the sample and reagents again and the strands split again; cool it down and the primers bind; heat it up a little more and additional bases in the tube bind in that same predictable pattern. The yield after two PCR cycles is, therefore, four copies of the areas of the original DNA strand in which we are interested.

Of course, we don't start with one cell's worth of DNA (about 6 pictograms) because a minimum of roughly 16–32 cells is needed (approximately 100–200 pictograms) to get reproducible results. For optimal results, many commonly used DNA kits recommend that 0.5–1.0 nanogram of DNA (500–1,000 pictograms or about 80–160 cells) should be used. Because sperm cells carry only half the genetic information of other nucleated cells, twice as many (or 160–320 sperm cells) are needed to achieve a target of 0.5–1.0 nanogram of DNA. Newly designed kits seek to generate reproducible results while using a smaller amount of sample DNA.

This process of heating and cooling takes place in a thermal cycler and is done approximately 30 times (this will vary, depending on the kit used by the laboratory and their validated protocol). At the end of the process, there are literally billions of copies of areas of interest.

For a technical explanation of the PCR process, go to https://amplification.dna.gov/m01/01/ (free registration required). Once we have a large number of copies, the generated DNA pieces or fragments are separated by size. The most commonly used method for separation is via the use of a *capillary electrophoresis (CE)* instrument. A *capillary* (shaped like a very thin straw) is inserted into a tube or well and draws out a small amount of the amplified product mixture. The DNA travels up and through the straw in a predictable manner — smaller DNA fragments moving faster than larger DNA fragments. Once a piece of DNA reaches the end of the capillary, it passes over a laser light. This excites the fluorescent dye incorporated during the PCR process and causes the bound dye to fluoresce (light up). A camera captures and measures the emitted light, which is reproduced in the corresponding dye color in an *electropherogram* (see Figure 4). It takes approximately 20–30 minutes for all of the DNA to pass through the capillary.

Not only does each locus contain DNA fragments of varying sizes (for example, a 5 at THO1 is shorter than a 9.3 at THO1), but the DNA fragments for each locus also vary in size. D3 alleles, for example, range from about 120 base pairs to 150 base pairs. By contrast, CSF alleles range from about 305-350 base pairs. Because shorter fragments of DNA travel faster than longer fragments, all D3 alleles come across the laser light well before any of the alleles from CSF.

Some loci do overlap in size. If, for example, there are three loci that overlap, they are tagged with different dye colors to avoid confusion in interpretation of the electropherogram. When the alleles come across the laser light, the color as well as the length of the fragment is recorded. The color of the alleles, along with their length, indicates which alleles go with which other alleles as well as what locus they come from.

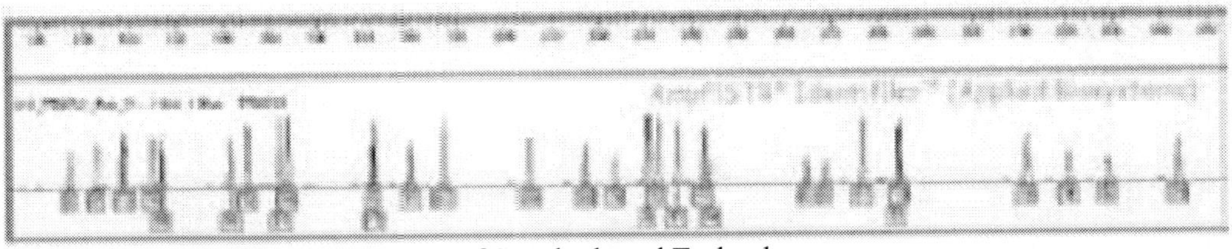

Source: John Butler, National Institute of Standards and Technology.

Figure 4. Tagging Loci With Different Colors.

Section 2. Where Is DNA Evidence Found?

Where is DNA evidence found? The short answer: essentially everywhere. As forensic scientist Keith Inman put it, "The world is a messy place."

The obvious places are blood, semen, hair pulled from the body, skin and saliva. Many times these stains are obvious, but DNA exists in other places. DNA can be found on cigarette butts, the lips of beer bottles, and envelopes that were licked before they were sealed. DNA can be found on surfaces that were touched, such as a counter, a phone or a gun. It can be found on clothing — again, both the obvious, such as the crotch of a pair of underwear, and the less obvious, such as the collar or underarms of a shirt, the waistband of a pair of shorts, socks, and the headband of a baseball cap. It can be found on pens, particularly those that have been used for any length of time or put in someone's mouth. A used toothbrush is an excellent source of DNA, as can be chewing gum or spit on the ground. A rather famous case in Cook County, Ill., was solved by obtaining DNA from chicken bones — not chicken DNA, but the DNA of the person who ate the chicken off the bones.

Serology/Body Fluid Stain Testing

The term *serology* is used by many forensic laboratories to refer to the initial examination of items of evidence to test for the presence of biological materials and/or to recover portions of samples for DNA testing. Serologic testing can be used to indicate or identify the presence of a particular body fluid — such as blood, saliva, semen or urine — in the investigation of a crime. A serologist may also visually examine hair, teeth, bone, tissue and skin cells by using a microscope. Biological stains in the dried state are reasonably stable and can be detected months or years after being deposited.

Understanding serology is critical, as proper serologic testing may identify the type of body fluid comprising a stain from which a *DNA profile* was generated. If, for example, a stain on an item of clothing is at issue, knowing whether the male DNA present in the stain is from saliva versus semen can radically impact the context of the case and the nature of the defense. For this reason, defense counsel must be aware of the differences between presumptive and confirmatory/conclusive tests (see below) and must understand which were used in a particular investigation.

Lawyers often focus on the DNA evidence in a case, neglecting to pay attention to the type of biological evidence from which the DNA profile was generated. The positive identification of a stain as originating from a particular body fluid may be probative evidence in itself and important in developing the case theory regarding what happened and how it happened. For example, a DNA sample, taken from the nightdress of a child living in an extended-family home and matching a cohabiting male suspect, may provide profiles with different degrees of probity, depending on whether the source of the DNA is saliva or semen.

Presumptive Versus Confirmatory Tests

Serologic tests can be classified as *presumptive* or *confirmatory*.

Presumptive tests are often used for bulk or rapid screening of evidence. A positive presumptive test suggests the presence of a particular body fluid. Because of the possibility of false positive reactions, however, a presumptive test is not conclusive.

To positively determine the presence of a particular body fluid, a *confirmatory test* — with a high degree of specificity for the body fluid in question — must be performed. For some body fluids, such as vaginal secretions, most labs do not use a confirmatory test.

If a confirmatory test was not conducted in a case, defense counsel should determine why, whether it might still be requested, and how this could impact the theory of the case. Most often, a confirmatory test is not done because there is no currently available test that can definitively identify a body fluid, or the test has not been validated by the lab. Sometimes, however, confirmatory testing is not done because of lab protocol, limited sample size or lack of resources. Be aware that there are documented examples of examiner or laboratory neglect resulting from the failure to conduct a confirmatory test or to correctly report what serologic testing was conducted.

See Figures 5 through 9 for examples of various analyses used in presumptive and confirmatory tests for epithelial (skin) cells, blood, semen and sperm cells.

Blood

The common presumptive tests for blood are phenolphthalein (Kastle-Meyer reagent; see Figure 6), leucomalachite green (LMG), o-tolidine and tetramethyl benzidine (TMB), which cause color changes that indicate the potential presence of blood. These tests are considered presumptive because human blood, animal blood, oxidizing agents (such as rust) and plant extracts can all show a positive reaction.

Another presumptive test called *luminol* (3-aminopthalhydrazide) is typically used to screen large areas for the presence of blood that is not visibly detectable. In the presence of blood, luminol glows in the dark; however, a common household item such as bleach may also give a positive reaction.

There are currently two types of confirmatory tests for blood — *crystal tests* and *antibody-antigen tests.* Crystal tests, such as the Takayama test, are specific for hemoglobin, a protein found in blood; however, these tests cannot determine whether the blood is human or animal. An example of the antibody-antigen type of confirmatory test for blood is the widely used ABAcard® HemaTrace® test for the Identification of Human Hemoglobin. The HemaTrace® test (an immuno-chromatographic lateral-flow test strip) is used as a *confirmatory test* for human blood, although it gives a positive reaction with the blood/hemoglobin of other higher primates and has been reported to react with ferret blood. The *Rapid Stain Identification of Human Blood (RSIDBlood)* test is another immuno-chromatographic lateral-flow test-strip device. The RSID-Blood test detects a component of the human red blood cell membrane called *human glycophorin A antigen;* it purports to be specific for human blood and has no cross-reactivity with other species. More extensive *validation studies* are ongoing.

Semen

Semen is a mixture of sperm cells *(spermatozoa)* and seminal fluid. *Seminal acid phosphatase (SAP)* is an enzyme present in high concentrations in semen. SAP (or AP) is a presumptive test for semen that causes a color change indicating the potential presence of semen. Other body fluids (such as vaginal fluid) also contain AP in lesser concentrations and can give positive results.

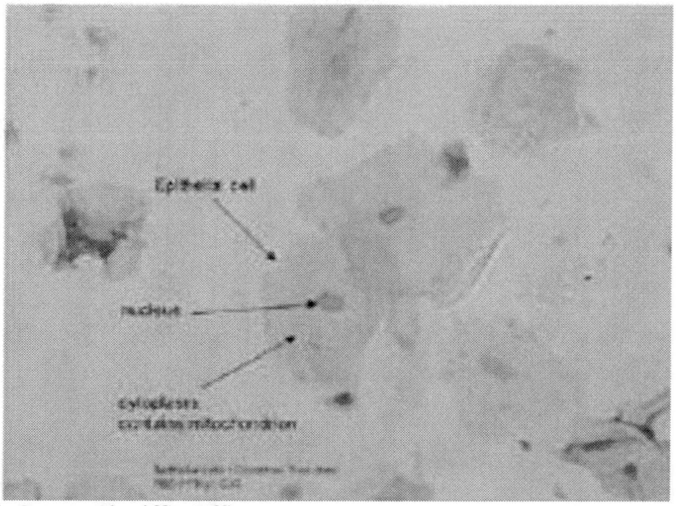

Source: Palm Beach County Sheriff's Office.

Figure 5. Epithelial Cell With Christmas Tree Stain.

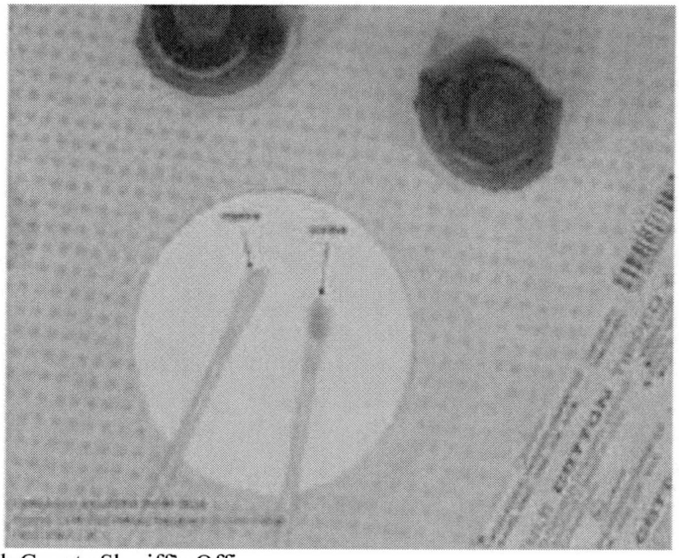

Source: Palm Beach County Sheriff's Office.
Note: Positive test is pink, and negative test has no color change.

Figure 6. Kastle-Meyer Presumptive Test for Blood.

Visual observation of sperm cells under a microscope confirms the presence of semen. The *Christmas tree stain* is an example of a histological stain used to color the sample for ease in visualization. With this stain, under a compound bright-field light microscope, sperm heads are a deep neon-like pink/red with pale pink, almost colorless tops *(acrosomal caps)* (see Figure 8). When sperm tails are present, the area that connects to the sperm head (the neck and mid-piece) stains a pale bluish green, whereas the remainder of the tail stains pale green. Epithelial cells appear as pale bluish green with red to purple nuclei (see Figure 5). An expert should be consulted regarding the significance of the presence or absence of sperm tails.

Another way to confirm the presence of semen is to test for *p30*, also called *prostate-specific antigen (PSA),* a male-specific protein. This can be done with the use of an immuno-chromatographic lateral-flow test strip such as the ABAcard® p30 test for the Forensic Identification of Semen (see Figure 7), via an *enzyme-linked immunosorbant assay (ELISA)* test, or by using another immunoassay procedure. Another confirmatory test for semen is the RSID-Semen test, which detects two *semenogelin* proteins, other male-specific proteins found in high quantities in seminal fluid. Tests for male-specific proteins can be particularly useful in detecting seminal fluid from males with no sperm cells or a reduced number of sperm cells (e.g., vasectomized, *azoospermic* or *oligozoospermic males).*

Given current DNA methodologies, some laboratories use the *differential DNA extraction* process to confirm that sperm cells are present in a sample. The differential extraction process uses the differences in the properties of sperm cells and epithelial cells to attempt to separate the two cell types. The ability to obtain a single (or major) male DNA profile in the designated male (sperm) fraction, particularly when a single (or major) female DNA profile is obtained in the corresponding female (nonsperm/epithelial cell) fraction, can be interpreted to definitively state that sperm cells were present in the tested sample.

Saliva

Saliva can be found in a variety of places at crime scenes. Some of the most common sources are discarded cigarette butts and drink containers. Saliva is also commonly found in sexual assault cases that involve oral contact; it is often mixed with other body fluids such as vaginal secretions or semen.

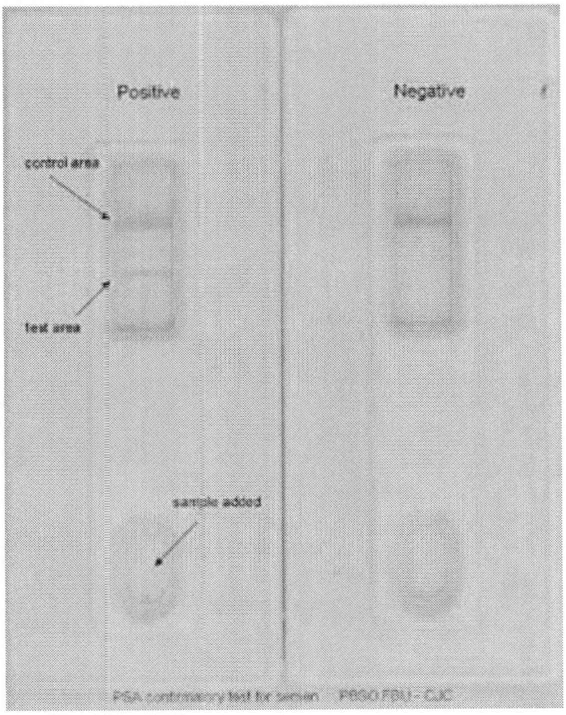

Source: Palm Beach County Sheriff's Office.

Figure 7. Prostate-Specific Antigen (PSA) Confirmatory Test.

Saliva

Saliva can be found in a variety of places at crime scenes. Some of the most common sources are discarded cigarette butts and drink containers. Saliva is also commonly found in sexual assault cases that involve oral contact; it is often mixed with other body fluids such as vaginal secretions or semen.

The most commonly used presumptive tests for saliva involve the detection of alpha-amylase (_-amylase). Significant levels of alpha-amylase are strongly indicative of the presence of saliva. Examples of amylase tests include the Phadebas, radial immunodiffusion, and RSID-Saliva assays.

Currently, there is no confirmatory test for the positive identification of saliva. Research is ongoing in an effort to develop an affordable confirmatory test.

Urine

When the need arises to test for urine, presumptive testing for the presence of urea and/or creatinine — two substances found in large amounts in undiluted urine — can be conducted. Another test for urine is the RSID-Urine assay that tests for Tamm-Horsfall protein, which is also found in urine in large amounts.

Currently, there is no confirmatory test for the positive identification of urine. Research is ongoing in an effort to develop an affordable confirmatory test.

Epithelial (Skin) Cells

Our epithelium covers most of the internal and external surfaces of our bodies. Any cells from the epithelium are referred to as epithelial cells. When discussed in a forensic setting, epithelial cells typically originate from the outer surface of the skin or a body cavity such as the vagina, mouth or rectum.

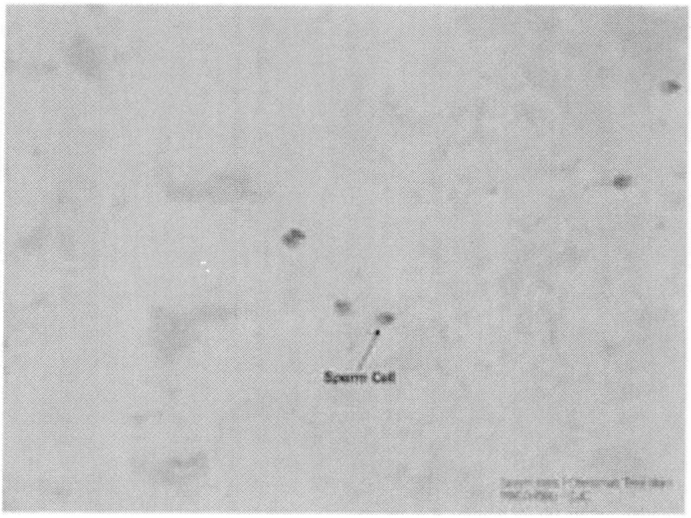

Source: Palm Beach County Sheriff's Office.
Note: Sperm heads appear in pink.

Figure 8. Sperm Cells With Christmas Tree Stain.

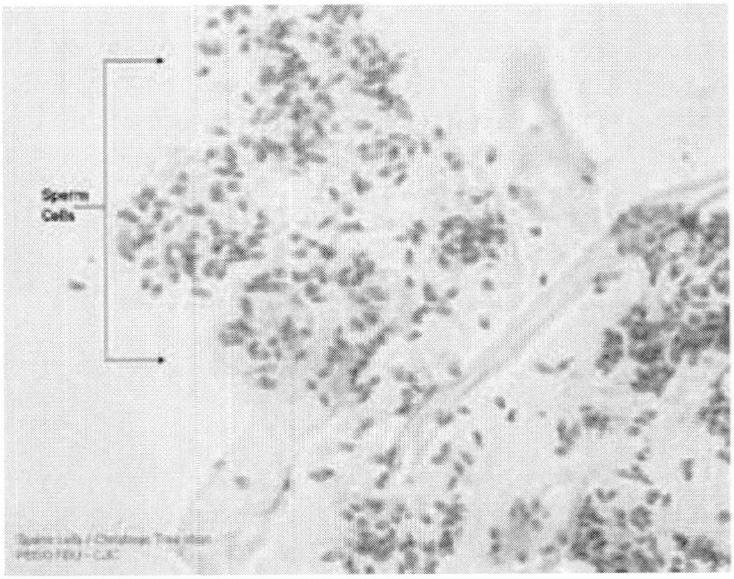

Source: Palm Beach County Sheriff's Office.

Figure 9. Sperm Cells With Christmas Tree Stain.

From a practical perspective, DNA testing is commonly conducted on items believed to contain skin cells shed from the external surfaces of our bodies. Further, it is not uncommon to hear someone refer to a DNA profile as being obtained from "sweat." In reality, any DNA taken from areas on which someone has left sweat is from the person's epithelial cells — there are no commonly used screening tests to indicate the presence of sweat.

The microscopic appearance of epithelial cells from different areas of the body can vary. However, given that most forensic biology examinations involve reconstitution of dried body fluid deposits, which can affect the microscopic appearance of a cell, most forensic laboratories do not seek to identify the origin of epithelial cells by using a microscope.

Serology Validation Studies

It is unlikely that the testing lab will possess internal validation studies for the classic presumptive and confirmatory serologic tests, such as Kastle-Meyer, acid phosphatase, Takayama, and so on, because these are long-established testing processes that would have been in use well before validation requirements for testing procedures were put in place. However, the lab must be able to reference each test they are using to the scientific literature.

The lab should have performed internal validation studies on the more recently developed and adopted tests, such as the ABAcard® HemaTrace® and p30 assays, and the RSIDTM range of products, to demonstrate their robustness, reproducibility and reliability. These validation studies should be available for review.

Serology Interpretation Guidelines: A Must for Each Test

The lab must have interpretation guidelines for every test result (both presumptive and confirmatory). The interpretation guidelines are found in the laboratory's protocol/procedure documents. Interpretation guidelines may vary from lab to lab; however, the lab must demonstrate that its body fluid test worked properly at the time it was used for the samples at issue in a case. This is often done by demonstrating the appropriate reactions of positive and negative control samples. A positive control contains a known sample of the body fluid that is being tested for (e.g., blood) and must give a positive test result. A negative control is a sample that is known to not contain any of the body fluid being tested for and must provide a negative test result. Defense counsel should ensure that the lab followed its defined interpretation guidelines during testing procedures.

Section 3. What Are the Basic Steps in DNA Typing?

Following the processing of items of evidence to obtain samples for DNA testing, there are five basic steps to follow during DNA typing:

- Extraction of DNA from cellular material.
- Quantification of the amount of human DNA recovered.
- Amplification of the areas of DNA being tested using PCR.
- Separation and typing of *amplified DNA* fragments (typically using capillary electrophoresis, or CE).
- Review, interpretation, comparison and reporting of typing results.

Section 4. What Are the Categories of DNA and DNA Tests?

Nuclear DNA

Nuclear DNA (nDNA) is found in the nucleus of a cell, packaged in *chromosomes.* The nucleus of a human cell contains 23 pairs of chromosomes (46 total), half of them inherited from each parent. One notable exception is that each individual sperm contains only 23 chromosomes (1 pair); forensic scientists look at multiple sperm, which collectively provide a testing result that represents the full complement of 46 chromosomes. Every individual — except for identical twins — has a unique autosomal short tandem repeat *(STR)* nDNA profile.

Short Tandem Repeats

Autosomal nDNA testing uses STR technology as the basis for local, state and national DNA databank entries. STR testing examines regions of the DNA molecule that tend to *repeat* themselves in *short,* adjacent, or *tandem* segments. Autosomal STR technology has a very high degree of discrimination.

"Multiplex" STRs are groups of genetically independent STR markers that can be examined at the same time. The conventional or most commonly encountered type of STR testing is conducted on loci found on autosomal (non-sex determining) chromosomes. Accordingly, this testing is sometimes referred to as autosomal STR analysis, testing or typing.

See https://forensic.dna.gov/ module4/1/011/ for a graphic of an STR sequence (free registration required).

Mini-STRs

Mini-STRs correspond to shorter sequences of DNA than those occurring in conventional autosomal STRs. Mini-STRs target some of the same regions as the 13 CODIS STRs but have been redesigned to amplify smaller portions of the DNA strand, which are used when conventional autosomal STR testing is not possible because of DNA *degradation.*

Amelogenin (Gender Typing) Testing

When the *amelogenin* locus (found on sex chromosomes in the nucleus) is tested, the gender of the sample donor may be determined. Generally, females type as XX and males type as XY. Amelogenin data may aid in interpretation of a mixture containing both male and female DNA. See https://forensic.dna. gov/module6/3/005/ for an image depicting the difference between amelogenin typing results from a female and a male (free registration required).

Low-Quantity Template DNA Testing

Low-quantity template DNA testing, also referred to as *low copy number (LCN) DNA testing,* is sometimes defined as typing of samples that contain less than 0.1 nanogram (100 pictograms) of sample DNA. Recall that existing protocols typically recommend using 0.5–1.0 nanogram of DNA to generate a complete autosomal STR DNA profile. Another definition is that LCN typing is the analysis of any results below the stochastic threshold for normal interpretation — in other words, if the results fall outside of the laboratory's defined peak height ratio norms, then the typing result in question is defined as being LCN. LCN typing has also been defined as typing a sample with less than 0.2 nanogram of DNA or as an increase in amplification cycles.

Use of the term *low copy number* has become more precise over time. It is important to note that there is a lack of consensus in the scientific community regarding the use and meaning of this term. From a practical perspective, this means you should not assume how LCN is being used — and it may be important to put its use in historical perspective.

Given the lack of clarity of the term *low copy number,* there is a move to use the more precise terms of either *low-quantity template samples* or *low-level DNA samples*. Low-quantity template samples are those that exhibit stochastic effects. Stochastic effects are defined by the *Scientific Working Group on DNA Analysis Methods (SWGDAM)* as the observation of intralocus peak imbalance and/or allele drop-out resulting from random, disproportionate amplification of alleles in low-quantity template samples.

Y-STRs

Y-chromosome DNA is inherited from the paternal parent. Essentially, fathers pass down their Y-STR DNA profile to their male offspring, from generation to generation, without a change in the profile (barring mutation).[1] Every male in a paternal lineage — fathers, sons, brothers, uncles, and first, second, third and fourth cousins as well as widely dispersed male relatives — will share the same Y-STR profile. In simple terms, all members of the same paternal lineage have the same Y-STR profile. Mutations in a male lineage do occur, resulting in different Y-STR profiles, but they are rare.

Just as maternal lineages can be tracked with *mitochondrial DNA (mtDNA),* paternal lineages can be tracked with Y-chromosome markers.[2] A Y-STR profile can be thought of as a "single genetic locus" — as contrasted with the numerous independent loci available for traditional autosomal STR DNA typing. Because Y-chromosome DNA does not undergo recombination (shuffling of alleles) at each generation in the manner that autosomal STR loci do, its discriminatory power pales when compared with results of traditional nuclear autosomal STR DNA typing.

Y-chromosome DNA is found in the nucleus of the cell, along with the autosomal (non-sex) chromosomes from which the more commonly encountered traditional STR profiles are derived.

The Y chromosome is found only in males, which limits its application but also makes it particularly attractive to scientists and law enforcement in cases involving male and female DNA mixtures that cannot be *deconvoluted,* and in cases where a male profile is believed to be present but traditional testing detects only female DNA.

Because of these types of issues, Y-STR DNA-typing kits have been developed that detect only Y-chromosome markers. This allows a crime lab to isolate male DNA that might otherwise have been overwhelmed by the presence of female DNA or might have gone undetected altogether. As such, Y-STR DNA evidence is introduced most commonly in sexual assault cases.[3]

A core set of locations on the Y chromosome comprises a "minimal Y-STR *haplotype,*" which has served as the basis of forensic applications since 1997.[4] In 2003, SWGDAM recommended the use of both the minimal haplotype loci and two additional Y-STR markers known as DYS438 and DYS439.[5] Since that time, commercial kits have been developed that allow for co-amplification of 12 Y-STR loci (PowerPlex® Y) and 17 Y-STR loci (Yfiler®); both kits include the minimal haplotype loci and the additional loci recommended by SWGDAM.

As indicated above, the STR genetic markers on the Y chromosome can be used to obtain the genetic profile of the male donor(s) in mixtures of body fluids from males and females. In mixture cases, when the concentration from the female donor is very high compared with the male contributor, the standard *autosomal STR analysis* may fail to detect the DNA profile of the male donor(s).

When this occurs, Y-STR analysis can be used to target the Y chromosome, and the DNA from the female contributor is ignored. Because Y-STR DNA typing is considered a tool to augment autosomal STR results, crime labs will typically conduct Y-STR typing on samples either simultaneously with or subsequent to the traditional autosomal STR DNA typing.

For example, the fingernail scrapings from a female victim who has scratched her male assailant may benefit from Y-STR analysis because, typically, most of the DNA recovered from under the victim's fingernails will be her own, and the male perpetrator's component will generally comprise a tiny fraction of the DNA present. Unlike mixtures of sperm and nonsperm cells, it is not possible to perform a *differential DNA extraction* (a procedure that attempts separation of the sperm cells from the nonsperm cells) of mixed-source epithelial cells in fingernail scrapings. These types of cases are well suited to Y-STR testing.

Other cases in which Y-STR analysis may be useful include (a) sexual assaults involving saliva/ saliva and saliva/vaginal secretion mixtures and (b) cases in which the interval between the incident and the collection of *intimate samples* from the victim is greater than two days. In the latter case, too few sperm cells may remain to obtain a sperm-fraction DNA-typing result that does not also contain the victim's DNA profile (because of the much larger proportion of donor DNA and limitations of the differential extraction process).

Mitochondrial DNA

Mitochondrial DNA (mtDNA) is found in the *cytoplasm* of the cell, the area that surrounds the nucleus. The mitochondrial genome is different from the nuclear genome and is distinct in its size, variability and method of inheritance. Perhaps the most significant distinction is the manner in which mtDNA is inherited: Rather than inheritance of randomly combined alleles from both parents, mtDNA is passed unchanged from the mother to all of her offspring. In the simplest of terms, an exact copy of her mtDNA is passed from a mother to her child. A man's mtDNA is inherited from his mother; he does not pass on his own mtDNA type to his children.

This maternal inheritance pattern has two important implications in forensic testing. The first implication is an advantage: The mtDNA of a single maternal relative, even distantly related, can be compared with the mtDNA from skeletal remains and other forensic samples. The second implication is a disadvantage: mtDNA is not a unique identifier. Because maternal relatives share the same mtDNA type, the source of a biological sample can never be conclusively identified with mtDNA. And, consequently, the discriminatory power of mtDNA pales in comparison to that of nuclear autosomal STR DNA typing, as all members of a given maternal line will share the same sequence — barring mutations — from generation to generation, future and past.[6]

As indicated above, mtDNA differs from nuclear DNA in its storage location in the body. mtDNA is found outside the cell's nucleus in energy-producing organelles called mitochondria, where a person's entire mtDNA genome occurs in abundance — up to thousands of individual, complete copies.[7] Compare this to the singular set of nuclear DNA found in the nucleus.[8] Because it is present in such dramatically higher quantities, mtDNA can often be detected when nuclear DNA cannot.[9]

A person's mtDNA sequence has approximately 16,569 base pairs in total, as compared with nuclear DNA's approximately 3.2 billion base pairs.[10] mtDNA also differs from nuclear DNA insofar as most people are *heteroplasmic* with respect to their mtDNA genomes, meaning that they may have more than one mtDNA genome. A person's mtDNA sequence can differ at various locations within the body, from tissue type to tissue type, or even within the same tissue.[11] Certain bodily tissues, such as hair — often used for forensic mtDNA analysis — tend to exhibit higher levels of mtDNA variation than others.[12] Some research suggests that the occurrence of *heteroplasmy* in an individual increases with age, whereas other data appear to contradict that finding.[13]

Nonetheless, while the causes of mtDNA heteroplasmy are not fully understood, its existence has been widely observed and is not disputed.[14]

mtDNA profiling was originally developed outside of the criminal justice context as a research tool for population studies in various scientific disciplines. Its relative integrity across generations proved to be an effective means of deriving information about groups identified by a common mtDNA sequence.[15] Law enforcement adapted the practice of mtDNA profiling — initially found useful for the very fact that large groups of people share the same profile — as an inculpatory tool to aid in criminal investigations.

The nature of mtDNA places limits on its forensic adaptation, which a number of U.S. courts have acknowledged.[16] The relatively small mtDNA genome consists of two primary regions, one of which is a noncoding region and thus deemed suitable for forensic use. The noncoding region of the mtDNA genome regulates replication of the mtDNA molecule and is known as the control region.[17] The control region is typically 1,125 base pairs in length.[18]

To distinguish one person's mtDNA from another, forensic scientists look at specific locations within the control region that are known to be highly variable among humans, as they do with nuclear DNA. In the context of mtDNA, however, only two such regions are currently used[19] for forensic identification because of their observed variability from person to person: *Hypervariable* Region I (HVI) and Hypervariable Region II (HVII). Together, these two regions encompass a total of only about 610 base pairs.[20] An individual's mtDNA "profile," or type, is a list of the differences between the sequences observed in the two regions and those in a reference sequence known as the *revised Cambridge Reference Sequence (rCRS).*

These 610 bases, as previously noted, provide something far short of the discriminatory power of nuclear DNA, which has 13 routinely examined distinct and genetically independent locations that follow Mendelian genetic inheritance laws, enabling them to be used to form the statistical basis of compelling forensic identification evidence. Comparatively speaking, the typically 610 base pairs available for mtDNA analysis are tantamount to a "single genetic locus."[21]

While both nuclear autosomal STR DNA and mtDNA are routinely used for identity testing in missing and unidentified persons cases, analysis of mtDNA is considered most useful in traditional forensic cases when nuclear DNA is insufficient in quality or quantity to obtain a useful STR typing result.

One reason for this is the multiple copies of the mtDNA genome in cells discussed above, which means there will be more copies of the mtDNA to begin with. Another reason is that the mtDNA genome is smaller than the nuclear DNA genome, so it is more impervious to degradation. Shed body, head and pubic hairs with no cellular material attached to the root and aged skeletal remains are examples of samples analyzed for mtDNA because nuclear DNA of sufficient quantity or quality is often not recoverable from these biological materials. See https:// forensic.dna.gov/module8/1/003/ for an interactive exercise that explores the differences between nuclear and mitochondrial DNA (free registration required).

Section 5. Emerging Technologies

Single Nucleotide Polymorphisms

Single nucleotide polymorphisms (SNPs) (often pronounced "snips") are the most common type of genetic markers found in humans. SNPs can play an important role in establishing ancestry and, outside forensics, SNPs are often used to identify genes involved in complex diseases. SNPs can also be used to obtain results from even smaller and more degraded DNA samples than with STRs.

Although the future utility of SNPs is uncertain, it seems unlikely — due to the limited variation of SNPs and difficulties with mixed-sample interpretation — that this method will replace STRs for routine DNA analysis. See https://forensic.dna.gov/module15/1/003/ for an example of an SNP sequence (free registration required).

Plant and Animal DNA

Forensic DNA methods can conceivably be applied to criminal cases for virtually any nonhuman source imaginable. Domestic cat and dog DNA have been used in criminal cases, usually to link a suspect to a pet. Specialized STR analysis has been used when animal blood or hair with adequate root material is available. STR analysis provides nearly the same discriminatory power in domestic animals as it does in humans, making an STR match very powerful when animals are involved.

Here are some examples of how plant and animal DNA can be used in the criminal justice system:

- Fur, feathers, bone, blood, urine, feces and saliva may link an animal to a poacher or help prove illegal importation of animal products, such as pelts or tusks.
- Meat products may be traced to cattle with mad cow disease.
- Pods, seeds, leaves, bark and roots of illegal plants or controlled substances (such as marijuana) may be present at a crime scene. Data collected from plants constitute the newly emerging field of forensic botany. See *Arizona* v. *Bogan*, 183 Ariz. 506, 905 P.2d 515 (Ariz. App. 1995).

See also https://forensic.dna.gov/module15/ 3/002/ for information about the use of DNA analysis and microbiology in the criminal justice system (free registration required).

Section 6. How to Find Resources and Stay Current

Attorneys can keep current on forensic DNA through a number of resources, including books and websites for the beginner or seasoned practitioner. Regional and national meetings for attorneys offer training sponsored by public defenders' offices and professional associations. Attorneys familiar with forensics can also attend forensic science meetings and trainings such as those sponsored by the National Institute of Justice (www.nij.gov), the National Academy of Sciences (NAS, www.nasonline. org), the American Academy of Forensic Sciences (www.aafs. org), and the National Clearinghouse for Science, Technology and Law (www.ncstl.org) — all good resources to help stay abreast of the most recent research and findings.

Some additional resources include:

- The President's DNA Initiative (www.dna.gov). Website has free online training and is an excellent resource for laypeople.
- The National Legal Aid Defender Association (www.nlada.org). Website has downloadable forensic DNA articles, transcripts and protocols. Most materials are available to the public, although more confidential materials are accessible only to criminal defense attorneys.
- Butler, J.M., Forensic DNA Typing: Biology, Technology and Genetics of STR Markers, 2nd ed., Burlington, MA: Elsevier Academic Press, 2005. A leading treatise on forensic DNA (including diagrams).
- Two authoritative studies conducted by National Research Councils (NRCs) of the NAS, including helpful language on the limitations of forensic DNA:
- DNA Technology in Forensic Science, Washington, DC: National Academy Press, 1992, commonly referred to as NRC I.

- The Evaluation of Forensic DNA Evidence, Washington, DC: National Academy Press, 1996, commonly referred to as NRC II.
- Buckleton, J., C.M. Triggs, and S.J. Walsh, eds., Forensic DNA Evidence Interpretation, Boca Raton, FL: CRC Press, 2005. An excellent resource for understanding calculations used in forensic DNA evidence and explains more complex problems in DNA statistical analysis than most textbooks; offers approaches used and approved internationally, although less frequently by U.S. crime laboratories.
- Kreeger, L.R., and D.M. Weiss, Forensic DNA Fundamentals for the Prosecutor: Be Not Afraid, Alexandria, VA: American Prosecutors Research Institute, 2003. Useful for defense attorneys wanting to know what to expect from prosecutors in DNA cases.
- Forensic Bioinformatics Annual Conference: Three-day seminar during August in Dayton, Ohio. Less experienced attorneys can learn how to litigate a DNA case, including introductory training in DNA testing, and comprehensive lessons on specific issues in cases involving DNA.
  An advanced track is available for more experienced attorneys (www. bioforensics.com).
- Moenssens, A.A., C.E. Henderson, and S.G. Portwood, Scientific Evidence in Civil and Criminal Cases, 5th ed., New York: Foundation Press, 2007.
- Payne-James, J., R.W. Byard, T.S. Corey, and C. Henderson, Encyclopedia of Forensic and Legal Medicine, Oxford, England: Elsevier Academic Press, 2005.

Section 7. Forensic DNA Lab Report Basics

Although there are national standards for reporting DNA analysis results, laboratories differ regarding information included in their reports. Basic elements that commonly appear include the following (those in **bold** are required under Standard 11.2 of the *Quality Assurance Standards for Forensic DNA Testing Laboratories,* effective July 1, 2009):

- Administrative information on the case, such as agency, **case identifier** (e.g., file number), evidence item numbers, victim name and suspect name.
- **Date the report was issued and a signature and title, or equivalent identification, of the person accepting responsibility for the content of the report.**
- **A description of the evidence examined.**
- **A description of the technology** or technique used.
- **Locus or amplification system** used.
- Testing **results and/or conclusions.**
- **A quantitative or qualitative interpretive statement.**
- **Disposition of evidence**.

Reports also may contain the results of any non-DNA tests (serology, for example) that were performed to locate and characterize the type of biological evidence.

Some laboratories include genotype data in their reports, usually in the form of a table. As noted above, reports must contain an interpretive statement, which will address whether a DNA profile from an evidence sample can or cannot be associated with:

- A known individual (suspect, victim or third party).
- Other evidence samples (scene samples or sexual assault evidence).
- Databank samples (offenders, *forensic unknowns* or missing persons).

If one or more of the known samples are consistent with any of the evidence samples, the report must provide a statistical frequency or frequencies for the most probative finding(s).

In addition to the laboratory analytical report, other documents related to a case may be available. For example, the National DNA Standards require laboratories to maintain a case file with all records related to case analysis and generated by examiners. Case files commonly include:

- A chain of custody for items received by the laboratory.
- Sketches or photographs taken in the laboratory.
- Examination ("bench") notes of any testing by the analyst.
- Laboratory logs or standard forms related to testing.
- Strips, photographs or copies of *autoradiographic film* or electropherogram data.
- Communication between the analyst and others involved in the case.

In addition to information in the case file, other lab documents may include:

- *Equipment calibration* and maintenance records.
- Analyst training and proficiency test records.
- Unexpected results or corrective action reports.
- Quality assurance reports and audits.

Possible DNA Report Conclusions

An *inclusion,* or *DNA match,* occurs when a known sample is compared with an evidence sample and the profiles are the same. An inclusion or a DNA match may also occur when all of the alleles in a known sample are also found in a DNA mixture profile. The significance of the inclusion or match will depend on the statistical data obtained. Some labs report this finding as "cannot be excluded."

When a known sample is compared with an evidence sample, the donor of the known sample is excluded as a source of the evidence if the profiles are different. This is referred to as an *exclusion* (or a DNA nonmatch).

When an individual is excluded as the potential source of DNA, it does not necessarily mean the individual was not involved. For example, a true perpetrator who left no detectable biological material will be excluded as a source of DNA. Conversely, if an individual is a potential source of DNA at a crime scene, it does not necessarily mean that the person was involved in the crime. (See Section 2 of this chapter regarding alternate explanations for the presence of DNA.)

Sometimes, no conclusion can be drawn as to whether a known individual is included or excluded as the potential source of DNA evidence. *Inconclusive or uninterpretable results* may be due to complicating factors such as multiple contributors, contamination, degradation of samples, or misinterpretation or misrepresentation of the results.

Note: A defense attorney should seek an independent expert to review a laboratory's finding of inconclusive or uninterpretable results to determine whether, in fact, the opinion is supported by the data, particularly when no results that support the defense theory of the case have been reported.

Sometimes testing of a sample is attempted, but no results are obtained. This could indicate:

- Absence of DNA in the sample.
- Insufficient DNA in the sample.
- Extensively degraded DNA.
- Presence of a substance that inhibits the PCR process (PCR inhibitor), such as denim dyes, carpet glue or certain types of latent print powder.
- Improperly conducted or incomplete testing.

Note: Where initial testing produces no evidence of DNA, consider sending the evidence for independent testing.

A threshold amount of *amplified DNA* must be observed before a laboratory will report an *allele* (a different form of a genetic marker at a particular locus) or *genotype* (the individual's inherited allele or alleles at a specific genetic marker, or locus). The threshold (called a *threshold value*) can differ among laboratories and is based on internal validation studies used to establish guidelines. Laboratory guidelines determine whether — and under what conditions — data under the threshold are reported. A laboratory may have more than one threshold value based on its validation studies. For example, it is not uncommon for a laboratory to have different threshold values for reporting *homozygous* and *heterozygous* results.

Single-Source Profile

DNA from one contributor is commonly referred to as a *single-source DNA profile.* A single-source profile could be derived from:

- A reference sample (victim or suspect).
- An elimination sample (first responders, EMT personnel, consensual sex partners, or anyone who might have had authorized access to the crime scene).
- A crime scene or other evidence sample (blood stain, chewed gum, cigarette butt).

More Than One Source

Mixtures of DNA from more than one contributor are commonly encountered. A mixture could be due to:

- Actual contribution by multiple donors during the crime.
- Presence of DNA on the item prior to the crime.

- Testing of intimate samples (e.g., vaginal swabs or breast swabs).
- DNA added by handling an item after a crime but before recovery (e.g., handling of a gun used in a crime by a person(s) other than the police).
- Contamination during crime-scene processing and sample handling (collection, packaging or testing).

Any biological material (blood, semen, saliva, urine, hair, sweat and skin cells left behind after contact) can be mixed and found in combination with any other biological material.

Detecting Small Amounts of DNA

DNA can be detected in minute amounts. Laboratory reports may classify certain profiles as belonging to a major contributor or a minor contributor, although labs may differ in how they report more than one profile detected in a sample. See http://static.dna.gov/flash/6.3.003_ Steering Wheel.swf online for a report referencing a major DNA profile obtained from a small amount of "touch" DNA.

Note: Counsel is advised to consult with an expert regarding the interpretation and significance of classifying a contributor as major or minor.

End Notes

[1] Butler, J.M., Forensic DNA Typing: Biology, Technology, and Genetics of STR Markers 201 (2d ed., 2005).

[2] Butler, supra note 1, at 201.

[3] Id. at 201-03.

[4] Id. at 207.

[5] Id.; see Mulero, J.J., et al., "Development and Validation of the AmpFISTER YFiler PCR Amplification Kit: A Male Specific, Single Amplification 17 Y-STR Multiplex System," 51(1) J. FORENSIC S CI. 64 (2006); Shewale, J.G., et al., "Y-Chromosome STR System, Y-PLEX 12, for Forensic Casework: Development and Validation," 49(6) J. FORENSIC S CI. 1278 (2004).

[6] The mutation rate for mtDNA, however, is significantly higher than that for nuclear DNA.

[7] The average has been estimated at approximately 500 copies for most cells. See Satoh, M., and T. Kuroiwa, "Organization of Multiple Nucleoids and DNA Molecules in Mitochondria of a Human Cell," 196 Exp. C ELL R ES. 137 (1991).

[8] There are technically two "copies" of nuclear DNA in the nucleus of a cell — one from the mother and one from the father — but each "copy" comprises only half of the set required for forensic typing.

[9] Butler, supra note 1, at 241.

[10] Butler, supra note 1, at 242.

[11] See Buckleton, J., S. Walsh, and S. Harbison, "Nonautosomal Forensic Markers," in Buckleton, J., C.M. Triggs, and S.J. Walsh, eds., Forensic DNA Evidence Interpretation at 304 (2005).

[12] Grzybowski, T., "Extremely High Levels of Human Mitochondrial DNA Heteroplasmy in Single Hair Roots," 21 E LECTROPHORESIS 548 (2000).

[13] Buckleton et al., supra note 11, at 305.

[14] See Buckleton et al., supra note 11, at 303; D'Eustachio, P., "High Levels of Mitochondrial DNA Heteroplasmy in Human Hairs by Budowle et al.," 130 FORENSIC S CI. I NT. 63 (2002) ("Major unresolved issues include the molecular mechanisms responsible for the occurrence of heteroplasmy to different extents in different tissues.").

[15] See, e.g., Wallace, D.C., "Mitochondrial Disease in Man and Mouse," 283 S CIENCE 1482 (1999); Shriver, M., and R. Kittles, "Genetic Ancestry and the Search for Personalized Genetic Histories," 5 NATURE R EV. G ENET. 611 (2004); Cann, R.L., et al., "Mitochondrial DNA and Human Evolution," 325 NATURE 31 (1987).

[16] See, e.g., Vaughn v. State, 646 S.E.2d 212, 214 (Ga. 2007) (observing that "mtDNA analysis is more applicable for exclusionary, rather than identification, purposes," but admitting evidence nonetheless); Wagner v. State,

864 A.2d 1037, 1045 (Md. App. 2005) ("mtDNA analysis provides significantly less ability to discriminate among possible donors than does nuclear DNA analysis and has been said to be a test more of exclusion than of identification."); State v. Scott, 33 S.W.3d 746, 756 (Tenn. 2000) ("Because it is not possible to achieve the extremely high level of exclusion provided by nuclear DNA, mtDNA typing has been said to be a test more of exclusion than one of identification.").

[17] Human Mitochondrial DNA — Amplification and Sequencing Standard Reference Materials 1-2, National Institute of Standards and Technology, Pub. No. 260-155 (September 2003).

[18] See Anderson, S., et al., "Sequence and Organization of the Human Mitochondrial Genome," 290 NATURE 457-65 (1981).

[19] There is a third highly variable region (HV3) that has been studied, but it is not currently used by forensic laboratories.

[20] Holland, M.M., and T.J. Parsons, "Mitochondrial DNA Sequence Analysis: Validation and Use for Forensic Casework," 11 FORENSIC S CI. R EV. 21, 24 (1999).

[21] U.S. Department of Justice, "Mitochondrial DNA Analysis at the FBI Laboratory," FORENSIC SCI. C OMM UN. (July 1999).

3. DISCOVERY: GETTING TO KNOW A CASE WITH DNA EVIDENCE

Section 1. From Crime Scene to Laboratory

Crime Scene Collection

Depending on the type of case and the investigating body, documentary evidence regarding the crime scene will be available to the defense, such as:

- A written report.
- Photographs.
- Diagrams, including the location of evidence.
- A videotape.

Such documentation may reveal not only what was collected but also what was not collected. For example, an ashtray with cigarette butts in it may provide DNA information: Were the cigarettes collected? Is there more than one brand of cigarettes in the ashtray? Is there more than one ashtray? What about the beer bottles? Is there evidence that wasn't collected that could have provided additional DNA information?

Note: It is not the existence of DNA per se that makes the collection important. It is the existence of a potential third party's DNA that makes collection important. This, of course, will be fact specific and will not be a valid argument in every case.

Pay attention to the cleanliness or dirtiness of a crime scene. Are the ashtrays overflowing? Have the dishes been washed lately? The DNA profile found on a single coffee cup in an otherwise clean kitchen sink is likely far more relevant than a DNA profile found on a beer bottle in a house with six dozen beer bottles strewn about the living room.

Also consider potential transfer issues; for example, if the victim was stabbed 100 times and there is blood from floor to ceiling, blood on the victim's driver's license on the dresser may merely be due to circumstance and not particularly relevant.

Evidence Collection

Good resources providing an overview of evidence collection from a law enforcement perspective are the brochure, *What Every Law Enforcement Officer Should Know About DNA Evidence* (https://www.ncjrs.gov/ pdffiles1/nij/bc000614.pdf), and the online training courses, What Every First Responding Officer Should Know About DNA Evidence (https://letraining. dna. gov, free registration required) and What Every Investigator and Evidence Technician Should Know About DNA Evidence (https://letraining.adv. dna.gov, free registration required). The online publication and training courses discuss various collection protocols, although law enforcement agencies typically have their own protocols for evidence collection. Also, officers or forensic technicians who collect crime scene evidence may receive ongoing training, so collection protocols may change over time. All such information should be gathered during discovery.

In general, evidence collection techniques that minimize contamination include:

- Wearing gloves when collecting evidence and using personal protective equipment (PPE), such as booties and a face/particle mask, as needed. In the crime scene environment, the PPE works both ways — ensuring the safety of the person collecting the samples and protecting the samples from contamination with DNA from the collector.
- Changing gloves between sample collections, or as needed, to avoid cross-contamination.
- Ensuring that the area(s) used for evidence collection and packaging of evidence are as clean as possible.
- Inspecting the evidence collection materials before their use to ensure that they are new (unused) and clean.
- Ensuring that reusable implements such as pens and clipboards have been cleaned/ decontaminated before starting any evidence collection.
- Ensuring that reusable containers (e.g., bottles with chemicals for blood screening tests) and instruments (e.g., the alternate light source and the trace materials recovery vacuum) have been cleaned/decontaminated before starting any evidence collection.
- Using disposable instruments for sample collection or cleaning collection instruments thoroughly before and after handling each sample.
- Avoiding touching the area on an item of evidence where it is probable that DNA exists.
- Avoiding talking, sneezing or coughing over evidence.
- Avoiding touching the face, nose, mouth or exposed skin when collecting evidence.
- When possible, ensuring that each piece of evidence is dry (or will readily dry, such as swabs put into swab boxes with airholes) and in its own separate paper (not plastic) bag, envelope or other appropriate container. When it is not possible to completely dry evidence before transporting, it can be temporarily stored wet in a plastic bag or container; however, the item should be immediately removed and dried according to agency protocol as soon as possible.
- Sealing envelopes and bags with evidence tape (not by licking them) or using envelopes with self-adhesive backing. Staples should be avoided because they may puncture the skin and could lead to bleeding.

Compare the police department's list of collected items with the notes made at the crime lab that document what was received by the crime lab. Pay attention to the condition of items as they are received at the lab. For example, if an item arrives wet, the lab will document that condition. If an item arrives with hair stuck in the tape sealing the envelope, the lab notes should reflect that. If a series of envelopes arrive at the lab with red stains on the outside, this will also be documented.

These types of notations could indicate potential issues with collection procedures used at the crime scene. Needless to say, laboratory personnel cannot control contamination that occurs at the crime scene or the conditions in which the evidence is stored or transported before arriving at the lab. However, they should be aware of any potential issues with collection and packaging that could affect the usefulness of the evidence for subsequent examination and reporting.

The Crime Scene

Crime scenes come in all shapes and sizes. A temperature-controlled town home with the blinds drawn may be the ideal crime scene, affording officers the luxury of collecting evidence undisturbed. Outdoor crime scenes, such as the middle of a cornfield in winter, the woods or the desert, present challenges. Changes in the weather, such as the onset of a storm, as well as changes in lighting may limit law enforcement's ability to observe and collect all relevant evidence.

Once the crime scene is accessible, attorneys should visit it and other relevant scenes whenever possible. Being at the crime scene(s) provides additional information, possibly identifying other evidence that could be or should have been collected.

Chain of Custody: Proper Preservation Techniques

A *chain-of-custody* record (either written or electronic) that follows the evidence from crime scene to courtroom should accompany the evidence. This record — which may also document proper preservation techniques — is discoverable.

The officer or forensic technician who collects the evidence documents where the item was recovered and who put the evidence in the container. The chain-of-custody documentation should contain the names of everyone who had custody of the evidence at any point, including:

- The person who collected the item of evidence and initiated the chain of custody.
- The person who brought the evidence to the police department/crime lab.
- The person who logged it into the evidence room.
- The person who logged it out of the evidence room and brought it to the lab.
- The person who received it at the lab.
- The person(s) who received it from the lab after final disposition.

Note: Each agency involved in the chain of custody may also keep separate chain-of-custody documentation for each item of evidence. These documents should also be requested and reviewed to determine if there was a possible break in the chain of custody or an indication of inadvertent introduction of contamination.

Because chain-of-custody documentation will reveal not only who had the items but also how and where they were stored, a review should consider both the chain of custody and the conditions under which the DNA evidence is collected and stored before it is submitted to the lab and while it resides at the lab. DNA evidence is vulnerable to deterioration when subjected to sunlight, heat or humidity.

Thus, evidence should not be stored in the trunk of a police vehicle or on the dashboard in hot weather. It is also reasonable to ask if the transporting vehicle had air conditioning. If you are not familiar with the notations used by an agency on chain-of-custody documents, you may need to request additional information to clarify the exact location and conditions under which an item of evidence was stored.

List of Evidence Items

Not all evidence collected by law enforcement will necessarily go to the crime lab. Likewise, not all evidence that goes to the crime lab will be tested for DNA evidence. Depending on the case, a lab analyst — alone (typically based on the laboratory's protocol) or with the assistance of law enforcement, the prosecution or a lab supervisor — will determine what items of evidence to test. Resources, potential probative value of the evidence, and the law enforcement or prosecution's theory of the case are all factors contributing to decisions about which evidence to test. For example, certain types of DNA testing may not be performed on a particular item if preliminary testing indicates that DNA testing may not be successful.

Photographs and Video

There should be photographs and/or video of the crime scene. In addition to recorded images taken by law enforcement and/or coroner/ medical examiner staff, there also may be local news footage. Reviewing all of the images and footage may provide reasons to collect additional crime scene evidence.

Section 2. Lab-Directed Discovery

Defense counsel should be thorough when requesting discovery and should make detailed requests for files. Under most discovery rules, defense counsel is entitled to (a) all information regarding the prosecution's efforts to make available the scientific test reports and relevant raw data used in the case as well as (b) all information describing the prosecution's efforts to maintain and preserve the evidence. The jurisdiction should inform defense counsel if an *evidentiary sample* will be or was entirely consumed during the testing process. Therefore, in all cases in which there is DNA evidence, defense counsel should request the production of full discovery and the preservation of all DNA evidence. If full discovery is not provided or if evidence is not preserved, defense counsel's recourse is to seek relief from the trial court, either through a motion to compel discovery or through a motion to preclude the state from using the DNA evidence at trial.

In response to a defense discovery request, the prosecutor may initially provide a generic discovery response that is independent of the specific case at issue. Followup discovery requests may be necessary for the prosecutor to provide more specific discovery materials. In addition to paper and electronic discovery (discussed below), defense counsel should ask to visit the laboratory at which the DNA tests were performed and meet with the lab personnel who worked on the case.

A thorough review of a comprehensive discovery packet of the laboratory analyses may reveal weaknesses in the chain of custody, scientific procedure or analysis of the DNA data. Defense counsel should learn the local protocols used in the case and scrutinize the underlying data to see if they support the conclusions drawn in the DNA reports. Refer to Chapter 6 for in-depth information regarding data review and interpretation.

What Should Be Requested from the Lab?

1. A disk containing *raw data,* including but not limited to the sample files, project files, injection lists, sample sheets and injection logs.

Start by reviewing the injection lists, sample sheets and injection logs, noting the time and date stamps on all runs to check the order in which samples were run and to make sure that controls were not substituted with those from a different day. This could reveal a mistake or, in the most extreme circumstance, indicate laboratory fraud; in either case, conclusions based on the evidence and reference profiles would be unreliable. If it was an honest mistake, the lab should be able to provide data for the actual controls run on that day. If controls did not perform properly or were not run, conclusions based on the evidence and reference profiles would be considered invalid. A laboratory that follows the *Quality Assurance Standards for Forensic DNA Testing Laboratories* (the *QAS*) — any laboratory that uses *CODIS (Combined DNA Index System)* — is prohibited from using any testing results that are not supported by properly performing controls (Standard 9.5, July 1, 2009).

Note: If the injection logs and sample sheets do not match up, hire an expert to find out why. To make sure you know what you are looking for, you may want to hire an expert for at least one case — or attend a training class that addresses this topic.

Defense counsel should determine if the *allele calls* of the evidence samples were made before or after the reference samples were processed. The lab's protocol may be to run and call the *alleles* on the evidence samples before knowing the profiles of the reference samples. This is particularly important for samples with complicated mixtures, partial profiles or low peak heights, represented by low *relative fluorescent unit (RFU)* values close to the *analytical threshold.* Context bias may occur if the analyst knows the profiles for the reference samples before interpreting the profiles for the evidence samples.

The compact disk handed over during the discovery phase will also include electronic files containing data on the DNA fragments separated during *capillary electrophoresis,* which should include the runs of positive and negative controls, reference samples and evidence samples. The disk should also include the raw data before they were processed for *peak heights* and allele calls.

Graphs *or electropherograms* are generated that are based on the analyst's review and interpretation of the raw data. The reported *DNA profile* for each sample is depicted on these graphs. Software programs for forensic *STR DNA analysis* are in common use (e.g., Gene-ScanTM and GenotyperTM (which must be used in tandem), Gene-Mapper®, PowerTyperTM, and TrueAllele®). The laboratory's protocol will specify which program(s) have been validated and are being used.

The criteria used to call alleles vary across laboratories. To analyze the raw data using different criteria, counsel or a defense expert must be able to operate the macros on the computer using a computer program. Independent analysis of the raw data may reveal potentially *exculpatory peaks* that might have been missed because they were below the calling threshold used by the laboratory. For example, the minimum peak height that a lab considers might be 50 RFUs, but there could be a true peak that is exculpatory at 45 RFUs. It takes only a one-allele difference in a full single-source profile to exclude a suspect.

Note: Independent analysis of the data allows the defense to look for data outside the threshold value(s) relied upon by the testing laboratory.

The disk will also include the electropherograms — including all formats of loci data output — or other images that have already been processed by the laboratory to include alleles and peak heights. These should be compared to the final report to ensure that there were no transcription errors. Also, look for signals that were dropped from the reported profiles because they were considered to be *artifacts* or *noise*.

2. Copies of the DNA typing results.

The electropherograms are frequently printed out by laboratory analysts. These printouts may be color copies, which are ideal, or they may be black-and-white photocopies, which are somewhat harder to read. Regardless, the printouts are often easier to view than the electronic files because they do not require a computer or a license to run the computer programs that generate the electropherograms. In fact, many DNA analysts make allele determinations based on the printouts, not the electronic files. Analysts will often write notes on these printouts, such as notations on peaks that were considered artifacts, descriptions of the baseline, and calculations of *peak height ratios* to distinguish *stutter* from *real allele peaks* and distinguish *heterozygous* peaks from peaks of different contributors. These notes are also likely to be initialed by the analyst making the calls. The defense should consider whether an abundance of artifacts possibly indicates unreliable data or artifacts that are masking true alleles. Defense counsel should obtain copies of all printouts retained by the laboratory.

3. Copies of real-time, slot blot or other quantitation data.

Knowledge about the quantity of DNA obtained is very important and is acquired using different methods, the most common current method being *real-time PCR,* also called *qPCR.* Each quantitation method contains calculations for estimating how much DNA was extracted from each sample (some labs also determine how much human male DNA is present). This information allows the defense to determine how much human DNA was used and how much remains from each sample, which could be retested. If the DNA was consumed by the laboratory, knowing how much DNA there was originally can help determine whether consumption was indeed necessary.

4. Case notes, including handwritten bench notes, chain-of-custody record, and descriptions of the evidence.

There may be multiple forms or logs that comprise the chain-of-custody record. The defense should determine that each item and sample was accounted for at all times. Also check whether any items were signed into two places at the same time, which would indicate that the laboratory was not keeping track of the evidence properly.

The defense should investigate whether the *reference samples* were stored with the evidence samples, which could create a vulnerability to contamination in storage. The defense should also determine whether the evidence examinations, DNA extractions, PCR setup and DNA typing setup were conducted at separate times or in separate spaces, as required by the lab's protocol, to ensure that the integrity of the evidence was preserved and that the DNA was not vulnerable to contamination. For further discussion of DNA contamination issues, see Chapter 6, Section 2.

Normally, as indicated above, many items of evidence are collected and only some of them are tested. The items tested are usually chosen according to the investigating agency's or prosecution's theory of the case. The defense should consider its own theory and check to see if there are other collected items that were not tested but might support its theory, had they been tested for DNA. The defense then might consider independent testing or an argument that failure of the laboratory to test the items indicates a reason to doubt the prosecution's case.

5. Serology reports.

Because a positive *serology* report — a report that indicates or identifies the presence of a body fluid or biological material — provides context for the DNA, it is important to determine if serologic tests were *presumptive* or *confirmatory*. The most common serologic tests check for the presence of human blood, semen and saliva. *Presumptive tests* reveal less information because it is not human specific, and there may be a number of substances that result in a false positive result. For example, some presumptive tests for blood may yield a positive result from rust or some plant extracts.

In developing its theory, defense should consider where physiological fluids should be expected or not expected. The full serology report may help bolster the defense theory or contradict the government's theory. Although DNA may be identified on certain evidence samples, knowing what type of biological material made the stain may inform the defense's theory of the case. Consider a case in which oral copulation with ejaculation is alleged and DNA on the lip area swab matches the defendant. In the absence of serology results, DNA deposited from consensual kissing could be misinterpreted as supporting the state's case. In such a case, serology testing can confirm the presence of seminal fluid. Serologic results could be critical to the defendant's case if semen was not found.

See Chapter 2, Section 2, for more on serology.

6. Correspondence between lab personnel and any law enforcement, prosecutorial or other state/county/jurisdictional officials.

It is important to look for sources of *expectation bias,* or a strong belief or mindset toward a particular outcome. Often, lab personnel will be given more information about the case than is needed to conduct the forensic testing, and the seriousness of the case could sway the scientist. It is also possible that law enforcement inadvertently or directly states or hints to the forensic analyst what the desired results should be. Needless to say, this information might be developed as an area of cross-examination.

7. All documents routinely kept in the type of case file referenced.

This request will ensure that the defense has all of the other information from the laboratory case file that was not already specifically requested.

8. Documents related to the case that were referenced regularly but are kept in a place other than the case file.

This request will ensure that the defense has reference material that is not included in the specific case file but may have been consulted in the case.

9. A copy of all documentation regarding corrective action when casework errors are detected pursuant to the QAS (July 1, 2009 revision), Standard 14.

All laboratories that follow the *QAS* — specifically, all laboratories that use CODIS and, typically, all laboratories accredited by ASCLD/LAB, ASCLD/LAB-International, FQS or FQS-I — must be in compliance with this standard. Disclosure of any casework errors that occurred during a reasonable period of time (e.g., 6 months), either before or after the case-specific testing was done, may open up areas of cross-examination. See Chapter 5 for more information.

10. The quality assurance review (administrative and technical reviews).

Standard 12 of the *QAS* (July 1, 2009) requires that a second qualified technician or supervisor review the reports, notes, data and other documents and information related to the case to ensure there is an appropriate and sufficient technical basis for the scientific conclusions. This is often referred to as a technical case file review, conducted by the technical reviewer. All case reports issued by a forensic DNA lab must be subjected to both a technical review and an *administrative review.* The administrative review is conducted to ensure consistency with lab policies and editorial correctness and may be conducted by the technical reviewer.

Standard 12 also specifies that the technical case file reviewer must be currently qualified, or recently tested for proficiency, in the methodology(ies) used for technical reviews. Any discrepancy between the initial analyst's and the quality assurance reviewer's conclusions should be examined more closely to determine if there is a basis for exclusion of the DNA evidence or the presence of exculpatory evidence. The defense may wish to call on both the analyst who made the initial conclusions and the technical reviewer (or, in rare instances, the administrative reviewer, if different from the technical reviewer) when a significant discrepancy involving the initial/reporting analyst's report conclusions departs significantly from laboratory policies.

11. Data from the original case file if the case is based on a convicted offender match, a case-to-case match or a forensic match.

In a *"cold hit" DNA match* — and cases in which the evidence is linked to DNA evidence in another case — complete copies of all files should be obtained and reviewed for any discrepancies in protocols, procedures and analyses, including potential contamination issues. Be aware that if the cases were examined at different times, the laboratory may have used different laboratory protocols that would explain the discrepancies.

12. Documentation stating that the laboratory has searched for the casework DNA profile(s) in its staff DNA database, including the results of that search.

The defense should request documentation that the DNA profiles of all laboratory employees, especially those who worked on the case, have been compared (typically via a database search) against the casework profile(s) and what the results of that search were. Defense should also request documentation that the DNA profiles of all crime scene investigators and law enforcement agents involved in the case have been compared with the casework profile(s) and what the results of that search were. This is particularly important in cases where the evidence samples contain mixtures. It is a good practice to provide a list as part of your discovery request of lab employees, crime scene investigators and case investigators that you want to ensure have been compared with the casework profile(s).

13. Documentation from the three most recent laboratory accreditation assessments/ audits, including DNA Quality Assurance Standards audits and ongoing communications from the most recent assessment/audit.

As part of laboratory *accreditation,* the laboratory undergoes both external and internal *assessments* (also called *audits*) against the accrediting body's defined criteria. The purpose of laboratory accreditation is to have a systematic, independent and documented review of essentially all aspects of a laboratory's processes to objectively evaluate them. By virtue of laboratory accreditation, labs seek to demonstrate that they are providing a quality product and that their customers can rely on the results and reports issued by the laboratory.

As part of the ongoing accreditation process, regularly scheduled assessments/audits are conducted to determine whether the required accreditation criteria are being fulfilled and result in the generation of documentation. In a manner of speaking, the accrediting body's assessment/audit documents, the laboratory's internal assessment/audit documents and the DNA Quality Assurance Standards audit documents may include criticisms of the laboratory. Each assessment or audit has the potential to result in proposed corrective actions or findings of identified problems, to which the laboratory has time to comply and fix (or appeal). Because remediation of identified deficiencies can take some time, it is important to ask for and review the communications/documentation related to the status of open corrective actions/findings resulting from the last assessment/audit.

Minor issues that do not affect the quality of the product are identified all the time during laboratory assessments and are used to improve the laboratory's services and system. Of particular importance are any findings or proposed corrective actions that are defined as ones that could affect the quality of the results being issued by the laboratory. For example, a finding or corrective action regarding a problem with the failure of laboratory examiners to follow the defined protocol for using a single strikethrough with initials and date would not be considered to be a finding/proposed corrective action that affects the quality of the testing results. On the other hand, a proposed corrective action or finding that multiple instances were found during review of case files of reported conclusions deviating from the laboratory's defined interpretation guidelines would be considered more significant — as this could possibly affect the reliability and quality of the lab's reports.

Note: An *accrediting body* can receive information suggesting noncompliance of an accredited laboratory with standards, either by the laboratory self-reporting or by the accrediting body receiving a written allegation. If a complaint has been received against a laboratory, the accrediting body will notify the laboratory. It is acceptable to ask the laboratory, as part of discovery, for information regarding any pending allegations against the laboratory of serious negligence, misconduct or noncompliance that are being investigated by its accrediting body.

14. References to summaries of developmental validation studies for each of the DNA testing methodologies used in the case, if conducted by an organization other than the DNA testing laboratory, and to summaries of the laboratory's internal validation studies (and developmental studies, if applicable) of DNA testing methodologies used in the case and the supporting documentation.

In-house or *internal validation* studies provide the defense attorney with knowledge of how the laboratory determined the parameters for its protocols used to provide accurate and reliable DNA profiles. Prior to using any forensic DNA procedures on casework samples, the methodologies must be validated by the lab, and these studies must be documented and summarized.

These validation studies provide the basis for the interpretation guidelines and quality assurance parameters for the laboratory's DNA unit, including those for mixture studies.

They can be reviewed by an expert to assess the appropriateness and reliability of the laboratory procedures to accurately determine DNA profiles from mixed samples and define the range of detectable *mixture ratios*.

Note: Copies of the validation summary reports (both developmental and internal) can be provided easily; it is not burdensome to comply with this request. Labs may be unwilling or unable to create copies of the supporting documentation for their internal validation studies because of the labor required to produce such copious amounts of paper. However, most labs should allow defense counsel and the defense expert to visit the lab to review such documentation.

15. Proficiency testing results of analysts, technical reviewers and technicians in the case.

Standard 13 of the *QAS* (version effective July 1, 2009) requires that all DNA analysts, technical reviewers and technicians must participate in external proficiency testing twice per calendar year. The results of proficiency examinations of analysts, technicians and technical reviewers who tested evidence or reviewed data in the case may reveal problems, opening up potential areas for cross-examination. Poor testing results or deviation from the requirements as set forth in the *QAS* may be a reason to try to preclude the analyst from testifying at trial.

While it may be considered burdensome by a laboratory to produce copies of all proficiency testing records for the personnel involved in the analysis/review of the case, the laboratory can easily produce documentation that provides a summary of the external proficiency testing history of each relevant DNA analyst, technician and technical reviewer. Information on a proficiency testing summary document should include:

- The external proficiency test identifier (including the name of the test provider as well as the identifier for the test in question, e.g., CTS #10-574, which refers to a proficiency test sold by Collaborative Testing Services [CTS] in 2010, with their designated test number of "10-574").
- The date the proficiency test was assigned to the employee, received by the laboratory, or a similar metric.
- The manufacturer's due date.
- The date the proficiency test was completed by the employee and/or the test results were submitted to the manufacturer. To be counted as an external proficiency test under the *QAS,* this date must *always* be before the date the manufacturer released the expected results of the testing. Each manufacturer's Web site will contain information regarding the due date for submission of results and the expected testing results (once released).
- A record of whether or not the proficiency test was successfully completed and/or if any discrepancies between the employee's answer sheet and the manufacturer's answer sheet were noted. The laboratory may also indicate on the summary if any corrective action was required as a result of participation in the proficiency test.

The *QAS* provides very clear instructions regarding the acceptable length of time between participation in proficiency tests. Since July 2004, the term *semiannual,* when applied to the DNA proficiency testing interval, has been interpreted as an event that takes place two times during one calendar year, with the first event taking place in the first six months of that year and the second event taking place in the second six months of that year, with the interval between the two events being at least four months but not more than eight months. Any summaries of proficiency testing participation should be reviewed to ensure that the analyst, technical reviewer and technician's testing results are in compliance with this requirement. It is important to note that the proficiency testing standards have evolved over time, so you may need to put historical entries in the context of the standard in effect at the time the testing was conducted.

Note: Proficiency testing participation *summaries* can be provided easily by a laboratory for each analyst, technician and technical reviewer; it is not burdensome to comply with this request. Many labs will be unwilling or unable to create copies of the supporting documentation for the proficiency tests these employees have participated in because of the extensive amount of labor required to produce such copious amounts of paper. However, most labs should allow defense counsel and the defense expert to visit the lab and review the relevant proficiency testing documentation in house. Should this opportunity arise, you will want to obtain a list of the proficiency test providers used by the laboratory ahead of time. Consult with or bring an expert who can ensure that a thorough audit of the proficiency testing documentation is conducted by comparing this with the summaries provided by the laboratory, the *QAS* and the test results provided by the manufacturers.

16. Reference for, or a copy of, the documented population database(s) used to generate the statistical interpretation of autosomal (STR) loci testing results (commonly displayed as allele frequency tables) and a summary or notation of what statistical method was used; if applicable, a summary or notation of the documented interpretation guidelines/ procedures for reporting of statistics for mixtures; and, if applicable, for analyses involving nonautosomal testing, such as mitochondrial or Y-STR DNA testing, a summary or notation of the laboratory's documented statistical interpretation guidelines for such testing.

Allele frequency values (typically found in tables) provide the basis for the statistics used to describe the frequency of occurrence (rarity) of the STR evidence profile in a given population. A laboratory must use a documented population database(s) for which the calculated allele frequencies are available for review (for the most commonly used databases, see http://www. cstl.nist.gov/strbase/population/PopSurvey. htm#ReferenceListing).

In addition, Standard 9.6.2 of the QAS (July 1, 2009) specifies which statistical formulae, as per the recommendations of the National Research Council (NRC) in its 1996 report titled The Evaluation of Forensic DNA Evidence, can be used by a laboratory for autosomal STR statistical interpretations. A laboratory can deviate from the NRC II recommendations if the court in its jurisdiction has directed that a different method be used. Because of this, the statistical formulae used should be provided readily by the laboratory.

Of note is the fact that there can be genetic frequency variations within populations — for example, the frequency of occurrence of the THO1 9.3 allele is significantly different in samplings taken from African-American and Caucasian-American populations. This sort of variation is normal and expected; however, it creates a topic to pursue with the DNA analyst on cross-examination. The analyst should be able to clearly describe what segment of the population was sampled to create the database(s) relied upon by the laboratory to conduct the statistical analysis. If the laboratory's statistics are generated based on frequencies obtained by a convenience sampling of convicted offenders, the analyst should be able to explain why the sampling of STR markers in a prison population does not deviate significantly from a sampling of the same STR markers in the general population and how they know this. In addition, the analyst should be able to clearly discuss why it is not necessary to generate statistics from a subsection of the world population (to which the defendant can be assigned) to be able to provide information to the jury regarding the relative rarity of the evidence profile.

17. Laboratory protocols used for all analytical procedures, including evidence-handling procedures; serology/evidence screening testing procedures; reagent preparation and use methods; sample preparation methods; extraction and quantitation methods; autosomal STR methods and/or other relevant DNA testing methods; instrument calibration, maintenance and operating methods; software operation methods; data analysis, interpretation and reporting methods; processes for monitoring of analytical procedures using controls and standards; and administrative and technical review procedures for case files and reports.

Case notes should be compared with laboratory protocols to ensure that there were no deviations.

18. The quality assurance program manual.

The defense should compare the case notes with the quality control procedures outlined in the quality assurance/system manual to ensure that there were no deviations.

19. All reports for the case issued by the laboratory.

At the conclusion of testing, the laboratory issues a report. In many jurisdictions, this report, absent aggressive discovery requests from the defense, is the first — and often the only — document provided by the prosecution. Beyond simply obtaining this report, it is important to ensure that it has not been supplemented or modified/ amended over time. All reports (including preliminary, supplemental and amended reports, if issued by the laboratory) should be obtained during discovery.

Note that Standard 11 of the *QAS* (July 1, 2009), which deals with reports, specifies that the laboratory must retain sufficient documentation for each technical analysis to:

- Support the conclusions in its report.
- Enable another qualified individual to evaluate and interpret the data.

Accordingly, the materials requested under items 1 through 5 in this section should contain all of the case notes and *analytical documentation* related to the case and generated by the analysts, as required by the laboratory's written procedures for taking and maintaining casework notes that support the conclusions drawn in the laboratory reports.

20. Any other information in the form of documentation, or encompassed in some other manner, that the lab has in its possession or control, or knows of and can access regarding the case.

This particular request is designed to ensure that there is no additional information that was not specifically requested. The request to the lab may not be worded this vaguely, but the attorney should be aware that, after initial discovery is received, they may want to consult with their own expert to determine whether additional items should be requested.

Section 3. Brady and DNA Cases

The discovery obligation imposed on the prosecution includes the duty to preserve[1] potentially exculpatory evidence, and the separate duty to disclose evidence or information that might be exculpatory as to guilt or to punishment. Both duties have constitutional foundations; both may be made more rigorous by application of each state's constitutional protections, statutes or rules of procedure.

The duty to disclose potentially exculpatory evidence is rooted in the guarantee of due process. In Brady v. Maryland,[2] the U.S. Supreme Court held that "the suppression by the prosecution of evidence favorable to an accused upon request violates due process where the evidence is material either to guilt or to punishment, irrespective of the good faith or bad faith of the prosecution."[3] The requirement extends to the disclosure of impeachment evidence[4] and applies regardless of whether the individual prosecutor is aware of the evidence, as long as it is in the possession of those acting on behalf of the prosecuting entity.[5] Prosecutors have "a duty to learn of any favorable evidence known to the others acting on the government's behalf in the case."[6]

The list of evidence that might meet the Brady standard in a case involving DNA evidence is substantial and can include:

- Flaws in the collection process or chain of custody.
- Lab-related evidence:
- Prior incidents of laboratory error.
- Failed proficiency tests by the lab technicians or analysts.7
- Inconclusive results.[8]
- Evidence of contamination.
- DNA evidence from other crimes that might exonerate the accused in the case at hand.[9]

Unresolved at this time is whether the defendant has a constitutional right to demand that DNA profiles from a crime scene that do not match the defendant's profile be uploaded into and checked against local or national databases to potentially identify another suspect.[10] Several states allow such access by law.[11]

End Notes

[1] See § 32.04[3], infra.

[2] 373 U.S. 83 (1963).

[3] 373 U.S. at 87.

[4] United States v. Bagley, 473 U.S. 667, 676 (U.S. 1985) ("Impeachment evidence, however, as well as exculpatory evidence, falls within the Brady rule.").

[5] Kyles v. Whitley, 514 U.S. 419, 437 (1995).

[6] Ibid.

[7] See, e.g., State v. Proctor, 348 S.C. 322, 332 (S.C. Ct. App. 2001) (holding that the defense is entitled to proficiency tests under Brady).

[8] In some factual settings, the inconclusive test may not be exculpatory. See, e.g., People v. Kazarinoff, 2004 Mich. App. LEXIS 3379 *15-16 (Mich. Ct. App. 2004).

[9] See, e.g., People v. Rathbun, 2007 Cal. App. Unpub. LEXIS 6877 (California Unpublished Opinions 2007) (explaining, "[a]ppellant's theory is that if DNA profiles similar to his were identified, they could either indicate that a different party committed the crimes, or cast doubt on the prosecution's interpretation of the samples adduced as [the] appellant's," but finding a failure to prove that the prosecution had such evidence).

[10] The argument in favor of such a requirement is that the cooperation between local law enforcement (the agency prosecuting the accused) and state and national law enforcement authorities (the operators of DNA databases) brings them within the framework of Kyles v. Whitley, supra, n. 6. For a discussion of when the Brady disclosure obligation extends to law enforcement agencies outside of the prosecuting jurisdiction, see United States v. Risha, 445 F.3d 298 (3rd Cir. 2006).

[11]. 725 Ill. Comp. Stat. § 5/116 5 (2005) (allowing such searches by court order); Ga. Code Ann.§ 24-4-63 (2005) (providing similar access upon a showing that "access to the DNA data bank is material to the investigation, preparation, or presentation of a defense at trial or in a motion for a new trial"). Other statutes seem to permit such access without specifically identifying criminal defendants as those with rights to request such searches. Haw. Rev. Stat. § 844D-82 (2006); N.C. Gen. Stat. § 15A-266.8 (2005); Cal. Penal Code § 299.5(g)-(h) (West 2005); N.J. Stat. Ann. § 53: 1-20.21 (2006).

4. DNA Evidence: Evaluation, Assessment and Response

Section 1. Evidence

Evaluate the Context of DNA Results

They've "got DNA." What does this mean? A variety of tests exist to detect the type of biological stain from which the DNA profile was obtained. For example, sperm cells can be identified under a microscope; blood and saliva may be detected through *presumptive testing.* (For a more in-depth discussion on *serology,* see Chapter 2, Section 2.)

Determine how DNA Evidence Is Relevant

This seems like a basic question, but it warrants some discussion. As a general rule, a date or time stamp cannot be put on the deposition of a sample from which a DNA profile was obtained. Additionally, a DNA profile itself does not explain the circumstances under which the DNA was deposited at a specific location. Here are two scenarios that explain how the presence of DNA may be irrelevant to proving a case:

- If a woman says she was raped by her husband, and he says that the sexual contact was voluntary, the fact that his semen is in her vagina may not be significant.
- If a man's DNA profile is found at the scene of his wife's murder, and the crime scene is the man's own home, the fact that his DNA was found at the scene may not be significant. It is expected that our DNA is in our homes, in our cars, on our clothing and at our offices. DNA can and does transfer.

Review DNA Test Results with the Client Early, as It May Affect the Case Resolution

Reviewing the DNA test results with the client is crucial and should be done as early and as thoroughly as possible. It is unlikely that a criminal defendant will have a working knowledge of DNA and how DNA profiling works. Therefore, discussions between lawyer and client serve two purposes: (1) to determine if there is an innocent explanation of how the DNA may have been deposited on an item in question and (2) to ensure that the client understands the significance of the DNA results. Counsel should be prepared to talk to the client about the risks and benefits of seeking additional testing, including the capability of detecting *touch DNA.*

It is not enough to tell the client over the phone that there is a lab report finding his or her DNA at the crime scene. This conversation must take place in person. Take the time to talk about each sample and where it was found and collected. A cigarette butt found on the corner of Fourth and Main may not be particularly damning.

There are times when defense counsel must have a serious conversation with the client about the strength of DNA test results. Competent and effective defense counsel must educate the client regarding the power and accuracy of the DNA results. The client should be made aware that — in the absence of an identical twin (with *autosomal* STR analysis), contamination or a highly unlikely coincidental match — the jury may see the client as the source of the DNA evidence and there may be no credible defense. One straightforward way to explain this is to review the underlying data in person, showing the client a comparison, locus by locus, of the alleles found on a particular piece of evidence to the alleles of his or her DNA profile. This review can make this clearer to the client. Defense counsel has an obligation to give the client the best advice. Under certain circumstances, this may be to encourage a plea.

Consult with Someone Familiar with Forensic DNA

Consulting is invaluable in DNA cases and can assist counsel in understanding all of the issues, including how to communicate in ways that laypeople can understand. If a jury is unable to understand an issue, it does not matter how important that issue is. It is crucial to be able to clearly communicate problems with the government's interpretation of the evidence.

Hire an Expert (Not Necessarily to Testify)

An expert should be able to look at the evidence and tell you what it means in easily understandable language. An expert may say that there is no issue regarding the DNA evidence, that there are issues but they are not relevant, or that there are important and relevant issues.

An expert may serve solely as a consultant to assist counsel in understanding the evidence and in preparing the cross-examination of the government's expert. An expert may also be used to help explain the significance of evidence to the client.

In addition to using a consulting expert, counsel may wish to hire an expert to testify. Defense counsel should first review the expert's résumé and prior testimony, and consult with experienced attorneys who are familiar with the expert's work and courtroom effectiveness. Counsel should also investigate a prospective expert's background (e.g., see "Digging Up Dirt on Experts" at www.ncstl.org). Public defender offices, innocence project offices and criminal defense attorney organizations are excellent sources for referrals. Transcripts of expert testimony are available online to certified defense attorneys through the National Association of Criminal Defense Lawyers at www.nacdl.org.

Once the testifying expert has been vetted and selected, counsel must prepare a very specific list of questions and review them with the expert before trial. Counsel should go over the direct examination and anticipate cross-examination questions with the expert at least once before trial. During the preparation, the expert should be encouraged to identify any vulnerabilities or limitations. This is not the time to wing it.

Choose the Right Kind of Expert

Typically, the question in a DNA case is not "Do I need an expert?" but rather "What kind of expert do I need?"

The answer lies in the specifics of the case. The first challenge is to locate an expert to review the electronic data and paper case file to determine whether there are issues with the equipment, test kits, controls, testing methods (e.g., longer-than-usual injection times), interpretation of the results or statistics used. If the statistics are questionable or flawed, a statistician or population geneticist is a much better choice as a testifying expert than a lab analyst.

Conversely, if the injection time deviates from the lab's validated protocols, a lab scientist with an understanding of the importance of following protocols and the impact of an extended injection time will be a better choice.

Choose an Expert with the Right Qualifications

Some take the position that only a person who has worked in a crime laboratory is qualified to testify about forensic DNA testing. Others assert that crime lab experience is not necessary to consult on a DNA case. All science is reproducible and verifiable, when done properly, regardless of the context of the testing.

The ability of your selected expert to properly and thoroughly review the forensic biology case files — both the paper and electronic data — is critical. Complete review of a case file is a tedious process that requires attention to detail as well as the ability of the expert to consider how the provided documentation fits into the larger picture of the laboratory's policy documents, its accreditation requirements, and the *Quality Assurance Standards for Forensic DNA Testing Laboratories* (the *QAS,* July 1, 2009). Use of an expert who has worked in a crime laboratory may be valuable, provided the individual has the requisite skills to conduct this type of data and policy review.

When choosing a qualified expert for your DNA case, consider those who have previously conducted evidence screening and DNA testing on forensic samples and have testified in DNA cases for the defense. For example, if the issue is a statistical one, the head of the local university's biostatistics department may make a fine choice as an expert. The main considerations in selecting an expert are to find someone who can properly and completely review the case files and who can also effectively communicate the defense's theory of the DNA evidence to the judge or jury.

When choosing an expert, set up a meeting (teleconferencing or in person). Include a layperson in the meeting who has no forensic DNA experience and can act as a test person. If the expert can communicate the issues in a manner the test person can understand, he or she will likewise be able to explain the issues to a judge or jury.

Section 2. Funding for the Defense DNA Expert

Ake v.Oklahoma[1] guarantees an indigent defendant reasonable funding for expert assistance:

> [W]hen a State brings its judicial power to bear on an indigent defendant in a criminal proceeding, it must take steps to assure that the defendant has a fair opportunity to present his defense. This elementary principle, grounded in significant part on the Fourteenth Amendment's due process guarantee of fundamental fairness, derives from the belief that justice cannot be equal where, simply as a result of his poverty, a defendant is denied the opportunity to participate meaningfully in a judicial proceeding in which his liberty is at stake.[2]

Although *Ake* involved access to funding for psychiatric assistance, its reach has been extended to the forensic sciences, including DNA testing.[3]

The right to assistance is not absolute; rather, it is case dependent and requires showing the centrality or significance of DNA evidence to the prosecution's case: "[A] defendant must show the trial court that there exists a reasonable probability both that an expert would be of assistance to the defense and that denial of expert assistance would result in a fundamentally unfair trial."[4] The assistance need not be limited to serving as a witness; it is enough if the expert's assistance is needed to prepare defense counsel for understanding and addressing the prosecution's DNA evidence.[5]

Because DNA evidence has become more common and is likely to be admissible under either *Frye* or *Daubert* standards, showing the need for expert assistance may require disclosing issues at the heart of the defense's case theory. Therefore, counsel should seek funding by *ex parte* motion.[6] In every case, defense counsel will need to assess the strength of the DNA evidence as well as possible noncriminal explanations for how the suspect's DNA was found at the scene. There are cases in which the need for expert assistance may be particularly strong, including:

- Cases with partial matches.
- Cases where the results are reported as uninterpretable or inconclusive.
- Cases involving interpretation of mixtures.
- Cases involving defendants from population subgroups that may affect the statistical significance of the DNA match.
- Cold cases, or cases where evidence collection or storage raises concerns of *degradation* or contamination.
- Cases involving less frequently used, newer or emerging forms of DNA analysis and statistical interpretation (such as Y-STRs, mtDNA, mini-STRs, SNPs and *low copy number DNA testing* [also called *low-level* or *LCN testing*]).

The motion seeking funds for DNA expert assistance should detail that: (1) forensic DNA evidence is central to the prosecution's case and (2) expert assistance is needed to determine any, some or all of the following:

- The meaning and significance of the prosecution's evidence.
- Whether retesting or testing other crime scene evidence could be beneficial.
- Whether and how the prosecution's evidence is subject to attack.
- That the defendant is indigent and entitled to assistance under Ake, and any applicable state constitutional or criminal rule provision(s).

The motion should also identify the cost of expert assistance by listing each expert contacted, his or her hourly fee, and an estimate (by each expert) of the cost of initial case review and consultation. Leave should be sought for filing a supplemental motion if, after consultation, additional expert assistance is necessary. In doing so, be cautious about local practices in revealing content of ex parte communications.

Note: A motion seeking funds to hire an expert may reveal defense strategy. For this reason, counsel should seek to file the request ex parte and under seal. If there is any risk that the motion will be disclosed to the jurisdiction's funding authorities or others, it should be a bare-bones pleading that can be supplemented at a hearing before the court in camera.

It is critical that the motion and any supplements — oral or written — be made part of the record. If proceeding ex parte, the motion papers, any transcript of an ex parte hearing, and the judge's order should be filed under seal. If the motion is denied, counsel must take whatever steps are necessary to ensure that the issue is preserved for pretrial, interlocutory appeal, or appellate review if the trial results in a conviction.

Section 3. Evidence Consumption

As an initial matter, labs are obligated, under the *QAS* (July 1, 2009) Standard 7.2, to retain a portion of the evidence for subsequent defense testing: "Where possible, the laboratory shall retain or return a portion of the evidence sample or extract."

It is important, however, to understand what constitutes exhaustion of a sample. Based on the guidance provided by the *QAS,* if a portion of the DNA extract remains, even if the entire initial evidence sample has been subjected to the *DNA extraction* process, the sample has not been consumed in analysis. An illustrative example is the procedure commonly used for the processing and extraction of DNA from cigarette butts. When a lab conducts DNA testing on a cigarette butt, it will typically opt (based on its protocol) to "take the best evidence" for DNA processing. Because of this, it is commonplace for all of the outer paper from the anterior area of a smoked cigarette butt to be put through the lab's DNA extraction protocol. The end product of the DNA extraction process is a DNA extract, with the volume of this extract varying depending on a number of factors, including the laboratory's protocol. Only a portion of the generated DNA extract should be used to estimate the quantity of human DNA present and to generate the DNA typing result(s). Accordingly, a portion of the DNA extract will remain — the volume of which should be clearly discernible upon review of the case file. The lab will preserve this remaining portion of the DNA extract in the manner required by its protocol. This remaining DNA extract is considered the portion of the initial sample that remains for repeat testing, if necessary, or for possible subsequent testing on behalf of the defendant. One potential exception to this definition of evidence consumption is when a laboratory defines exhaustion of a sample more stringently in its protocol — if so, the laboratory must adhere to its own definition.

There are times when the DNA testing process will consume the entire sample. In this instance, if required by the jurisdiction's court or by laboratory protocol, the lab must notify either the prosecuting authority or the relevant law enforcement agency before consuming the entire sample. The case notes should clearly indicate whether a sample was consumed in analysis and, in some jurisdictions, the lab report will also indicate this. In cases where there is no suspect, some labs will proceed with testing without first notifying anyone. In cases where there is a suspect (either charged or not yet charged), labs will notify either the prosecuting authority or the relevant law enforcement agency before consuming the sample. When dealing with an older case, it is important to put the lab's and forensic community's practices into the proper historical perspective. As with a number of other current policies, the requirement to retain or return a portion of the evidence sample or extract has evolved over time and was not always the required practice.

Defense counsel should be aware of notification requirements in consumption cases. For example, some jurisdictions mandate that the defense attorney or the public defender's office be notified before evidence consumption when a suspect has not yet been charged. Other jurisdictions put the onus on the defense to request notification from the lab before any testing that might consume evidence.

The basis for notification has to do with the rights of the defendant. Because the evidence will not be available for retesting, the defendant is given the opportunity to ensure that the testing is performed in a reliable manner.

One option in cases where the evidence will be consumed is for the defense and the prosecution to agree on an independent lab (other than the jurisdiction's lab) at which to perform the testing. An alternative may be to grant the defense the right to have its expert observe the initial testing.

Neither option is ideal. Agreeing to a joint laboratory puts the defense in the position of endorsing the results, which makes it more problematic to challenge them if the results are not favorable to the defendant. If the defense provides an expert to observe the lab's testing and no errors are observed, the defense may have created an additional witness to vouch for the accuracy of the lab's test results. Counsel may wish to consider a court order, before observing the testing, that would prohibit the prosecutor or the prosecution's witnesses from testifying to the presence of a defense observer.

A different problem presents itself if the defense observer sees a lab analyst mishandling a sample or deviating from protocol. Most labs have rules for what observers can and cannot do. For example, the laboratory policy may dictate where in the lab nonemployees can go, or whether guests must provide a known sample for comparison. Court rulings should be obtained before testing to address what should be done if certain situations arise during the testing process. For example, should observers remain silent and record their observations of deviations from protocol? Should observers object to the conduct? Do observers have the power to prevent an analyst from proceeding a certain way? The answers to such questions should be obtained before the testing process. *Note:* It is advisable that, at a minimum, the defense attorney seek an independent expert to review a laboratory's case file when there is a possibility that the evidence was consumed in analysis.

End Notes

[1] 105 S. Ct. 1087 (U.S. 1985).

[2] 105 S. Ct. at 1092.

[3] See, e.g., Dubose v. State, 662 So. 2d 1189, 1199 (Ala. 1995); Polk v. State, 612 So. 2d 381 (Miss. 1992) (applying reasoning of Ake to hold that due process considerations require defendant to have access to a DNA expert); Polk v. State, 612 So. 2d 381: Superseded by Mississippi Transportation & Communication v. McLemore, 863 So. 2d 31, 2003 Miss. LEXIS 532 (Miss. 2003); 863 So. 2d 31, 39; Moore v. State, 889 A.2d 325, 336 (Md. 2005); Husske v. Commonwealth, 448 S.E.2d 331, 335 (Va. App. 1994), cert. denied, 487 U.S. 1210; Giannelli, Paul C., Husske v. Commonwealth, 448 S.E.2d 331: Opinion withdrawn by, vacated by, different results reached on rehearing at, en banc: Husske v. Commonwealth, 21 Va. App. 91, 462 S.E.2d 120, 1995 Va. App. LEXIS 700 (1995); "Ake v. Oklahoma: The Right to Expert Assistance in a Post-Daubert, Post-DNA World," 89 CORNELL L. R EV. 1305 (2004).

[4] Moore v. State, 889 A.2d 325, 339 (Md. 2005).

[5] Moore v. Kemp, 809 F.2d 702, 712 (11th Cir. 1987); Moore v. State, 889 A.2d 325, 339 (Md. 2005) (citing Kemp approvingly and collecting cases adopting the Kemp formula).

[6] See, e.g., Gianelli, "Ake v. Oklahoma The Right to Expert Assistance in a Post-Daubert, Post-DNA World," note 3; Shane, B., "Money Talks: An Indigent Defendant's Right to an Ex Parte Hearing for Expert Funding," 17 CAP. DEFENSE J. 347 (2005); Winbush, K.J., "Right of Indigent Defendant in State Criminal Prosecution to Ex Parte In Camera Hearing on Request for State-Funded Expert Witness," 83 AM. L. R EV. 5th 541 (2000). Nationally, courts are split as to whether the defense is entitled to proceed ex parte. See Moore v. State, 889 A.2d 325, 341 (Md. 2005) (collecting cases and approving of the ex parte process).

5. DNA BASICS: LABORATORY ISSUES

Section 1. Standards for Labs, Personnel and Procedures

Getting to know the crime lab involves knowing what types of documents and records exist and what requests to make. One important document that provides a set of requirements is *The Quality Assurance Standards for Forensic DNA Testing Laboratories* (the *QAS*) (version effective July 1, 2009) (http://www.fbi.gov/about-us/ lab/codis/qas_testlabs). In addition to this set of standards, the FBI also promulgates *The Quality Assurance Standards for DNA Databasing Laboratories* (version effective July 1, 2009) (http:// www.fbi.gov/about-us/lab/codis/qas_databaselabs) for laboratories that process DNA databank samples.

The *QAS* standards describe "the quality assurance requirements that laboratories performing forensic DNA testing or utilizing the Combined DNA Index System (CODIS) shall follow to ensure the quality and integrity of the data generated by the laboratory." Likewise, *The Quality Assurance Standards for DNA Databasing Laboratories* describe "the quality assurance requirements that laboratories performing DNA testing on database, known, or casework reference samples for inclusion in the Combined DNA Index System (CODIS) shall follow to ensure the quality and integrity of the data generated by the laboratory." In the versions of these quality assurance documents referenced before July 1, 2009, the *QAS* standards were guidelines; they are now requirements.

Section 2. QAS Requirements for Laboratories

A Quality Assurance Program

As can be noted from a review of the list of *QAS* standards, DNA testing labs are required to have a *quality assurance* program. Each lab is required to document the details of its quality assurance system in a manual that includes or references the following elements: goals and objectives, organization and management, personnel qualifications and training, facilities, evidence control, validation, *analytical procedures, equipment calibration* and maintenance, reports, review, *proficiency testing,* corrective action, *audits,* safety and outsourcing (Standard 3.1.1).

Although you can rely on an expert hired to review the DNA laboratory report and associated documents, it is recommended that you be familiar with the basics of the lab's quality assurance system to ensure that any feedback received from your expert is on target.

A Testimony Monitoring Program

The *QAS* standards (July 1, 2009) require labs to "have and follow a program that documents the annual monitoring of the testimony of each analyst" (Standard 12.7). Testimony monitoring is also a requirement of the *accrediting bodies.* Each lab can adopt whatever approach they deem appropriate to ensure that each DNA analyst's testimony is monitored annually. Examples of commonly used approaches are: monitoring by a lab supervisor, review of the testimony transcript, or a customer survey-type process that obtains feedback from customers such as the prosecuting attorney, defense attorney and/or judge.

Organization and Management Documentation

In the *QAS,* Standard 4.1.5 addresses the organization and management of the laboratory. This *QAS* standard requires labs to "specify and document the responsibility, authority and interrelation of all personnel who manage, perform or verify work affecting the validity of the DNA analysis."

Facilities and Evidence Storage/Control

The lab facilities' setup requirements are found in Standard 6.1, and the requirements for documentation of an evidence control system to ensure the integrity of physical evidence are found in Standard 7.1.

Section 3. QAS Requirements for Laboratory Procedures

Validation Studies

Standard 8 of the *QAS* addresses the details of the requirements for *validation,* which involves the extensive and rigorous evaluation of methods and procedures before acceptance for routine use in casework. The DNA lab can only use methodologies that have been validated (Standard 8.1). As part of the validation process, the procedures will have been tested both within the normal limits of the method and at the outer edge of the method's capabilities. There are two types of validations — *developmental* and *internal. Developmental validation* refers to the testing of new DNA testing systems that precede the use of the novel methodology for forensic DNA analysis. *Internal validation* refers to testing done within the lab that has been reviewed and approved by the technical leader before using the methodology for forensic casework applications.

The purpose of reviewing the summaries of the lab's validation studies is to assist you in ensuring that testing is done in a manner that is both reliable and reproducible. The validation studies can show the limitations of the system and when the system is expected to work well. All of the paperwork from the validation studies must be kept at the laboratory and be available for review.

Based on the *QAS* standards of July 1, 2009, all labs must perform a mixture study during their internal validation before adding any new DNA typing methodology for casework. The design of a typical mixture study includes a 1-to-1 mixture of DNA from two people. The test result should show even peaks within the mixed profile — in other words, if each individual contributing to the mixture has two different alleles that are being tested at one DNA location, four peaks should be present that are of relatively the same height, or intensity. A second portion of the study would examine what the *electropherograms* would look like, for example, in mixtures of DNA from the same two people in ratios of 1-to-2, 1-to-5 and 1-to-10. Using the validation testing to show what the varying ratios of DNA are expected to look like is extremely valuable. Knowing at what mixture ratio the results no longer appear as a mixture is also helpful information.

Serology validation studies — where they exist — can also be helpful. For example, if a lab has recently put a new presumptive test for seminal fluid online for casework, it will have conducted an internal validation study. The validation study summary should show what other fluids were tested in addition to seminal fluid and should specify if any other body fluids or materials tested positive with the *presumptive test* for seminal fluid.

Analytical Procedures and Equipment

Each lab must have and follow a set of *analytical procedures* (*QAS* Standard 9) that specifies *reagents,* sample preparation, extraction methods, equipment and controls that are standard for DNA analysis and data interpretation. In addition, the lab must have and follow written interpretation guidelines (Standard 9.6).

QAS Standard 10 addresses the detailed requirements for equipment calibration and maintenance schedules. You will want to ensure that the lab is following its documented program for conducting performance checks and calibration of its instruments and equipment as well as its planned maintenance processes.

It is important to understand which lab equipment and instruments (such as pipettes and thermal cyclers) have been deemed critical — meaning that they require calibration or a performance check before use and periodically thereafter.

Writing Reports and Reviewing Files

Detailed report-writing requirements are found in *QAS* Standards 11.1 and 11.2. Standard 11.3 addresses the confidential nature of reports, case files, DNA records and databank databases. This standard requires that the lab must have and follow written procedures to ensure the privacy of the reports, case files, DNA records and databases. The lab must also have and follow written procedures for the release of these documents and information.

Standard 12 discusses the acceptable methods for reviewing case files and reports. Labs are required to conduct and document both an administrative and a technical review of all case files and reports to ensure that the report conclusions are supported by the data and that the conclusions — given the supporting documentation — are reasonable and within the constraints of current scientific knowledge. The documentation of the report and case file reviews should be included in the case file notes. Evaluation of this documentation can assist counsel and experts in determining whether, in the lab's estimation, the testing was performed correctly and within the bounds of its procedural requirements. In addition to the laboratory's analytical report, other documents related to a case must be maintained and reviewed. For example, *QAS* Standard 11.1 (July 1, 2009) requires laboratories to maintain a case file with "all *analytical documentation* generated by analysts related to case analysis. The laboratory shall retain, in hard or electronic format, sufficient documentation for each technical analysis to support the report conclusions such that another qualified individual could evaluate and interpret the data." Case files commonly include:

- A *chain of custody* for items received by the laboratory.
- Sketches or photographs taken in the laboratory of items examined.
- Examination ("bench") notes by the analyst of steps taken in testing.
- Laboratory logs or standard forms related to testing.

- Electropherogram data (in older case files, there may be strips, photographs and/or copies of *autoradiographic film*).
- Communication information between the analyst and others involved in the case.

Obtaining Transcripts of Analysts' Past Testimony

Reviewing past testimony transcripts of laboratory analysts can be helpful. For example, a scientist may have a particular way of explaining DNA transfer that he or she uses in every case. Knowing what this explanation is ahead of time can assist counsel in preparing cross-examination. Additionally, some scientists are better than others in explaining DNA evidence.

There are several ways to obtain transcripts. If analysts have testified in previous cases that resulted in convictions, the appellate office of the public defender is likely to have transcripts. Transcripts are also available through the National Association of Criminal Defense Lawyers. If the laboratory uses the review of transcripts to monitor personnel testimony, another option is to request copies of past testimony transcripts in the laboratory's possession for the relevant DNA analysts as part of discovery.

Outsourcing

The requirements for which laboratories can and cannot be used for outsourcing of DNA testing, as well as who can and cannot review the associated testing data, have also evolved over time. Standard 17 clearly states that all vendor laboratories performing forensic DNA analysis must comply with the QAS as well as the accreditation requirements of federal law. This is important to keep in mind, should you wish to have testing done on evidence that has not been tested by the DNA lab in your jurisdiction, or if you want to have retesting conducted.

Corrective Actions

Per Standard 14.1, "The laboratory shall establish and follow a corrective action plan to address when discrepancies are detected in proficiency tests and casework analysis. ... [D]ocumentation of all corrective actions shall be maintained in accordance with Standard 3.2." In most accredited laboratories, a corrective action is documented by a CAR — a corrective action report.

It is noteworthy that although some labs keep a central "corrective actions" file or logbook, other labs simply document corrective actions within the original case files. This information is discoverable and can illustrate how easily contamination can occur, even within a crime lab. For labs that maintain the corrective actions information within the specific case files, a court order demanding that the information be culled from the case files and compiled for discovery purposes can be obtained.

Audits

As dictated by Standard 15, DNA labs are required to conduct audits once a year to maintain compliance with the *QAS*. Per Standard 15.2, at least once every two years the laboratory must have an audit conducted by a team comprising qualified auditors from an agency(ies) other than its own. Standard 15.4 specifies that both internal and external audits must be conducted using the FBI DNA Quality Assurance Standards Audit Document (http://www.fbi.gov/about-us/lab/codis/audit_testlabs).

Section 4. QAS Requirements for Laboratory Personnel

Education, Training And Experience

The *QAS* standards also cover educational, training and experience requirements for laboratory personnel (Standard 5). These requirements are clearly defined for the DNA technical leader (Standard 5.2*), casework CODIS administrator* (Standard 5.3), analysts (Standard 5.4) and technicians (Standard 5.5).

Proficiency Testing

The requirements for participation in DNA proficiency tests have evolved significantly over the years. *QAS* Standard 13.1 (July 1, 2009) requires each person involved with casework files — analysts, technical reviewers, technicians, and other personnel designated by the technical leader — to undergo semiannual external proficiency testing in each technology performed to the full extent in which they participate in casework. In particular, addition of the technical reviewer to the list of those required to complete semiannual proficiency testing is new. Although most labs were including the technical review as part of the documentation for each proficiency test, this requirement now formalizes that process, given that all casework files must be technically reviewed before the release of a report. See Chapter 3 for recommendations regarding requesting proficiency testing records as part of discovery.

6. DNA Basics: Understanding and Evaluating Test Results

Section 1. With Your Expert's Guidance, Interview the Lab Analyst

What to Ask

Long before trial, defense counsel should meet with the lab analyst — in person, whenever possible — to discuss the DNA evidence. The preferred location for the meeting is at the laboratory. The meeting may also take place at the prosecutor's office, the courthouse or possibly the defense counsel's office. It is a good idea to bring a second person to the interview to listen and take notes on a separate copy of the file.

Before the meeting, counsel should review a complete copy of the lab's file with an expert. (For more information, see discussion on discovery in Chapter 3, Section 2.) The expert can discuss the significance of each document and alert counsel to any problems evident in the lab file. (For more information about meeting with experts, see Chapter 4, Section 1.)

When scheduling your meeting, ask the analyst to bring the lab's original copy of the case file to the meeting.

At the beginning of the meeting, it is important to confirm that you have all of the documents the lab possesses. This can be done by simply going through the stack of documents one page at a time and visually confirming that each stack has the same pages in the same order.

This can serve as a valuable ice breaker and ensures that defense counsel has a complete copy of the file.

Next, go over the lab analyst's curriculum vitae in detail. If the analyst has published any relevant articles, obtain copies and review them ahead of time. Ask for defendants' names and the locations of any previous testimonies so you can talk to the attorneys involved in those cases and order transcripts.

Review each page of the lab file in chronological order, starting from when the lab first received the evidence. This may require you to put the file in a different order than it was received during discovery — before your meeting with the analyst — to save time. Make sure you have a complete understanding of each document; do not proceed until you understand the information on each page and how that part of the analysis fits into the overall testing process. Because you have already reviewed the file with an expert, repeating the process will allow you to determine — well in advance of testimony — if the prosecution's expert disagrees with your expert or interprets things slightly differently.

During the page-by-page review, counsel can also incorporate questions critical to the defense theory without specifically highlighting a particular theory. Before concluding the interview, ask a close-out question such as, "Is there anything else important we have not talked about?"

Note: The purpose of the interview is to obtain information. It is not to argue the defense's case, give up defense theory, or divulge expert information that will assist the client. There may be times when counsel wishes to raise a particular, seemingly exculpatory subject with the lab analyst. However, before doing so, discuss possible tactical and strategic ramifications with an expert and your colleagues.

This meeting provides an opportunity to find out the lab's precise position on specific issues. It is not in the client's best interests to squander the opportunity.

Section 2. Interpretation and Reporting of Results

Single-Source Profile

Figure 10 shows an electropherogram of a 16-loci single-source sample. Generally speaking, a single-source DNA profile will have either one or two peaks (alleles) at all of the areas (loci) examined. If there are two peaks and the sample is from a single source, they will generally be of relatively equal height, or intensity.

It is important to note that not all *electropherograms* will have the loci printed above the peaks, as is seen in Figures 10 and 11. (Figure 11 is an enlargement of the results at three loci.) Furthermore, although it is most common to see electropherograms with boxes beneath each labeled peak containing three bits of information — the allele designation/call, the corresponding length of the DNA fragment, and the *relative fluorescence units (RFUs)* for the peak — you may also encounter electropherograms with only one or two of these data points. You will need to refer to the laboratory's protocol to determine what markings are required on the printed electropherograms.

Assuming no identical twin, the probability of two people — related or unrelated — sharing a matching autosomal STR DNA profile at 13 or more loci is highly unlikely. (For the statistical likelihood, see the discussion of random match probability in Section 7 of this chapter.)

That said, what about the single-source sample that does not include 13 loci? Each DNA profile should be examined to determine (a) whether there is a match and (b) whether that match is with a complete or partial profile.

Note: The fewer loci that yield results, the greater the percentage of the population that can be included as possible contributors.

Partial DNA Profile

At times, a *partial (incomplete) DNA profile* will be generated — that is, there will be no results, partial results (e.g., only one of the two alleles present at the locus has been labeled), or inconclusive data at one or more of the loci tested.

It is not uncommon for a partial (incomplete) profile to be generated. This can happen for a number of reasons, including the following:

- The sample size is very small.
- An insufficient amount of sample DNA was used in the polymerase chain reaction (PCR).
- The original DNA sample has started to degrade or break down, reducing the number of intact DNA molecules.
- The sample contains a PCR inhibitor, such as carpet glue or denim dye.

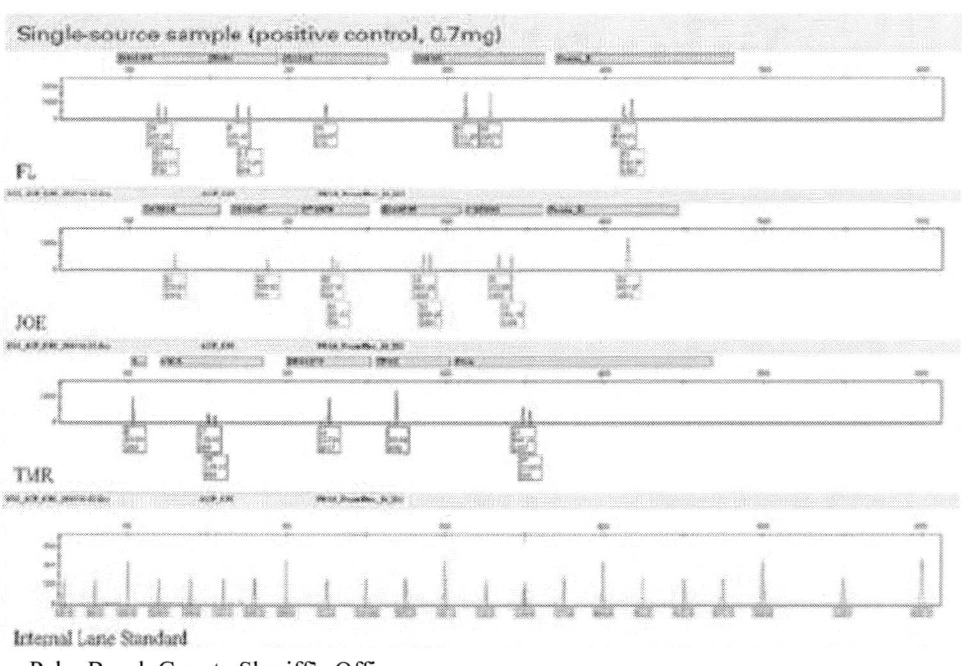

Source: Palm Beach County Sheriff's Office.

Figure 10. A 16-Loci Single-Source Sample.

When looking at an electropherogram, the smaller DNA fragments are located toward the left side of the image, and the fragment sizes get larger as you proceed from left to right.

Thus, referring to Figure 10, you'll note that the loci listed on the left — D3S1358 (in blue), D5S818 (in green) and amelogenin (in black) — are shorter in length than the loci on the right — Penta E (in blue), Penta D (in green) and FGA (in black). This principle holds true for all electropherograms, regardless of which testing kit is used.

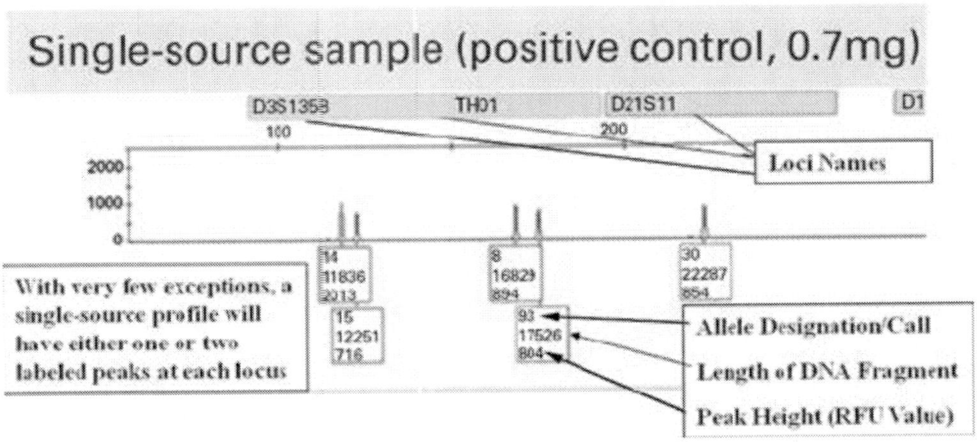

Source: Palm Beach County Sheriff's Office

Figure 11. Enlargement of Results at Three Loci, Single-Source Sample.

The presence of a partial DNA profile is clear when there are no labeled peaks under one of the electropherogram loci labels. Because longer pieces of DNA are more likely to be fragmented, or broken, during degradation of DNA and because the PCR process can be less efficient with longer pieces of DNA, you will be more likely to see a partial DNA profile with the loci results "missing" on the right side of the electropherogram. The presence of a partial DNA profile is less obvious when only one of the alleles at a locus is labeled.

The process of degradation occurs naturally over time, particularly when DNA is subjected to environmental factors such as sunlight, heat, water and/or bacteria. The DNA molecule begins to break down — not all at once, but gradually. Accordingly, it is not uncommon to see partial DNA profiles when cold case samples are examined. DNA degradation can also be seen in "new" cases where a sample may have been subjected to some sort of environmental insult before the sample is recovered/collected (e.g., blood deposited in the closed trunk of a vehicle that is subjected to direct sunlight during August in Louisiana for a period of time before collection). Degradation often signals itself by a characteristic pattern sometimes referred to as the "ski slope" effect. This ski slope pattern can be clearly seen in Figure 12, where the peak heights for the loci containing longer fragments of DNA get progressively smaller, or lower.

Unlike degraded DNA — in which the larger loci alleles tend to be lost first — in cases of inhibition that result in a partial DNA profile, random alleles may be lost.

When the lab report indicates that a partial DNA profile has been obtained, the DNA results should be examined by your expert to determine the following:

- The correct alleles have been identified and reported for the sample.
- There is actually a match.

- Whether the appropriate statistical formula was used in interpretation of the match.
- Whether the interpretation of the match follows the laboratory's guidelines.

Mixtures

Figure 13 shows a 1:1 mixture and Figure 14 shows a 6:1 mixture. As can be seen in both figures, it is clear that a mixture DNA profile has been obtained (a) when more than two alleles/ peaks are observed at multiple loci and/or (b) when there are only two alleles/peaks and there is a significant difference in the height of those peaks at multiple loci. The number of called alleles at the loci can be used to determine the most probable number of contributors to the mixture profile result.

Often, a main concern with a *mixture* result is whether or not the profile can be "resolved" to determine the DNA profile of one or more of the contributors. For example, in a mixture where 6 times more DNA is present from person #1 than from person #2, it may be possible to determine the alleles that would have been contributed to the mixture from this "major contributor," or person #1.

Conversely, in the same 6:1 mixture, it is typically not possible to discern with certainty all of the alleles that would have been contributed by the "minor contributor," or person #2.

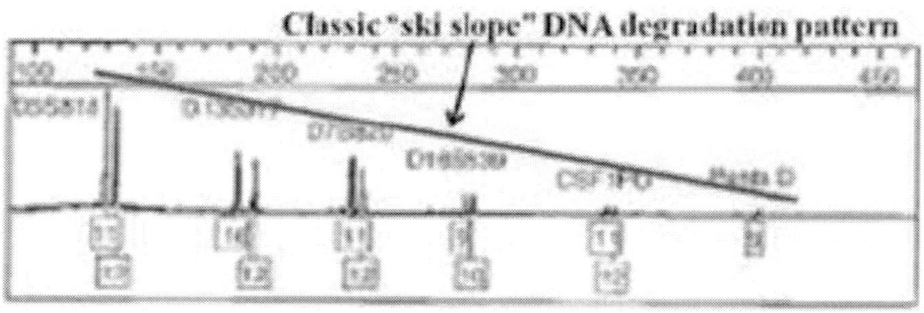

Source: Butler, Forensic DNA Typing, 2nd ed.

Figure 12. Portion of Electropherogram Depicting Degradation.

Labs use varying approaches in interpreting mixed samples. Historically, there has been a lack of consensus in the forensic science community as to which is the best approach because multiple approaches to mixture interpretation are possible and appropriate.

Laboratories also vary in the manner in which they conclude that a mixture is "resolvable." A *resolvable DNA mixture* is often identified as a mixture of DNA from two people in which at least one person can be definitively identified as a contributor. To complicate matters further, in samples where one of the contributors is known, many labs will use their knowledge of this known contributor's alleles to resolve, or *deconvolute,* the mixture. With the release of the *SWGDAM (Scientific Working Group on DNA Analysis Methods) Interpretation Guidelines for Autosomal STR Typing by Forensic DNA Testing Laboratories* (henceforth, *SWDAM Guidelines*), approved on January 14, 2010 (online at http://www.fbi.gov/about-us/ lab/codis/swgdam.pdf), some guidance has been provided regarding the way laboratories should establish their guidelines for mixture interpretation. Labs are now strongly encouraged to have the guidelines in place (listed in Table 1).

The intent is for each lab to clearly define how it is interpreting mixtures. However, it is anticipated that variations will continue to exist between labs regarding how mixtures are interpreted.

Note: If a mixture result implicates a client, it is strongly advised to consult with an expert to aid in interpretation.

When a mixture contains the DNA of three or more people — especially when all of the contributors are unknown and there is no clear major contributor to the mixture — teasing it apart into individual contributors is extremely challenging and may be impossible.

A *degraded DNA sample* or varying concentrations of DNA can further complicate interpretation of the mixture.

Note: If a complex mixture result implicates a client, it is strongly advised to consult with an expert to aid in interpretation.

Contamination

To a certain extent, many items that are collected in connection with investigations contain some amount of pre-existing (exogenous) DNA. By definition, this pre-existing DNA is a contaminant.

Accordingly, contamination can be said to exist when a sample of interest is deposited on an item that already contains DNA. In addition, contamination can occur when a sample comes into contact with an object that contains DNA before its collection. Contamination can also occur during the collection, examination or actual DNA analysis of a sample — this is typically the contamination of concern because it can be avoided.

Table 1. SWGDAM Guidelines for Mixture Interpretation (January 14, 2010)

Guideline	Summary of Guideline Intent
3.5.1	Establishment of guidelines based on peak height ratio (PHR) assessments to determine major and minor contributors.
3.5.2	Defining and documenting assumptions made in mixture deconvolution.
3.5.3	Defining other quantitative characteristics, such as mixture ratios, to assist in determining contributor profiles.
3.5.4	Establishment of guidelines for mixtures with a single major contributor and one or more minor contributors.
3.5.5	Establishment of guidelines for mixtures with multiple major contributors and one or more minor contributors.
3.5.6	Establishment of guidelines for mixtures with indistinguishable contributors.
3.5.7	Establishment of guidelines for determining whether separation of a known contributor's profile is applicable.
3.5.8	Establishment of guidelines for interpretation of potential stutter peaks in a mixed sample.

Source: Scientific Working Group on DNA Analysis Methods [SWGDAM] Interpretation Guidelines for Autosomal STR Typing by Forensic DNA Testing Laboratories, issued January 14, 2010.

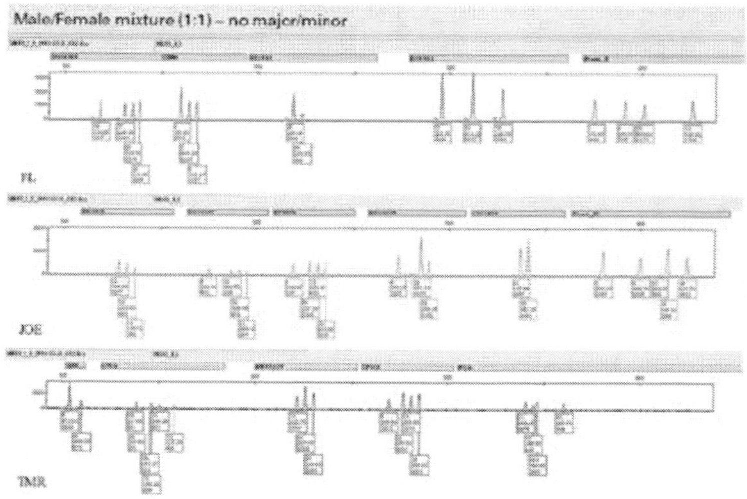

Source: Palm Beach County Sheriff's Office.

Figure 13. A 1:1 Mixture.

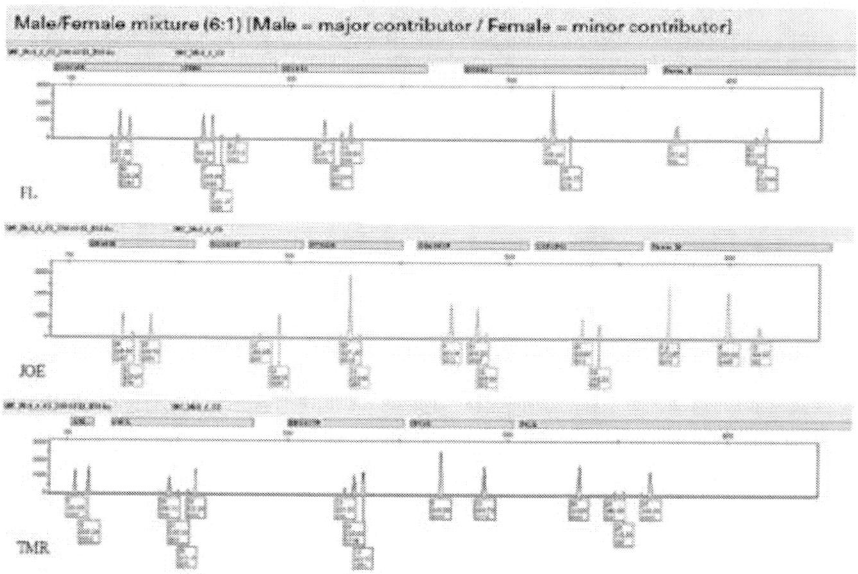

Source: Palm Beach County Sheriff's Office.

Figure 14. A 6:1 Mixture.

For example:

- If a bloody jacket comes into contact with a hair brush during evidence collection at the crime scene (or later) and blood gets on the hair brush, contamination occurs.
- If a scientist sneezes directly on a sample or open sample tube during testing, the sample may be contaminated.
- If a person collecting multiple items of evidence at the crime scene does not change gloves between items, the samples can become cross-contaminated.

- If the suspect's known sample is opened and processed before a crime scene sample on the same laboratory bench, there is a risk that the suspect's sample will contaminate the crime scene sample.

Some contamination sources are obvious, whereas others may never be verifiable as having occurred.

An expert can assist counsel in determining whether contamination could have taken place.

Interpreting the Data

As previously discussed, the *Quality Assurance Standards (QAS) for Forensic DNA Testing Laboratories* specifically require labs to have general guidelines for the interpretation of data, as do the *SWGDAM Guidelines* (January 14, 2010).

The laboratory's interpretation guidelines are based on its internal validation data and its experience with specific kits and instruments.

Based on the lab's protocol, the lab determines what alleles are callable for each sample; the next step is for the analyst to directly compare the evidence profile with the reference, or known, sample profile(s). Guideline 3.6.1 of the *SWGDAM Guidelines* states, "The laboratory *must* establish guidelines to ensure that, to the extent possible, DNA typing results from evidentiary samples are interpreted before comparison with any known samples, other than those of assumed contributors" [emphasis added].

The comparison of profiles will result in one of the following possible conclusions:

- The profiles match, and the known individual cannot be excluded or is included.
- The profiles do not match, and the known individual is excluded.
- The results are inconclusive or uninterpretable.
- The results from multiple evidentiary items are consistent or inconsistent with originating from a common source.

An inconclusive result is not the same as an exclusion. Exclusion means that the profile could not have originated from a source.

An inconclusive result means that the forensic data does not support an inclusion or an exclusion. Defense attorneys are especially encouraged to examine inconclusive test results to determine if there are alternate explanations.

If a person cannot be excluded on the basis of this direct comparison of evidence with known profiles, the next step is to perform a statistical analysis in support of any inclusion determined to be relevant in the context of the case. Specifically, Guideline 4.1 of the *SWGDAM Guidelines* states, "The laboratory *must* perform statistical analysis in support of any inclusion that is determined to be relevant in the context of the case, irrespective of the number of alleles detected and the quantitative value of the statistical analysis" [emphasis added].

The Source — A Single-Source Sample or Major Contributor

What does *source attribution* also referred to as an *identity statement,* really mean? Some labs feel comfortable reporting that a particular individual was "the source" of a DNA profile recovered from the crime scene. This conclusion is very different from stating that the defendant cannot be excluded as a possible source of the DNA profile obtained from the item of evidence. With source attribution, the lab is stating that this evidence DNA profile originated from this particular individual.

Source attribution is based on use of a mathematical equation called the uniqueness formula. Use of the uniqueness formula, to determine if an identity statement can be made, was recommended by the DNA Advisory Board and in the article, "Source Attribution of a Forensic DNA Profile," published in *Forensic Science Communications* (available online at http://www.fbi.gov/ about-us/lab/forensic-science-communications/ fsc/july2000/source.htm).

The uniqueness formula requires that both the size of the population being considered (typically, either the U.S. population or the world population) and the confidence interval be specified. The confidence interval essentially provides a range of values around a measurement that conveys how precise the measurement is. Typically, labs use confidence intervals of either 95% or 99%. To put this into perspective:

- When a confidence interval of 99% is used to calculate the uniqueness of a profile in the U.S. population of approximately 300 million, the value obtained is roughly 30 billion.
- Following NRC II guidelines of the *true match probability* being plus or minus 10-fold, the obtained value is multiplied by 10 as a conservative estimate, which results in a calculated value of approximately 300 billion.
- Based on the above uniqueness formula calculation, the generated *random match probability (RMP)* calculation (see below) for the evidence profile is compared with the uniqueness formula value. If the evidence profile frequency is greater than 1 in 300 billion — or less than 2.9×10^{-11} — the lab will make the source attribution statement in their report.
- The source attribution can be stated as follows: "We are 99% confident that, in a population of 300 million unrelated individuals, the STR DNA profile observed would occur only once (i.e., it is unique)."

This statement is based on the knowledge that the profile did occur once. Note that, without knowing whether it has actually occurred or been observed, any particular 13-locus STR DNA profile is unlikely to exist in a population of 300 million.

In a lab report, if the lab is using source attribution, the conclusion statement will be similar to the following:

[Suspect] (or his identical sibling) is the source of the DNA profile obtained from [item of evidence] to a reasonable degree of scientific certainty.

Because STR DNA technology cannot distinguish between identical twins, they would be expected to have the same DNA profile. Even if the other twin has not been accounted for and may be considered a suspect in the investigation or the true perpetrator, note that the calculated RMP values are still valid in relation to the probability of randomly selecting an unrelated individual in the population with the same DNA profile that was obtained from the evidence sample.

Should you have a case in which the suspects are related, be aware that the calculated RMP values will typically underestimate the expected frequency of the profile in related individuals. If your client's profile has been matched to the evidence profile, and if your theory of the case is that a relative of your client is the true source of the evidence profile, all efforts should be made to obtain a sample directly from this relative so that the generated profile can be directly compared with the evidence profile. This precludes the need to rely on a probability-based estimate of a coincidental match. If the relative's sample is not obtained, relatedness calculations should be requested if they are relevant and have not been conducted yet. *SWGDAM* Guideline 5.2.3 addresses which calculations for relatedness should be used.

Combined Probability of Inclusion or Exclusion for Mixture Profiles

Combined probability of inclusion (CPI) and *combined probability of exclusion (CPE)* calculations are commonly used by labs to indicate the statistical significance of mixture results. CPI is the percentage of the population that can be included in a mixture profile; CPE is the percentage of the population that can be excluded from a mixture profile. The CPI and CPE calculations are closely related: CPI is calculated by multiplying the probabilities of inclusion from each locus, and CPE is calculated by subtracting the value obtained from the CPI calculation from 1 (i.e., 1 − CPI). *Likelihood ratio (LR)* calculations are also commonly used (see the next section). The *SWGDAM Guidelines* do not state a preference for using one statistical method over another. However, labs are *required* to establish guidelines for selecting statistical formulas to be used when multiple formulas are applicable (*SWGDAM* Guideline 4.6.1) — in other words, it must be clear which statistical calculations are going to be used, and when.

In mixture calculations, the concepts of "restricted" and "unrestricted" come into play. In a *restricted calculation,* the relative peak heights at each locus are taken into account when pairing the alleles for the calculation. In an *unrestricted calculation,* all of the possible combinations of the alleles are deemed possible and are therefore used in the calculation. Figure 15, taken from the current version of the *SWGDAM Guidelines,* illustrates this point. In the example depicted in Figure 15, it is assumed that two donors and all peaks are above the *stochastic threshold.*

In a mixture profile with a distinguishable major contributor profile(s), the major contributor(s) may be suitable for statistical analysis, even in the presence of inconclusive minor contributor results. In general, with CPI/CPE calculations (where there are no assumptions regarding the number of contributors to the mixture), loci with alleles below the stochastic threshold may not be used for statistical purposes to support an inclusion. Because of the potential for allelic drop-out, there is a possibility of contributors possessing alleles not represented among the interpreted alleles, which is why those loci are not used in the calculation. An exception to this is the accepted application of a restricted CPI/ CPE to a profile with multiple major contributors, despite the presence of minor contributor(s) alleles below the stochastic threshold. *SWGDAM* Guideline 5.3.5 describes how this calculation would be conducted.

In a report, a CPI calculation looks something like this:

> The probability of a randomly selected, unrelated individual having contributed DNA to the mixture profile obtained from [evidence item] is approximately:
>
> 1 in 1.10 million for the U.S. Caucasian population.
> 1 in 456,000 for the African-American population.

1 in 525,000 for the southwestern Hispanic population.
2.92 million for the general Asian population. ~
These calculations were conducted using the combined loci in the Profiler Plus® and COfiler® DNA typing kits.

In a report, a CPE calculation looks something like this:

- The probability of randomly selecting an individual from the African American population that can be excluded as a contributor to this mixture is greater than 99.999%.
- The probability of randomly selecting an individual from the Caucasian-American population that can be excluded as a contributor to this mixture is greater than 99.998%.
- The probability of randomly selecting an individual from the southwestern U.S. Hispanic population that can be excluded as a contributor to this mixture is greater than 99.999%.
- This calculation was based on databases provided by the Federal Bureau of Investigation and was conducted using the 15 STR loci in the Identifiler® DNA typing kit.

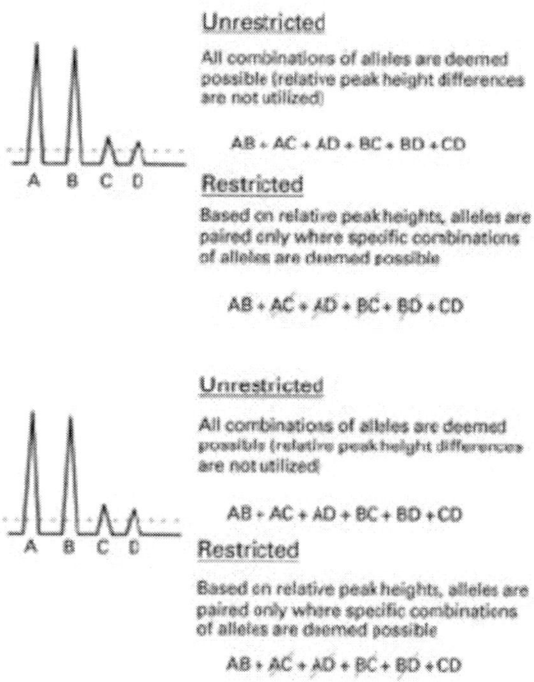

Source: Reproduced from SWGDAM Guidelines (January 14, 2010).

Figure 15. Restricted vs. Unrestricted Calculations.

The most commonly used formulas for CPE/ CPI are listed in Guideline 5.3 in the *SWGDAM Guidelines* (January 14, 2010). Examples of the use of CPE/CPI calculations for mixture profiles are also provided in Guideline 5.3.5.

Likelihood Ratios for Mixture Profiles

Although LR calculations are typically associated with mixtures, they can also be conducted on single-source evidence profiles. An LR is the ratio of two probabilities of the same event under different hypotheses. In forensic DNA testing, the numerator typically contains the prosecutor's hypothesis and the denominator contains the defense's hypothesis. This means that the obtained ratio indicates how much more likely the prosecution's theory of the case is (the defendant's DNA is contained within the mixture profile) compared with the defense's theory of the case (the defendant's DNA is not contained within the mixture profile). Note that the LR does not take relatedness into account — it is used to mathematically compare and contrast two possible theories for the evidence profile.

As with the previous types of calculations, LR calculations can also be restricted when relative peak heights are taken into consideration, or unrestricted when the LR is calculated without taking peak heights into consideration.

An LR calculation for mixtures is dependent on three things: the evidence profile, the reference profile(s) that have been compared, and the individual hypotheses. The lab has to "guess" the defense hypothesis. This usually means setting up the mathematical equation using two case theories: Either the defendant's DNA is in the mix or it is not in the mix. Because there are many testable hypotheses, most labs will select the most commonly encountered hypothesis to set up their LR calculations. This does not necessarily mean that they are opposed to or have dismissed other potential hypotheses.

When stated in a report's conclusion, the likelihood ratio looks something like this:

The DNA mixture profile obtained from [the item of evidence] is:

4.73 quadrillion times more likely to have originated from [suspect] and [victim/complainant] than from an unknown individual in the U.S. Caucasian population and [victim/complainant].

16.5 quintillion times more likely to have originated from [suspect] and [victim/complainant] than from an unknown individual in the African American population and [victim/complainant].

4.66 quadrillion times more likely to have originated from [suspect] and [victim/complainant] than from an unknown individual in the southwestern Hispanic population and [victim/complainant].

12.0 quadrillion times more likely to have originated from [suspect] and [victim/complainant] than from an unknown individual in the southeastern U.S. Hispanic population and [victim/complainant].

These statistics were generated using the STR loci in the Identifiler® System. The statistics assume unrelated individuals.

The significance of the LR calculations may generally be interpreted as follows:

- If the LR value is 1 to 10 times more likely, there is limited support for the prosecution hypothesis.
- If the LR value is 10 to 100 times more likely, there is moderate support for the prosecution hypothesis.

- If the LR value is 100 to 1,000 times more likely, there is strong support for the prosecution hypothesis.
- If the LR value is 1,000 or more times more likely, there is very strong support for the prosecution hypothesis.

Commonly used LR formulas are listed in SWGDAM Guideline 5.4.2 (January 14, 2010). Examples of the use of restricted and unrestricted LR calculations for mixture profiles are also provided in SWGDAM Guideline 5.4.2.

Expectation Bias

Expectation bias — having a strong belief or mindset toward a particular outcome — is a phenomenon that has been studied and published in the literature. For example, knowing the allele types of a potential contributor before analyzing the evidence may influence how the analyst interprets the evidence sample. All scientists should agree that expectation bias exists, but they will have differing opinions as to its relevance in forensic analysis. Expectation bias may be particularly relevant in mixture cases.[1]

Section 3. Technical Artifacts and Interpretation of Results

Artifacts are peaks or other abnormalities on the electropherogram that are not attributable to the DNA actually present on the evidence item. Artifacts can result from a non-allelic product that is generated during the amplification process, they can be associated with anomalies in the detection process, or they can be a by-product of primer synthesis. Technical artifacts have been documented and are routinely observed; laboratories are required to use protocols to distinguish between artifacts and real DNA peaks. An independent expert may disagree with a lab's conclusion that a peak is an artifact; instead, the expert may conclude that the peak accurately represents relevant DNA evidence or, conversely, that a peak that is called an allele by the lab is not a true allele.

The most common artifacts are stutter, spikes, nontemplated nucleotide addition, dye blobs, shoulders/split peaks, drop-out, drop-in, pull-up, and raised baseline/noise (or background).

Stutter

Stutter is a minor peak that is typically observed that is one repeat unit smaller than a primary STR allele and is believed to result from strand slippage during the DNA amplification process. Stutter is expected when STR *PCR* technology is used.

Here is how stutter occurs, according to the slipped-strand mispairing model: Recall that when the double-stranded DNA is heated up, the strands are separated into two halves to allow new DNA to be synthesized, after the temperature is changed to allow the DNA primers to adhere, or pair, to their corresponding areas on the DNA molecule, and after the temperature is modified again to favor extension of the DNA template strands. During the synthesis process, most of the bases re-pair as expected (As bond with Ts, and Gs bond with Cs), creating two paired strands of DNA where there was once one strand. Stutter occurs when, after the strand is heated and separated and the primers have been attached to the strands of DNA that are being copied (i.e., the template strands), one of the strands "breathes" or "bulges out" during the DNA extension process so that it is not lying down in a straight line, as is usually observed. This unpairing during the DNA extension process allows slippage of either the original template strand (*forward slippage*) or the strand that is being extended from the primer (*reverse* or *backward slippage*). The end result is that a shortened PCR product is created that is one less repeat unit shorter in length than the primary (real) STR allele.

Take, for example, a strand of DNA that has eight four-base-pair repeats. If the template strand bulges during the time the bases are being added to copy the DNA strand, the bases will continue to bond down the line, skipping the bulge. This results in a strand that has eight repeats on one side of the strand (with one four-base-pair repeat bulging away from the straight line) and, on the other side, only seven repeats. When the strand is heated again and breaks apart, there are eight repeats on one half and only seven repeats on the other half.

This seven four-base-pair repeat is now in the reaction tube and behaves just as the rest of the strands during new DNA synthesis. Thus, the seven-repeat copy (which was created because of strand slippage) continues to be copied along with the strands that contain eight repeats (the "true" original number of copies of the repeat unit of the template DNA put into the analysis tube). The seven-repeat copy will appear on the electropherogram as a minor peak.

Whereas Figure 16 demonstrates what normally occurs during the polymerase chain reaction *(PCR)* process, Figure 17 depicts how stutter occurs according to the slipped-strand mispairing model.

Why might this become a problem? In a casework mixed sample, the number and identity of the contributors can never truly be known — there is almost always the possibility of a small amount of pre-existing DNA being present before deposition of the biological sample of interest. Given this, a low-abundance, seven-repeat fourbase-pair product could be due to stutter created during the PCR testing process or to low levels of a seven-repeat fragment of DNA from a second contributor. There is often no way to know for sure, particularly when a complex DNA mixture is encountered that contains DNA from three or more people. The laboratory has guidelines — based on its established stutter percentage expectations (sometimes referred to as *stutter cut-off values*), its evaluation of *peak height ratios (PHR),* and its set *analytical threshold* based on signal-to-noise considerations — to help determine whether a peak should be declared a true allele and whether it is indistinguishable from stutter. The quantitative threshold for declaring a putative DNA peak as potential stutter is based on how single-source DNA samples behave in the lab during validation studies. The stutter percentage expectations for a laboratory are generated by quantifying the percentage of stutter product peaks during the review of single-source samples during validation. Using this method of evaluation, the percentage of stutter product formation seen for each allele at a locus is generated by dividing the observed stutter peak height by the corresponding allele peak height over multiple runs. Understand, however, that just because an allele meets the mathematical criteria for being "stutter," it is not necessarily stutter; it could be the original DNA from the deposited biological material. For example, by taking a look at the dilution studies done by the lab during its mixture validation studies (where two known samples of differing amounts are combined and then subjected to PCR), it is possible to see the presence of known donor alleles that now mathematically meet the stutter criteria as well as stutter alleles that mathematically meet the criteria for true alleles.

The latter is seen particularly when the minor peak appears between two real alleles (i.e., a minor peak of 11 repeats that is present when major, true peaks are present at 10 and 12, and either the donors are of relatively equal intensity or the donor of the 12 allele is the major contributor to the mixture). Although this is not the primary goal of the mixture dilution study, it illustrates how small amounts of DNA can sometimes be indistinguishable from stutter.

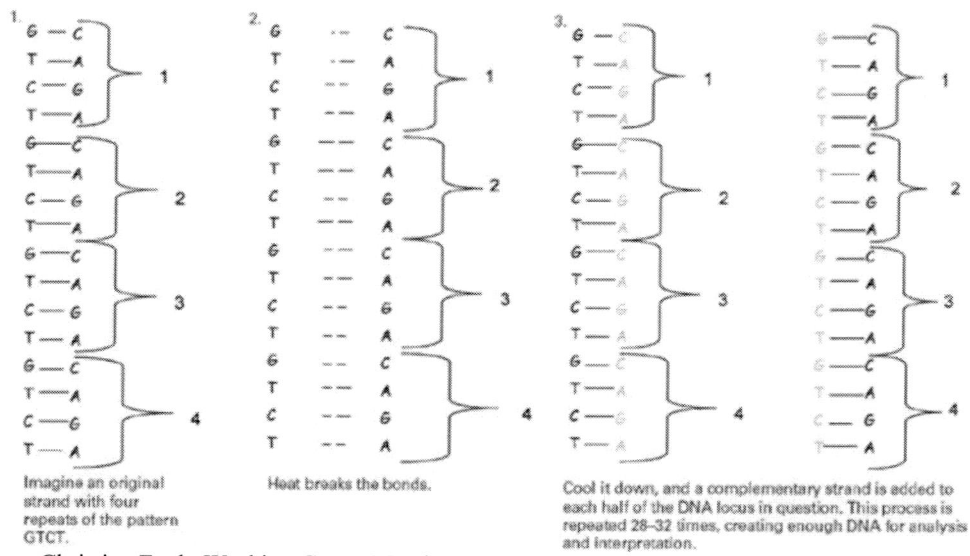

Source: Christine Funk, Working Group Member. LI
PCR = polymerase chain reaction. LI

Figure 16. Illustration of Normal Copying of Template Strand During PCR.

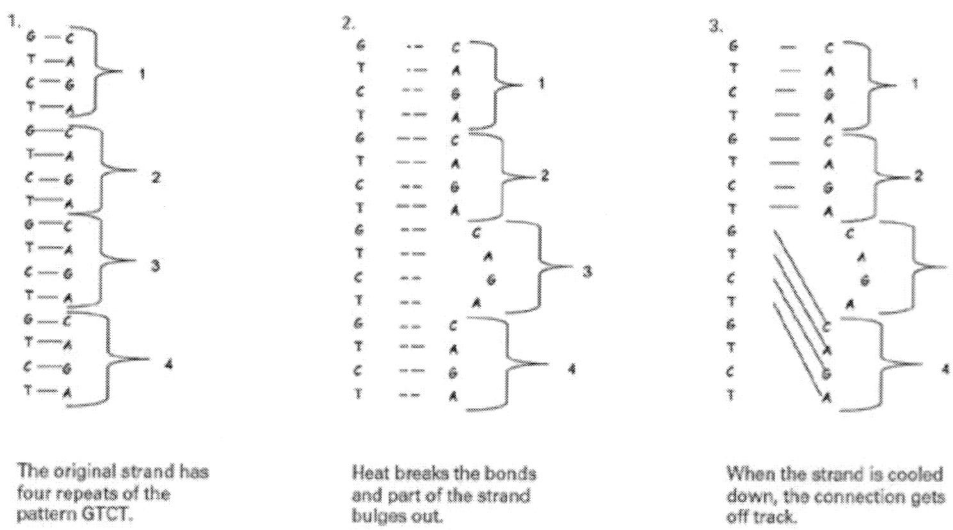

Figure 17. (Continued).

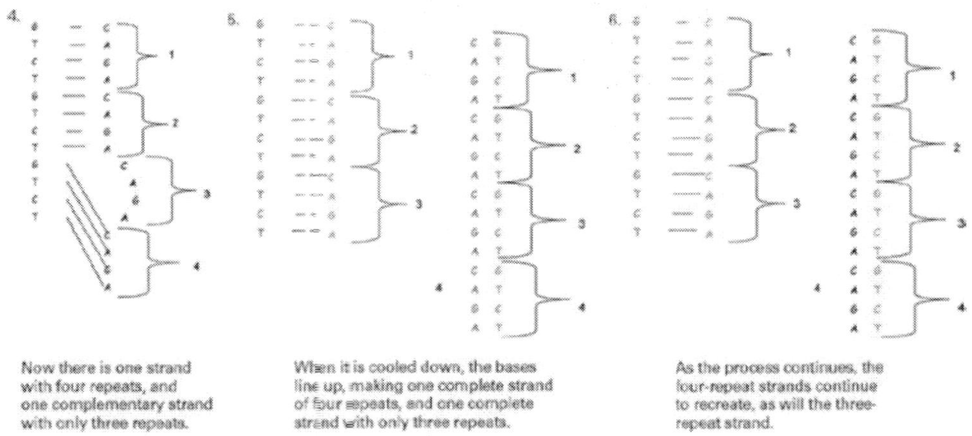

Source: Christine Funk, Working Group Member.

Figure 17. How Stutter Occurs.

Random Match/Man Probability (RMP) for a Single-Source Sample or Major Contributor

Many labs do not use the source attribution terminology. Instead, they report the result as an inclusion or a nonexclusion along with an RMP or other appropriate frequency estimate.

RMP is the probability of randomly selecting an unrelated person from the population who could be a potential contributor to the evidence profile.

Another way to think about the RMP is that the calculated number is the theoretical "chance" that, if you sample one person at random from the population, they will have the same DNA profile as the one obtained from the evidence sample.

It is important to be aware of what an RMP does not mean. In Forensic DNA Typing (2nd edition), Dr. John Butler provides some clear examples of what a RMP is not:

- RMP is not the chance that someone else is guilty.
- RMP is not the chance that someone else left the biological material at the crime scene.
- RMP is not the chance of the defendant not being guilty.
- RMP is not the chance that someone else, in reality, would have that same DNA profile.

RMP calculations must be conducted using DNA results obtained from evidence items — not from known sample profiles.

In addition, the lab should never use inconclusive or uninterpretable data in the RMP statistical analysis or any other statistical analysis. The lab should not calculate a "composite" statistic by attempting to multiply RMP values obtained at some loci with another type of statistic (LR or CPE/CPI) calculated at other loci. The lab can, however, calculate RMP for the major contributor to a mixture profile and can also conduct another type of calculation (LR or CPE/CPI) on the entire mixture profile.

RMP values are typically associated with single-source DNA profiles, but they can be calculated for mixture samples as well.

When applied to mixtures, this calculation is referred to as a *modified RMP,* which includes an assumption of the number of contributors to the mixture.

RMP values for single-source samples and for a single major contributor to a mixture are calculated using the formulas described in NRC II recommendations 4.1, 4.2, 4.3 and 4.4. Because the laboratory must document the population database and the statistical formulas used, that information should be easy to find in the laboratory's manual or in the case file. The most commonly used formulas are listed in *SWGDAM* Guideline 5.2. Examples of the use of RMP calculations for mixture profiles are provided in *SWGDAM* Guideline 5.2.2 as well.

Use of an RMP statistic in a report will be similar to the following:

The approximate frequency of the DNA profile obtained from [item of evidence] is:

1 in 3.3 sextillion in the Caucasian-American population.
1 in 75 sextillion in the African-American population.
1 in 38 quintillion in the Hispanic-American population.
1 in 22 septillion in the Asian-American population.

These statistics were generated using the 15 STR loci in the PowerPlex® 16 System. These statistics assume unrelated individuals.

The *SWGDAM Guidelines* (January 14, 2010) provide guidance for labs on the interpretation of potential stutter peaks in a mixed sample (Guideline 3.5.8). Specifically, Guideline 3.5.8 states that labs are expected to determine whether minor peaks in the stutter position are an actual stutter peak, an allelic peak, or indistinguishable as an allelic or stutter peak. Although this determination is based primarily on the height of the peak in the stutter position, and its relationship to the stutter percentage expectations established by the laboratory, the *SWGDAM Guidelines* clarify what the general expectations are and acknowledge that there will be some exceptions. The key is that the laboratory should declare what the minor peaks in the stutter position are — a stutter peak, an allelic peak, or indistinguishable — before any comparisons are made with any known samples, other than those of the assumed contributors.

Note: Stutter is another reason why a defense expert should evaluate the electronic data and electropherograms, particularly in mixture cases.

Spikes

Spikes are straight and narrow peaks, typically seen in relatively equal intensity, in all color channels on an electropherogram (see Figure 18). Spikes are generally due to alternating-current voltage fluctuations but can also be observed because of air bubbles or crystals of urea that cross the detector. Spikes are not reproducible from one run to another. Accordingly, when a spike is suspected, the sample is typically rerun because a spike does not normally appear in the same position twice.

It is important to be aware of spikes so that you understand why an analyst may disregard what appears to be a peak and why a sample may be rerun.

Dye Blobs

Dye blobs are artifacts made up of excess dye (see Figure 19). They can occur in any sample when dye, unattached to any DNA, comes through the *capillary* and passes over the laser light, which records the blob as an allele.

Dye blobs have physical characteristics — tending to be broad and often irregular in shape, with low RFU values, and corresponding to the spectrum of one of the dyes contained within the DNA typing kit. Dye blobs tend to appear in the same general location on the electropherogram, which makes them readily identifiable. However, when dye blobs have not been or cannot be removed, they may obscure true peaks or other relevant data, which can affect interpretation.

Drop-out

With low quantities of DNA or with degraded DNA, *allelic drop-out* can occur (see Figure 20). Drop-out occurs when alleles from a DNA profile "drop out" of the electropherogram because of a small quantity of an allele going undetected or an allele failing to amplify during PCR. This may involve one or both alleles at a particular locus. Typically, drop-out occurs at the larger loci first (the ones on the right-hand side of the electropherogram).

Note: Although drop-out is a documented phenomenon, it does not normally occur in robust samples and should not be used as a defense in these situations. An assertion of drop-out should be supported by objective criteria. In these situations, the defense is strongly encouraged to hire an expert to review the data.

Where drop-out is a legitimate possibility, the lawyer should be aware that it is theoretically possible that any allele could have dropped out at the locus. Other combinations of alleles not present in the reference sample could also be a possibility. For example, if the defendant is a 6,7 at THO1, the evidence profile shows only a 6 allele, and the scientist believes there was allelic drop-out, the scientist may still consider an inclusion based on the rest of the profile.

Typically, the associated statistical calculation will reflect the value (or lack thereof) of this determination by the examiner. However, the lawyer should understand that other alleles could have dropped out and that other combinations of alleles (e.g., 6,6, 6,7, 6,8, 6,9, 6,9.3 and 6,10) could exist in the sample.

Care should be taken when pursuing the possibility of allele drop-out if the results are consistent with the following: (a) the presence of a single contributor, (b) the locus/loci where allelic dropout occurred have higher molecular weights (are located on the right side of the electropherogram), (c) empirical data suggest that degradation was probable, and (d) the statistical calculation still strongly supports that the sample originated from your client.

Drop-in

Drop-in occurs when alleles not originating from the actual sample appear on the electropherogram. The source of *allelic drop-in* is often undetermined but may be due to low-level contamination in the laboratory that is introduced into the sample, sample container or reagents.

Some labs using robotic systems have ongoing difficulties with allelic drop-in. Drop-in alleles are typically not reproducible on subsequent reanalysis. Accordingly, if the sample size allows, the lab may opt to retype and/or re-amplify the sample.

Shoulders and Split Peaks

A *shoulder* (also called *minus A*) is a common artifact (see Figure 21). After the amplification process and during an incubation period, an additional adenine (A) base is added to the amplified DNA by the Taq polymerase used in the PCR process. This additional A base is expected to be on each amplified piece of DNA and is included in the sizing of each DNA fragment in the allelic ladder.

If too much of the sample DNA is added to the tube or well, there may not be enough time (or adenine bases) to add the extra base to each allele. Thus, a *shoulder peak* will be seen, which will be one base pair smaller/shorter than the actual allele. Shoulder peaks can also be seen if the PCR conditions are not optimized. In instances where the amount of sample injected onto the capillary is so great that it overwhelms the detection system, a split peak may occur (see Figure 22). The resulting peak ends up being fairly broad; it may appear that there are two peaks when there is only one. When samples are overloaded, the genotyping software cannot properly assign an accurate peak height to the off-scale data. These peaks are assigned an artificial height value that does not represent the true intensity of the peak. Therefore, peak height values for off-scale data should not be used in peak height ratio and stutter peak assessments. These split peaks are often called +A/−A artifacts because they may be one base pair larger or smaller than the true allele size used for comparisons. When split peaks are observed, the remedy is to retype the sample (or re-amplify it, in some cases, when too much template has possibly caused the problem).

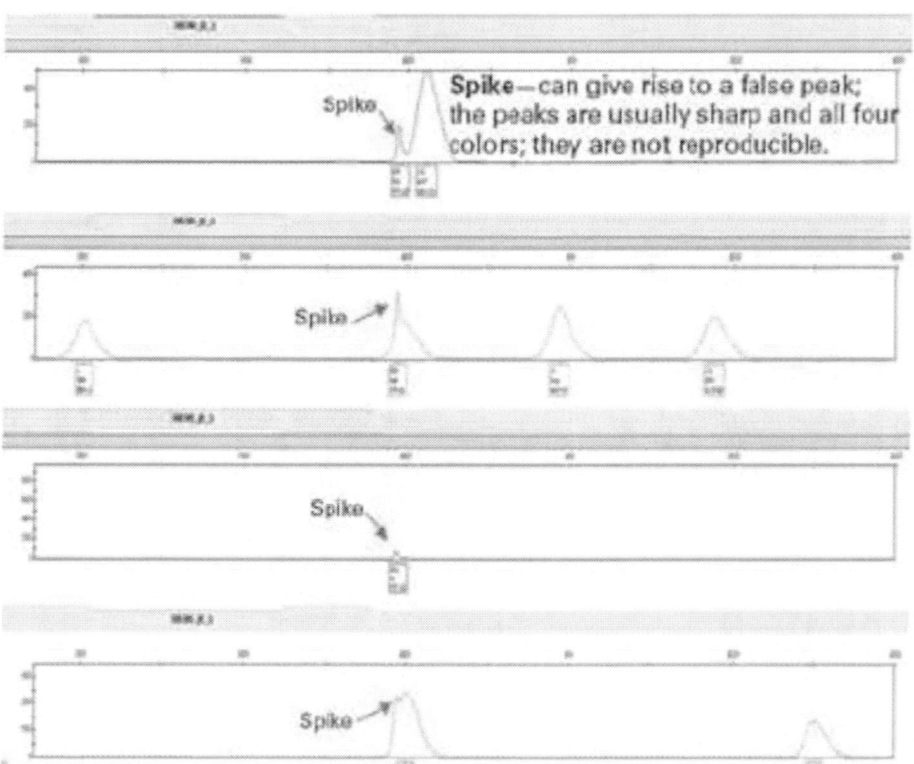

Source: Palm Beach County Sheriff's Office.

Figure 18. Spike Artifacts.

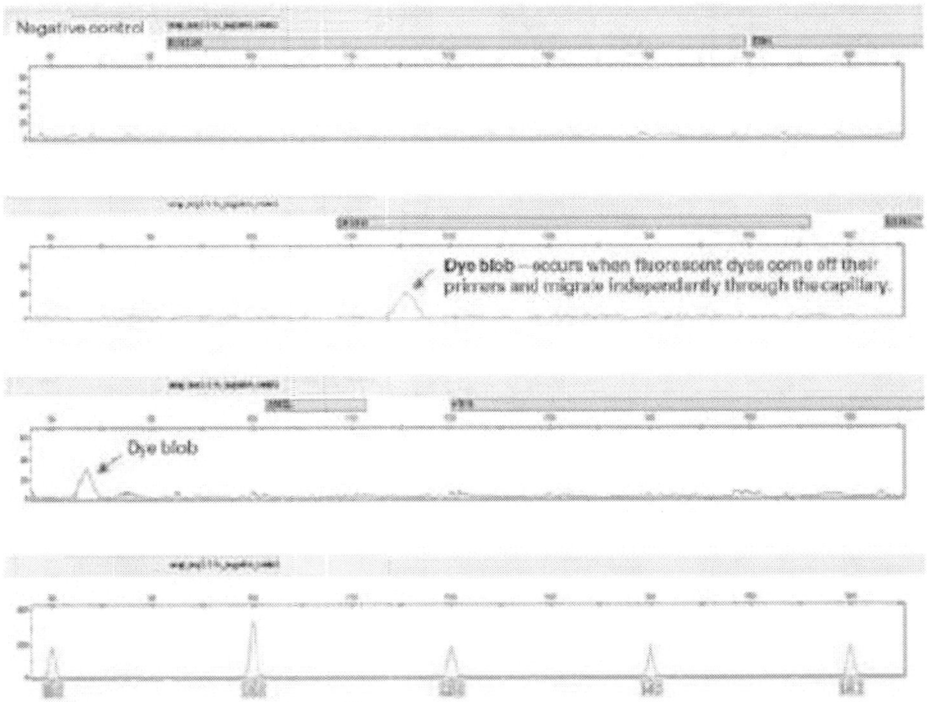

Source: Palm Beach County Sheriff's Office.

Figure 19. Dye Blob Artifacts.

Pull-up

Pull-up, also called *bleed-through* or *incomplete spectral separation,* occurs when the software cannot properly resolve the dye colors used to label the DNA fragments when the electropherogram is created from the raw data (see Figure 23) This is usually due to high concentrations of DNA in the sample; this can complicate data interpretation in complex mixture samples.

When pull-up happens, a peak corresponding to the length of one that actually exists in one color may appear to be recorded as a minor peak in one or more other colors. (When the dye colors are laid on top of each other, one can see the "peak-beneath-a-peak" phenomenon clearly.) These additional peaks are not actual DNA; however, they may look like DNA if they happen to line up where a true allele might be. Conversely, in a mixture with a minor contributor, pull-up may mask a true allele.

Peak Height Imbalance

Peaks are expected to be "balanced" within a locus (the same genetic marker) (see Figure 24) For example, if the DNA came from a single source and that individual is a 14,16 at a particular locus, the peaks should be of relatively equal intensity and the same height, and their measured RFU values should be approximately the same. However, there are certain conditions known to cause *peak height* imbalance, such as degradation, inhibition, low template DNA and mixtures. Although not typically considered to be an artifact, peak height imbalance can cause difficulty in data interpretation and is therefore mentioned here.

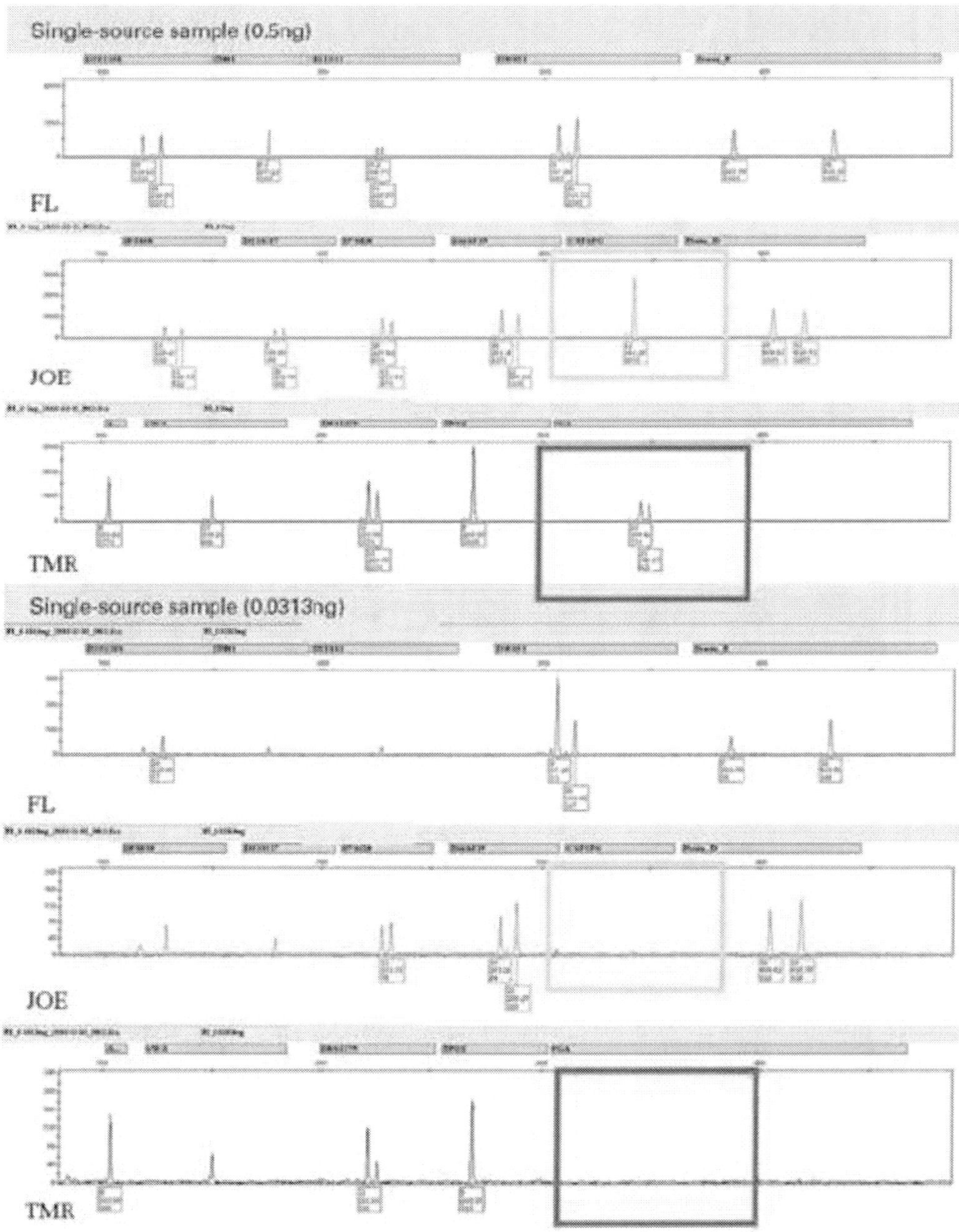

Source: Palm Beach County Sheriff's Office.

Figure 20. Drop-out Artifacts.

In the instance of a mutation that has occurred in or close to the area on a person's DNA where one of the primers binds (called a *primer binding-site mutation),* an actual peak height imbalance artifact can be observed. These individuals will exhibit more of one of their alleles at the affected locus than the other. This is not a problem when the same kit is used to type all of the samples that are compared; indeed, the imbalance between the two alleles at the locus is reproducible.

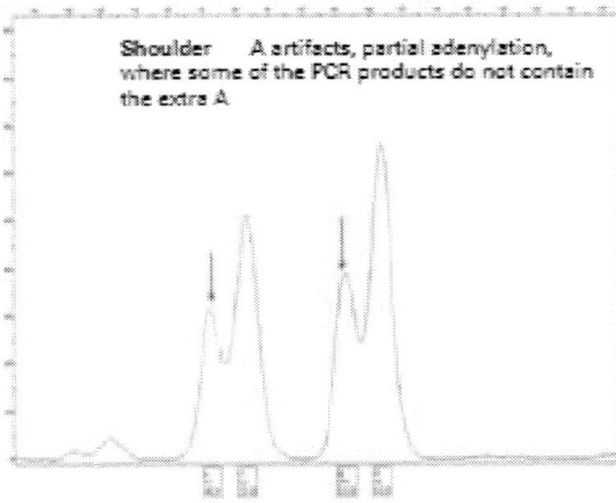

Source: Palm Beach County Sheriff's Office

Figure 21. Shoulder Artifacts.

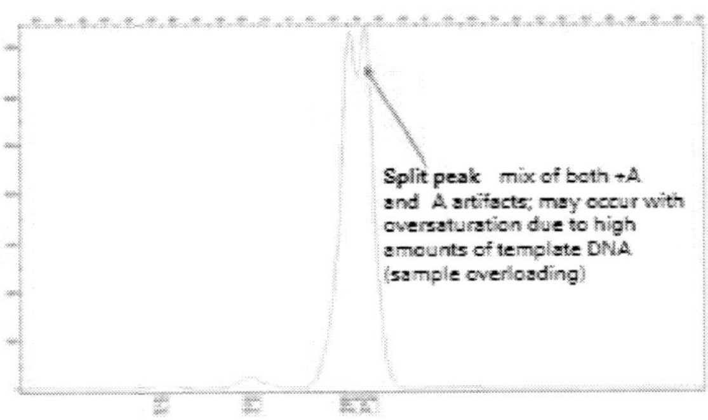

Source: Palm Beach County Sheriff's Office.

Figure 22. A Split Peak Artifact.

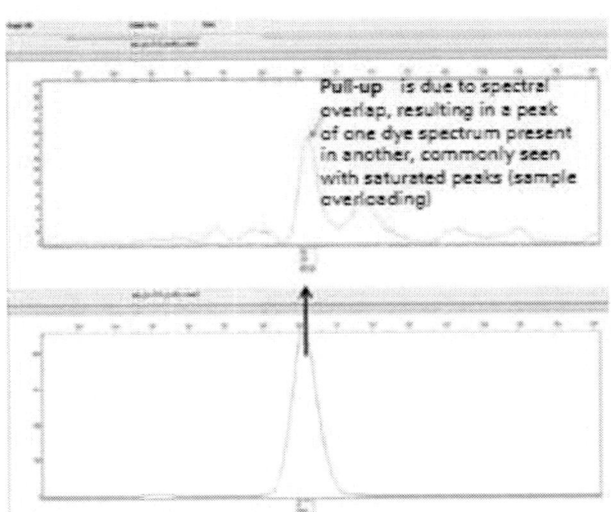

Source: Palm Beach County Sheriff's Office.

Figure 23. Pull-up Artifacts.

In some instances, a primer binding-site mutation may actually result in one of the alleles not being observed when one DNA kit is used versus another (i.e., D8S1179 typing results are 14,15 using one kit and 14,14 using another kit). The loss of an allele in this instance is a function of the primer used by one company corresponding to an area where there is a mutation, and the primer used by the other company falling inside the area where the mutation occurs, thus avoiding it. This lack of concordance between DNA typing kits has been documented a number of times and is the primary reason why the CODIS searching algorithm allows for one mismatch, or *wobble*, position (i.e., when searching using 13 loci, 22 out of 26 allele calls must match, rather than 26 out of 26.

Consult with the defense's expert to determine what significance, if any, peak height imbalance may have in a case.

Elevated Baseline or Background Noise

Noise is a natural by-product of the instrumentation and is always present on the electropherogram. It appears as a horizontal "squiggly" line (baseline) at the bottom of each color and can be seen when one has "zoomed in" on the baseline. Noise can be a problem (a) when the sample size is physically small, (b) the amount of the sample amplified contains less than the optimal amount of template DNA, or (c) the sample contains a mixture of DNA with one or more minor contributors. In these instances, if the noise becomes too high, it can result in a labeled "peak" that is close to the analytical threshold, which can be misinterpreted as an allele. Conversely, an allele may be misinterpreted as noise.

Each lab uses their validation data to set the analytical threshold. In some instances, the set analytical threshold is for all *capillary electrophoresis* instruments the lab owns; in other instances, analytical thresholds can vary for different instruments. The analytical threshold is based on signal-to-noise considerations, which allow the lab to distinguish potential allelic peaks from background noise in most instances. The *SWGDAM Guidelines* make it clear that the lab's analytical threshold cannot be established for purposes of avoiding artifact labeling because that would result in the potential loss of allelic data (Guideline 3.1.1.2).

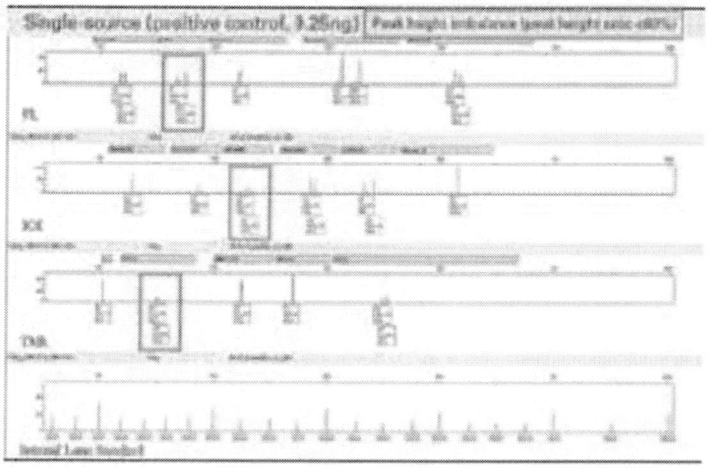

Source: Palm Beach County Sheriff's Office.

Figure 24. Peak Height Imbalance.

Section 4. What the DNA Results Do Not Show

Transfer

Although a defendant's DNA profile may be present in a sample, the testing itself cannot determine how or when it got there. For example, if the defendant's clothing was gathered and stored in the same bag as the victim's bloody clothing, it is possible that the victim's DNA transferred onto the defendant's clothing when it was collected — and not when the crime was committed.

Contamination

It is often difficult to determine whether a profile or part of a profile is due to the presence of contaminating DNA. Evidence may be contaminated:

- Before the crime was committed (commingling of items).
- During the crime.
- During evidence collection. If the contamination is due to the addition of the collector's DNA, this will often be detected, particularly when the lab has a staff database.
- During improper storage of the evidence.
- During laboratory testing.

Consent

DNA may identify the presence of an individual but not the circumstances under which the DNA was deposited. DNA deposited consensually looks exactly like DNA left without consent.

When

A DNA profile can confirm an individual's presence, but it cannot tell when the DNA was left in a given spot. For example, DNA found on a soda can at a crime scene may have been deposited at the time of the crime or at some earlier point.

Why

There may be a logical explanation for how an individual's DNA was deposited on an item at the crime scene. For example, the defendant's blood may be found on the victim because the victim attacked the defendant and a fight ensued.

Section 5. Alternate Theories of Defense

A DNA match between a defendant and an evidence sample does not mean that the prosecution's version of events is the only possible explanation for the DNA evidence. There are a number of possible defense theories. By understanding what DNA evidence is — and is not — the defense attorney can evaluate its impact on the defense theory.

Defense counsel must think critically about how to explain the DNA results to jurors so that they think about DNA from the defense's perspective. Counsel should consider how the DNA evidence is significant to the charges, how it relates to other evidence, and whether there is an innocent explanation for its presence.

Defendant is the DNA source

First, counsel may want to accept that the defendant is the source of the DNA. In some cases, the defense will be that the defendant was involved in the act with which he is charged, but his involvement was legally justified. Examples include consent and self-defense. In these cases, the existence of the defendant's DNA would be expected, and the DNA match actually corroborates the defense theory.

The defense may also argue that although the defendant is the source of the DNA, he was not involved in the act with which he is charged. Instead, his DNA became involved through another means, such as transfer, prior contact, laboratory contamination, prosecution or law enforcement malfeasance, or the possibility that another individual planted the defendant's DNA.

Defendant Is Not the DNA Source

Alternatively, when dealing with a partial DNA profile or a mixture DNA profile, the defense may be best served by arguing that the defendant is not the source of the DNA or a contributor to the mixture profile. The defense may argue that the match or inclusion is coincidental — particularly when paired with an argument that the government is inflating the match's statistical significance. For example, from the defense perspective, the lab analyst's use of a modified *RMP* in a mixture case would inflate the statistical significance of the match as compared with another method like the *CPI*. (See Section 2 in this chapter for a discussion of statistical calculations.)

The defense may claim that the report of a match or inclusion was false because of the analyst's subjective interpretations of low-level DNA. Counsel may also argue that a mixture demonstrates the existence of a third-party perpetrator or that there are reasons to exclude the defendant as a contributor to the sample. In such cases, the defense can actually use the DNA evidence to contradict the government's theory.

When arguing that the defendant is not the source of the DNA, it is important to ensure that no other significant evidence links the defendant to the scene. The defense argument that the DNA did not come from the defendant is most effective when there is:

- No eyewitness identifying the defendant (or there is an eyewitness who positively identifies someone else).
- No other inculpatory forensic evidence.
- No confession from the defendant.
- Additional evidence of a third-party perpetrator.
- Alibi or character evidence for the defendant.

The defense can also use serology to supplement DNA. Serology testing can help determine what kind of stain (blood, semen or saliva) was present. Counsel should investigate whether the type of sample can be used to corroborate the defense theory or dispute the prosecution's theory about what occurred. If the laboratory does not routinely perform certain body-fluid screening tests, you may be able to use this to your advantage by establishing, on cross-examination, that the analyst has no idea from which type of biological material the DNA profile was obtained.

Section 6. DNA and the Client

Defense attorneys must have conversations with clients to explain and discuss what DNA evidence means, any avenues for challenging it, the difficulties in challenging DNA evidence, and how the client can help in the defense.

These conversations can be difficult, but developing an early working relationship with the client will enable counsel to have substantive conversations about what DNA evidence may or may not show.

When first informing a client that evidence from the crime scene appears to match his or her DNA profile, consider having a discussion along the following lines:

> The lab results indicate that DNA found at the crime scene matches your DNA. With this evidence, it is going to be difficult to convince a jury that you were not at the crime scene. We have to really reconsider whether sticking with a misidentification defense (or other defense that does not address the DNA results) is the best strategy. Let's talk about your options in light of this evidence.

A client — invested in having his attorney believe his innocence — may continue to insist that the lab results are wrong. To address this, counsel can discuss the possibility — or improbability — of finding an expert to dispute the results and then focus on how a jury will be likely to view the DNA evidence.

The client may also want to have the evidence retested — or have the testing conducted on evidence that has not yet been tested. Counsel must explain the risks of retesting/testing (discussed in more detail in Chapter 3, Section 1), which include triggering the prosecution to test first, alerting the prosecution that the defense is testing, and having to turn the results over to the prosecution. The client must understand that a retest or newly requested testing can bear negative consequences if a result comes back that includes him/her as the main DNA contributor. As discussed earlier, the more counsel can avoid talking about the DNA results as being indicative of the client's guilt, the more successful the defense will be in having frank, substantive conversations about the risks and rewards of retesting and new testing of the DNA sample. Defense attorneys must also counsel clients on the issues surrounding DNA databases. First, the defense must inform the client if the offense he is currently accused of (arrest or conviction) may lead to his DNA being put into a DNA databank. You may also need to explain that his prior arrest or conviction in another incident may have resulted in his DNA profile already being in a DNA databank. Explain what a DNA databank is and what will happen with the DNA in the future — in particular, that future evidence samples from crime scenes will be checked against his and others' DNA profiles to see if there is a match. If counsel can challenge or petition for the removal of the client's inclusion in the databank — in some jurisdictions, after an acquittal — be sure to inform the client of his rights and provide help or a referral to get his DNA out of the databank. In addition, be certain to petition for removal of the evidence profile in addition to any arrestee sample that may exist. If the laboratory maintains suspect DNA profiles at either the local (LDIS) level or the state level of CODIS, or if the lab maintains a database separate from CODIS that contains profiles, be certain that the petition includes removal of the profile from those databanks/databases as well.

Section 7. Types of Statistics — What Do They Mean?

Where Do These Numbers Come From?

Frequency values used in statistical calculations come from databases. Some labs have their own databases, and some labs rely on work conducted by the FBI or other labs for their databases.

What Is in The Database?

DNA profiles. Typically, a lab has collected DNA samples and generated profiles from a minimum of 100–150 people from each of three or four population groups for inclusion in a portion of the database. Each person has two alleles at each locus; with 100 people, there would be 200 potential alleles in the database at each locus; with 150 people, there would be 300 potential alleles at each locus. These DNA profiles are examined, looking for their frequency of occurrence in the sampled portions of the population.

For example, if the 16 allele is observed, at the genetic marker D3, 15 times out of the 150 profiles (300 total possible alleles) in the database, 5% of the 16 allele is observed (15/300 = 0.05). Thus, the lab assigns the frequency of 5% to the 16 allele (in the frequency chart, this would appear as 0.05).

This is done for each allele at each locus. Based on work done in the field of statistics, it has been determined that a minimum sampling of this size — about 100 — can be used to infer the frequency of occurrence of each of these alleles in the entire population. Given that the STR loci comply with certain rules of population genetics *(Hardy-Weinberg equilibrium* and *linkage equilibrium* are discussed later in this chapter), these frequencies can then readily be extrapolated to the entire population. These basic calculations are relied on when determining a statistical frequency of the observed STR profile in court cases. See the section on use of the product rule, (page 68).

RMP and Source Attribution Statements

In addition to declaring a DNA match, inclusion or failure to exclude, lab reports must provide a statistical frequency that will give weight to the match.

An example of a typical RMP statement was provided in Section 2 of this chapter. Be advised that some labs, rather than give a statistical calculation in their report, may instead use language that declares the profile is unique within the U.S. or world's population, or declares that an individual is the source of the DNA. This is called *source attribution.* Each lab has its own protocols for report wording. They are discoverable and will typically be found in the lab's procedure manual and/or its quality assurance manual.

Hardy-Weinberg Expectations/Equilibrium (HWE)

How are these random match/man probability statistics calculated? For that matter, what is the basis for any of the calculations? A similar mathematical approach, using defined formulas, is used for all DNA profile statistical calculations, regardless of whether the profile is a single-source profile, a partial profile, or a mixture profile. Allelic frequencies from databases that have demonstrated adherence to Hardy-Weinberg expectations/ equilibrium (and no linkage) are used to calculate genotypic frequencies of each STR locus result. These genotypic frequencies are then multiplied together, using *the product rule,* to generate an estimated frequency of occurrence of the obtained DNA profile in the population to which the database corresponds.

Take, as an example, a person who has the heterozygous profile of 14,16 at the genetic marker D3. The corresponding values for the frequency of occurrence of those alleles can be obtained from the database frequency tables. To obtain the frequency of a person in the population having a 14,16 profile at D3, the frequency of having a 14 allele (as determined by the aforementioned database) is multiplied by the frequency of the 16 allele, and this number is then multiplied by two.

Why two? Because of something called *Hardy-Weinberg equilibrium,* or *Hardy-Weinberg expectations:*

$$1 = p^2 + 2pq + q_2$$

where: ~

p_2= frequency of AA genotype (homozygote) ~
$2pq$ = frequency of Aa genotype (heterozygote) ~
q_2= frequency of aa genotype homozygote). ~

It is important to note that HWE is applicable because we are talking about genetic markers — a system that follows Mendelian genetics. What is key is that during formation of gamete cells (egg and sperm), the markers segregate in pairs and sort themselves independently, which allows shuffling of genetic information for each gamete that is formed. In other words, the human population is considered a Mendelian population — it is a group of interbreeding individuals who share a common set of genes and genetic markers, called the *gene pool.*

Using the D3 profile example and the laws of HWE:

(frequency of 14,14 typing result)^2 + 2(frequency of 14 allele) × frequency of 16 allele [representing frequency of 14,16 typing result] + (frequency of 16,16 typing result)^2 = 1, over time.

What are the concerns about applying HWE to the calculation of statistics that assign levels of significance to casework DNA profiles that are generated? There are plenty of concerns. The laws of HWE require the following:

1. The population being tested is infinitely large. Clearly, the U.S. and the world populations do not meet this criterion.
2. There is random mating. This is not the case for humans, in general, because people tend to mate within their own geographic area, and with people who are visually similar to them or have the same belief system they have.

3. The population must be free from outside evolutionary forces such as mutation, migration and natural selection. Migration occurs constantly within the human population, as do mutations to our DNA, and natural selection occurs, naturally over time — meaning that if a particular DNA sequence of a gene brings some benefit to the population, this genotype will be favored during mating or proliferation of the species over time.

Why, in the face of such an apparently blatant mismatch between the concepts of HWE and the human population, is HWE applied to the assessment and application of statistics generally accepted within the scientific, population genetics, forensic biology, and mathematics communities? To shed some light on this question, each of these concerns, outlined above, is addressed:

- That the population must be infinitely large: In reality, for the purpose of generating statistics relative to casework genotypes, the population can easily be defined as infinitely large. Why is this, if we know that the U.S. and the world populations are finite in size? A minimum group size of 100–150 people has been repeatedly shown to be of sufficient size to demonstrate that HWE applies to the human population for STR loci, regardless of the geographic area sampled. What these population samplings show is that after testing a few hundred individuals, allele frequencies essentially "plateau" after about 200 data points. Accordingly, more extensive sampling of the population is not necessary or of any benefit.
- That the population must be randomly mating: There is nothing about someone's STR typing results that would have an effect on the choice of a mate in a population. Random mating implies that any individual of one sex is equally likely to mate with any individual of the opposite sex in the population. While geographic location, socioeconomic status or background, race, and physical characteristics such as body type, height and weight can and do influence the choice of mate, STR typing results are "invisible" to us and, therefore, the expectation of random mating applies to them. Often an analyst will not be able to successfully explain to a jury why STR loci can be considered to meet the HWE of random mating, so cross-examination on this point can be effective.
- That the population is free from mutation, migration and natural selection: Although the human population is clearly not free from these evolutionary forces, there is nothing about one STR allele that favors it over another. All of the STR loci do not code for proteins, as far as is currently known — human characteristics favored by mutation, migration and natural selection all code for proteins, the structure of which can and does lend an evolutionary advantage in some situations. Applying statistics to DNA typing results, the fact that the population frequency values of the samples continue to meet Hardy-Weinberg expectations supports the view that the human population is sufficiently free of these evolutionary forces to allow use of the frequency values that have been established. This can be the most difficult of the HWE concepts for an analyst to explain to a jury. Mutations of human DNA have resulted in variations of our population at each STR locus over time. This approach can be effective during cross-examination.

There are some departures from Hardy-Weinberg expectations that have been noted and are worth exploring further in some cases, for example, the observed inbreeding and kinship factors. The effects of these factors are the result of mating between closely related individuals. This results in an increase in the number of *homozygotes* and the decrease of *heterozygotes* compared with the general or randomly sampled population. Along these lines, a population subgroup's homozygosity can increase at a locus and concurrently decrease in heterozygosity. Population subgroups are considered to be small population groups (such as the Amish) that seek to mate solely within their own groups. Genetically speaking, this is similar to inbreeding, but it does not result in narrowing of genotypes to the same degree. Of note is that each of these population subgroups will demonstrate common ancestry through typing of different genetic markers. Of importance is that statisticians can measure the existence of population substructure that occurs in these groups with a value called a theta (0) correction — colloquially referred to as the *inbreeding coefficient.* The greater the theta value, the greater the corresponding substructure.

Homozygote "Correction" Factor

DNA labs use a correction factor, of sorts, in an effort to address concerns about the underestimation of homozygotes due to substructuring that has been documented in the population. NCR 11 recommends that the following formula be used to determine the frequency of occurrence of homozygous genotypes:

$$1 = p^2 + p(1 - p)0$$

where:

p = frequency of allele in the database,
0 = 0.01 for most populations, and
0 = 0.03 for small native populations.

Most DNA labs use a theta value (0) of 0.01 for most, if not all, of their calculations. This theta value, in conjunction with the modified formula for determining the frequency of a homozygous genotype, boosts the frequency estimate of homozygotes. This results in a more conservative value in the event that substructure is found to be a factor. Population geneticists argue that such an overestimation of frequency will always favor the defendant. Conditional subpopulation calculations may also be performed in accordance with NRC II formulas 4.10a and 4.10b, as per *SWGDAM* Guideline 5.2.1.4.

No theta correction is used for heterozygote genotypes because, as noted earlier, their frequency of occurrence is already overestimated if substructure exists in the genotype.

Minimum Allele Frequency

Because allele variants infrequently encountered in the population may not have been seen in the population sampling used to create the database, or the allele variants are underrepresented in the sampling, labs must have a statistically conservative mechanism for dealing with this situation. NRC II recommends that a minimum allele frequency value be assigned to any allele that is or was not observed, or was seen less than 5 times, in the samples constituting the database.

The minimum allele frequency is calculated using the formula, $5/2n$, where n equals the number of people in the database, such that $2n$ represents the number of possible alleles, given that number of people. Selection of a value of 5 minimum observations of an infrequently encountered allele ultimately results in a very conservative value for these alleles — 5 is an arbitrarily selected number. Using the example of a database of 150 people: the minimum allele frequency estimate of alleles not seen or seen less than 5 times = 0.017, or $5/(2 \times 150)$.

As with the correction factor for homozygotes, population geneticists argue for use of the minimum allele frequency value because this overestimation of frequency will always favor the defendant.

The Product Rule

The multiplication of the genotype frequencies across loci is called *the product rule.* Essentially, the product rule states that when two events are independently occurring, the chance that both will happen at the same time can be determined or estimated by multiplying the probabilities of occurrence of each event.

The use of the product rule in statistical calculations has been longstanding in the mathematics, statistics and population genetics communities.

For the product rule to be legitimately used in forensic DNA cases, it must be demonstrated that, at each STR locus, the allele inherited from the mother is inherited independently of the allele inherited from the father.

It must also be demonstrated that the STR loci used in DNA typing kits are inherited independently from each other. Repeated testing of the loci used in STR DNA typing has demonstrated that the loci are inherited separately and, therefore, no linkage between the loci has been found.

Note: The product rule is not used to generate frequency estimates for mtDNA or Y-STRs. (For more information on reporting results for Y-STR tests, see the discussion on the counting method, later in this section.) Some laboratories multiply RMP frequency and the Y-STR/mtDNA frequency. This emerging approach was being litigated as this publication went to press.

Testing of Databases: Hardy-Weinberg Equilibrium and Linkage Equilibrium

Most crime labs have been performing DNA testing for some time; it is likely that their databases have been previously examined to determine that they met Hardy-Weinberg equilibrium and linkage equilibrium criteria. Before a lab may use the product rule, its STR database must meet the HWE criteria.

Check to see whether the lab's database has been examined by a statistician or a population geneticist. If the database has been previously examined in the course of litigation, it is unlikely to be an issue.

The databases in common use by crime labs have been reviewed by an independent expert, and the results were peer reviewed and published.

Statistics for Partial DNA Profiles

A single-source DNA profile that does not contain complete information from all loci tested is called a *partial DNA profile.* A partial mixture DNA profile can also be encountered.

For a partial single-source profile, the RMP calculation is still used. The number will be more favorable to the defendant with a partial DNA profile because fewer loci are obtained; thus, fewer frequencies are multiplied together and the final statistic is less rare.

For a partial mixture DNA profile, the lab will use whatever calculation type their protocol requires — CPI, CPE, MRMP or LR. Again, the number will be more favorable to the defendant with a partial mixture DNA profile because results have been obtained at fewer loci.

Any partial DNA profile provides the best opportunity to refute the significance of the statistical calculation provided in the lab's report. Although partial single-source DNA profiles may still provide strong support for the value of a match, this will depend on how partial the profile is. Assigning allele calls in partial mixture profiles can be challenging, as noted in Section 2 of this chapter, which often translates into challenges in conducting the statistical calculation, depending on the formulas the lab uses.

These often give the most room for exploring the real value of the failure to exclude someone as a potential contributor to the profile. When attempting to do so, you must make sure the jury is aware of the types of statistical numbers that are usually encountered in forensic cases. Without such a benchmark for comparison, a combined probability of inclusion (CPI) number of 1 in 3,000, or even 1 in 100, may sound like it isn't "too bad."

Statistics for Mixture Profiles

The commonly used calculations for mixtures in U.S. crime labs are represented in the table from the *SWGDAM Guidelines* (January 14, 2010) reproduced below (see Table 2). What is clear from the Table is that many choices exist for calculating the significance of a mixture DNA profile. This naturally introduces the possibility that your expert may have a different recommendation for another method for calculating the statistic.

Strategic Considerations for Mixture Calculations

The RMP approach for mixture calculations expresses the DNA match in far more incriminating terms than other available approaches. Even use of a *modified random match probability statistic (MRMP)* helps to temper use of this statistical approach.

How do the labs justify the use of an RMP approach? They conclude that it is possible to separate out ("pull out") a major (or, in some cases, a minor) DNA profile in the mixed DNA sample and then conduct an RMP analysis on that major profile. Sometimes this is possible to do, as when a clear DNA profile from a major contributor can be discerned in the mixture; in those circumstances, use of an RMP statistic is appropriate. For reasons discussed elsewhere in this guidebook, this approach can be problematic when applied to a mixture DNA profile. It is not possible to discern the contributor's profile in the absence of known samples with which to compare it.

In general, defense counsel's default should be to promote use of the CPI/CPE method in any mixture case because the reported statistic is almost always more favorable to the defense.

However, labs *must* use whichever statistical approach their protocol dictates.

Even with the CPI/CPE method, counsel should consider how to convey statistical information to the jury. Psychological studies have shown that juries better understand the statistical value, and are less likely to inflate the discriminatory value of DNA evidence, when presented in the form reported by CPI. The way CPE results are worded can be problematic if your client is not excluded because lab results generally state that 99.998% (or greater, depending on the actual calculation) of the population can be excluded as a contributor to the mixture.

Therefore, in mixture cases, defense counsel should seek to have the significance of the evidence conveyed by their expert to the jury in the following form, based on the expert's CPI calculation: "1 in X number of (randomly selected, unrelated) people in the population would be expected to be included as a possible contributor to the evidence profile."

It is not reasonable to expect an analyst to conduct any statistical calculation on the stand, particularly one that is different from one routinely used by their laboratory.

Even if you ask for a break for the examiner to conduct the calculation, many lab protocols prohibit release of any data without a technical review. Furthermore, most statistical calculations are fairly complex, and computers are now routinely used to generate statistics. If you want to present different statistics to the jury, you should plan on using your own expert to present the testimony.

Table 2. Suitable Statistical Analyses for DNA Typing Results

Category of DNA Typing Result	RMP	CPE/CPI	LR (1)
Single source	/		/
Single major con- tributor to a mixture	/		/
Multiple major con- tributor to a mixture	/ (2)		/
Single minor con- tributor to a mixture	/		/
Multiple minor con- tributor to a mixture	/ (2)	/ (3)	/
Indistinguishable mixture	/ (1)	/	/

(1) Restricted or unrestricted; (2) restricted; (3) all potential alleles identified during interpretation are included in the statistical calculation.

Source: Reproduced from SWGDAM Guidelines (January 14, 2010).

Notes: The statistical methods listed in the table cannot be combined into one calculation. For example, combining RMP at one locus with a CPI calculation at a second locus is not appropriate. However, an RMP may be calculated for the major component of a mixture and a CPE/CPI for the entire mixture (as referred to in section 4.6.2).

Counting Method for Y-STR and Mtdna Testing

The *counting method* is used to report the significance of Y-STR and mitochondrial DNA testing results. Because of the mode of inheritance of Y-STR loci and mitochondrial DNA, the statistical approaches used for autosomal STR loci are not appropriate — simple Mendelian genetics laws do not apply to Y-STR and mitochondrial inheritance. The counting method calculates how many times a profile has been seen in the consolidated U.S. Y-STR database (http:// usystrdatabase.org) or another appropriate database identified by the lab.

Following is an example using the counting method approach for a Y-STR haplotype profile: "The Y-STR profile was seen three times in the database of ___ × ___ individuals" or, sometimes, "The Y-STR profile was not seen in the database."

Statistics for Related Individuals

The probability of randomly selecting an unrelated individual with a particular genotype is computed as an RMP. Individuals who are related to one another will share more alleles than unrelated individuals because they have a biological relative in common.

The same alleles will occur more frequently within a family than in the general population. Thus, for example, a homozygous profile that may normally be found at a 1% frequency in the general population may be found at a much greater frequency among relatives. For example, the expected frequency among full siblings for this same homozygous profile would be 25.5%.

Be mindful that once you ask, "What is the probability that a relative (full sibling, parent, child, etc.) would have a matching DNA profile," you are now asking a completely different question from "What is the probability of randomly selecting an unrelated individual with a matching DNA profile?"

Statistics for Database Searches

For a discussion on statistics for database searches, see Chapter 9, Section 6.

End Note

[1] See Risinger, D.M., et al., "The *Daubert/ Kumho* Implications of Observer Effects in Forensic Science: Hidden Problems of Expectation and Suggestion," 90 CALIF. L. R EV. 1 (2002); Krane, D.E., S. Ford, J.R. Gilder, K. Inman, A. Jamieson, R. Koppl, I.L. Kornfield, D.M. Risinger, N. Rudin, M.S. Taylor and W.C. Thompson, "Sequential Unmasking: A Means of Minimizing Observer Effects in Forensic DNA Interpretation," 53 J. F ORENSIC S CI. 1006 (July 2008).

7. DNA Basics: Pretrial Preparation

Section 1. Should the Defense Request Testing?

Once the DNA evidence has been evaluated by the defense's expert, a decision by the defense to conduct additional DNA testing should be considered. The defense theory of the case should be developed before a retest or other DNA testing is started. The following should be considered:

- What are you trying to prove with the defense DNA test?
- Is third-party guilt a possibility?
- Will DNA testing help refute the prosecution's DNA testing?
- Will DNA testing help prove a fact that is important to the case?

It is important to consult with the defense's DNA expert on the desirability of testing. It is also important to meet with the client and obtain input after discussing the DNA testing process and the option of conducting DNA testing with the client. For a discussion on talking to the client, see Chapter 4, Section 1, and Chapter 6, Section 5.

Once the decision to perform DNA testing is made, counsel should locate a DNA laboratory that is experienced in forensic DNA testing. The laboratory should also be available to testify if a favorable result is obtained.

Section 2. Evidentiary Issues

Three discrete ways DNA acquisition may be implicated in a criminal investigation or prosecution are as follows:

- Collecting a suspect's DNA before or during the investigation and matching it to crime scene evidence.
- Collecting DNA from an already charged defendant to determine whether it matches crime scene evidence.
- Collecting biological material from the defendant's body, clothing, car, or other possession that might be shown to belong to the victim.

Each method is likely to implicate the Fourth Amendment because retrieving biological material from the accused is a seizure, and subsequent testing is a search. This is made clear by application of the holding in *Skinner* v. *Railway Labor Executives' Ass'n.*[1] In *Skinner,* the U.S. Supreme Court made the following points:

- "The initial detention necessary to procure the evidence may be a seizure of the person, if the detention amounts to a meaningful interference with his freedom of movement."[2]
- "[C]hemical analysis of urine, like that of blood, can reveal a host of private medical facts. ... Because it is clear that the collection and testing of urine intrudes upon expectations of privacy that society has long recognized as reasonable ... , these intrusions must be deemed searches under the Fourth Amendment."[3]

Section 3. DNA Collection — Databanks of Convicted Person DNA

In 1989, Virginia became the first state to pass legislation requiring certain classes of criminal offenders to submit to DNA testing for the purpose of including the DNA samples in a DNA databank.[4] Virginia legislation also provided that DNA evidence was admissible evidence of identity in criminal proceedings.[5]

Over the next 10 years, all other states enacted laws requiring the collection of biological samples from certain criminals for inclusion in databanks. Some states require that DNA samples be taken from all classes of felons; other states also include individuals convicted of certain misdemeanors. All states require the collection of DNA samples from convicted sex offenders.[6]

A federal statute — known as the DNA Analysis Backlog Elimination Act of 2000 — was enacted on December 19, 2000.[7] The statute mandates the collection of DNA samples from individuals in custody or on probation, parole or supervised release if they have been convicted of certain qualifying federal offenses, such as murder, voluntary manslaughter, other homicide offenses, kidnapping, robbery or burglary.[8] Convictions for certain federal offenses relating to sexual abuse, sexual exploitation or other abuse of children; transportation for illegal sexual activity; certain felony offenses relating to sexual incest; and crimes of violence also trigger the DNA sample collection requirement.[9] To protect the privacy rights of individuals subject to this requirement, the federal statute limits the purposes for which DNA samples or results may be used and the circumstances under which they may be disclosed.[10] The federal statute also makes it a crime to obtain a DNA sample or result without proper authorization.[11]

State and federal statutes that require the collection of DNA samples from certain classes of convicted criminals have been the subject of various constitutional challenges, virtually all of them unsuccessful. In particular, Fourth Amendment challenges[12] have been rejected, regardless of the court's analytic approach. Absent action from the U.S. Supreme Court overturning these decisions — or unless their reach is found not to apply to certain categories of individuals, such as juveniles, arrestees, or people who have completed their probation, parole or supervised release — it will be difficult to challenge subsequent comparisons between an evidence profile and databank profiles from known individuals.[13]

Does a Databank Match Establish Probable Cause?

A match between DNA recovered from a crime scene and a defendant's databank (or otherwise lawfully obtained) profile— without more evidence — does not establish probable cause to arrest. Just as a fingerprint may be left at a crime scene in any number of innocent ways,[14] so, too, may a defendant's DNA be retrieved from evidence innocently left near the location, such as a cigarette butt. Probable cause can be found only with additional facts that make it likely that the *criminal* left the item.[15]

Section 4. DNA Collection — Taking DNA from an Arrested Person by Judicial Order

Some states require DNA testing of any person charged with a specified offense (e.g., a felony).[16] If the resulting DNA profile matches a crime scene profile, the following suppression issues should be examined:

- Was there probable cause for the arrest?
- Was the arrest for one of the designated offenses?
- Does the legislation mandating DNA testing of arrestees violate the Fourth Amendment or the state's constitutional protection against unreasonable searches and

seizures? Does it violate another provision such as a state constitution's privacy guarantee?

To date, courts are divided on whether mandatory testing of all arrestees (or all arrestees for designated offenses) is unconstitutional. This issue is likely to receive more attention: Federal legislation enacted in 2006 allows DNA testing of all individuals arrested for federal criminal felonies, and more states are moving toward arrestee databanks.

In *In re Welfare of C.T.L.,*[17] the Minnesota Court of Appeals forbad arrestee testing, deeming it a *search for evidence with no probable cause* determination:

> By directing that biological specimens be taken from individuals who have been charged with certain offenses solely because there has been a judicial determination of probable cause to support a criminal charge, [these statutes] dispense with the requirement under the Fourth Amendment that before conducting a search, law enforcement personnel must obtain a warrant based on a neutral and detached magistrate's determination that there is a fair probability that the search will produce contraband or evidence of a crime. Under the statute, it is not necessary for anyone to even consider whether the biological specimen to be taken is related in any way to the charged crime or to any other criminal activity.[18]

The Minnesota Court of Appeals also focused on the defendant's privacy interest with regard to his or her DNA.[19]

By contrast, the Virginia Supreme Court ruled in 2007 that such legislation is constitutional, finding that it is an identification procedure rather than evidence gathering. The court reasoned, "Like fingerprinting, the Fourth Amendment does not require an additional finding of individualized suspicion before a DNA sample can be taken."[20]

As a general matter, on the privacy interest dimension, there is a continuum from prisoners (who have little or no reasonable expectation of privacy) to parolees, then probationers. It can be argued that arrestees, especially ones who are ultimately acquitted or have the charges against them dismissed, should have the full privacy interests afforded by the Fourth Amendment.

Probable Cause

When there is no statutory authorization for across-the-board DNA "fingerprinting" of arrestees, judicial order must precede DNA testing of a pretrial defendant (absent consent, discussed later). In light of *Skinner*[21] and its prequel holdings, including *Schmerber* v. *California,*[22] the likely standard is a warrant or judicial order predicated on probable cause. As the U.S. Supreme Court explained in *Skinner:*

> In most criminal cases, we strike this balance in favor of the procedures described by the Warrant Clause of the Fourth Amendment. See United States v. Place, supra, at 701, and n. 2; United States v. United States District Court, 407 U.S. 297, 315 (1972). Except in certain well-defined circumstances, a search or seizure in such a case is not reasonable unless it is accomplished pursuant to a judicial warrant issued upon probable cause. We have recognized exceptions to this rule, however, "when 'special needs,' beyond the

normal need for law enforcement, make the warrant and probable-cause requirement impracticable."[23]

Although this language seems dispositive on the issue of whether the probable cause standard must be met, final analysis will depend on the level of intrusiveness found in a DNA search. In an earlier decision, the U.S. Supreme Court held that the fingerprinting of individuals might be allowed during lawful *Terry*[24] stops on the basis of reasonable suspicion.[25] Several states have upheld orders for DNA testing on the basis of a judicial determination that there is reasonable suspicion the individual is a suspect.[26] However, in those decisions, the testing occurred pre-arrest or pursuant to a grand jury subpoena.

Because of privacy concerns, the U.S. Supreme Court has characterized compelled intrusions into the body to seize blood samples as Fourth Amendment searches.[27] Accordingly, to obtain a warrant or court order for a blood sample, the government must show that probable cause exists to believe that the blood sample will produce evidence of the defendant's involvement in the crime.[28] Thus, it is likely that the probable cause standard will apply post-arrest. All of this is predicated on there being a lawful arrest or detention; if it is unlawfully seized, a DNA "fingerprint" must be suppressed as the fruit of the "poisoned" seizure.[29]

Testing Crime Scene Evidence Before Requesting a Defendant's Sample

Assuming the defendant is lawfully in police custody, a related concern is whether the prosecution must test the crime scene evidence to see if a *DNA profile* likely to be that of the perpetrator has been found *before* requesting the defendant's *DNA sample.* This is significant, because, once the police have the defendant's profile, it can also be checked against evidence from other crimes or uploaded into DNA databases. With these concerns apparently in mind, the ABA's *Standards for DNA Evidence* require as a prerequisite that there first be a judicial determination "that the sample will assist in determining whether the person committed the crime."[30]

The probable cause standard is the ultimate test for such searches. The U.S. Supreme Court has provided some guidance, however, stating that the government must establish not simply a "mere chance" that the extraction of blood will produce relevant evidence but "a clear indication that in fact such evidence will be found [through the compelled bodily intrusion]."[31]

Nonetheless, courts have typically issued such orders after either a DNA profile has been extracted from crime scene evidence[32] or the type of crime scene evidence (e.g., sperm) makes it clear that a profile will be obtained.[33] A mere "boilerplate" allegation — that it is "important in the investigation to take venous blood and saliva samples for comparison and/or elimination ... with no factual allegations whatsoever to demonstrate that the desired evidence would be found" — will be deemed insufficient.[34]

Courts analyzing the reasonableness under the Fourth Amendment of orders for blood samples, *buccal swabs,* hair samples and other bodily intrusions have closely scrutinized the government's claims for the need for such intrusions. Courts have required a clear showing that the intrusion will produce evidence.

For example, in *In re Lavigne,*[35] the Supreme Court of Massachusetts reviewed the government's request for a blood sample for use in its homicide investigation. The government did not offer evidence that it had relevant samples to which the defendant's blood could be compared. The *Lavigne* court held that the defendant was entitled to a hearing where the judge was required to "make findings as to the degree of intrusion and the need for the evidence of the blood sample."[36] The court noted that the government bore the burden of establishing a "nexus between the item to be seized and criminal behavior,"[37] and that an order compelling the defendant to provide a blood sample was unreasonable unless the government established such a nexus. The court ordered the return of the blood sample that had been taken from the defendant.[38]

Similarly, in *State* v. *Acquin,*[39] the government sought to seize the blood of a defendant in a murder case. The government was in possession of items believed to be the murder weapons, which bore reddish sticky substances that could have been blood. Relying on *Schmerber* and *Hayden,* the Supreme Court of Connecticut held, "In the absence of facts establishing, *at the very least,* that the 'substance' found on the alleged murder weapons was in fact blood," the court could not find probable cause to believe that the blood seized from the defendant would have a nexus to the crime charged.[40]

Section 5. DNA Collection — Taking DNA from an Arrestee Without a Warrant

Because a person's DNA profile remains constant, there is no exigency that would entitle the police, without a warrant, to seize the suspect's biological material for DNA testing once the suspect is in police custody.[41] However, exigent circumstances may arise when the defendant may have DNA evidence from the alleged crime victim on his or her person or clothing (for example, when a suspect is arrested within hours of an alleged sexual assault or homicide).

U.S. Supreme Court precedent permits warrantless seizures in such limited circumstances. In *Cupp* v. *Murphy,*[42] the Supreme Court approved the taking of apparent blood under the fingernails of a murder suspect, without a warrant, within hours of the crime because it was a "very limited search necessary to preserve the highly evanescent evidence they found under his fingernails."

Subsequent to *Cupp,* the Supreme Court held that clothing worn by a person arrested on the basis of probable cause may be seized and checked for crime scene evidence without police first needing to secure a warrant.[43] This principle has been applied to taking an arrestee's clothing to check for DNA.[44] The same applies when police want to swab the arrested person's genitals, hands, or other body parts in search of the alleged victim's bodily fluids. The analysis will turn on the proximity of the arrest to the time the crime occurred and whether there is any likelihood that the defendant could wash, shower, or otherwise remove the biological material before a warrant could be obtained.[45]

The taking of a DNA sample does not automatically confer the right to upload that profile into a local, state or national databank.

Section 6. Alternative Methods of Obtaining DNA Evidence — Consent

The doctrine governing consent searches is two-pronged: Consent must be voluntary[46] and the scope of consent is governed by the natural language used by the parties.[47]

Acquiescence to an explicit request for a DNA profile, if made without coercion, will be presumptively voluntary.

In the DNA context, three situations may raise concerns about the validity or proper scope of consent. First, police may ask for an item (e.g., a piece of hair or a hat) from which DNA can be extracted without telling the person of their intent to test it and obtain a DNA profile. A motion seeking to suppress the resulting DNA profile will be determined by (a) the language that was used by the police when asking for the item (e.g., "We want your hat to show it to witnesses")[48] and (b) whether surrender of the item eliminates any reasonable expectation of privacy.[49]

The second consent paradigm is one in which the police ask for a person's DNA, ostensibly to compare it with the evidence from a particular crime scene, but instead compare it to crime scene evidence from another crime or several other crimes. Courts addressing this practice have approved it, focusing on the loss of an expectation of privacy of the DNA profile. They have not discussed the separate issue of whether the scope of consent has been exceeded.[50]

This practice comes into play when a DNA profile is sought for comparison with evidence from one crime scene but is then uploaded into a local, regional or national DNA databank. Law enforcement has increasingly used *DNA dragnets* to collect genetic information that is sometimes entered into DNA databanks.[51] In a DNA drag-net, law enforcement officers investigating a particular offense ask the eligible population of a community for "voluntary" DNA samples. After the investigation, police might then request submission of the generated profiles of these innocent people to DNA databanks. It is important to note that, under the current regulations dictating what samples are suitable for submission to the national CODIS *(NDIS),* no suspect samples (volunteered or otherwise) can be submitted to NDIS. However; if the crime laboratory practices and state laws allow, they can be uploaded to the local CODIS *(LDIS)* and potentially the state CODIS *(SDIS),* if approved by the state CODIS administrator).

The decisional law analyzing such practices is limited, but some courts have found that relinquishing one's DNA profile to police ends *all* expectations of privacy.[52] As the Virginia Court of Appeals explained:

> [T]he overwhelming weight of relevant authority from our sister states indicates that society is unwilling to recognize as reasonable the subjective expectation of privacy infringed by the government when a DNA sample validly obtained from a suspect in one criminal case is used to analyze and compare the suspect's DNA in an unrelated criminal case.[53]

However, "voluntary" consent in DNA dragnet situations may be illusory. Individuals asked to give samples in DNA dragnets are often unaware of their right to refuse. It is important to be mindful of the potential that some police officers may have used coercive measures to obtain consent in DNA dragnets, such as threatening — implicitly or explicitly — that individuals who do not volunteer their DNA will become suspects.[54]

Police also may have reason to take a person's DNA, without his or her consent, in the context of a particular case, but they may not have the right to put that information into a DNA databank. For instance, in the "BTK" serial killer case, the police, with probable cause, searched a suspect's home and collected a genetic sample, which did not match the evidence sample.[55] A court granted the suspect's motion to have his DNA profile purged from law enforcement databanks.[56]

Section 7. Alternative Methods of Obtaining DNA Evidence — Abandoned Property

The Fourth Amendment has been held to permit police to seize abandoned property, that is, property that has been left in a public location in a manner accessible to others.[57] Unless in a jurisdiction where state constitutional protections are greater,[58] this means that police (or private parties acting at the behest of law enforcement) may seize any item believed to contain a suspect's DNA if that item has been "abandoned."

Cases of DNA seizures involving abandoned property include collecting a suspect's saliva:

- After he or she spits on a public street.[59]
- After police create a ruse and mail a letter purportedly from a law firm inviting the suspect to join in a lawsuit, and his or her saliva is on the return envelope.[60]
- From a soda can that police offered a suspect, who drank the soda and then threw the can in the trash.[61]

It is only when the allegedly abandoned item is still on the defendant's private property — or otherwise in circumstances where an expectation of privacy is retained — that seizure of the item must comply with Fourth Amendment protections.[62]

The U.S. Supreme Court has not yet made a determination of whether DNA left on a soda can, an envelope or some other collectible item may be collected under the doctrine of abandonment. Whether such seizures should be subject to Fourth Amendment scrutiny because of the vast amount of private information contained in DNA is still an open question. A person may have a higher expectation of privacy for the personal information contained in DNA than for the item that police found and used for forensic DNA testing. Thus, any subsequent testing of such surreptitiously found information and the potential for its inclusion in a local or state databank — even the initial collection of the item — may violate the individual's Fourth Amendment rights.

Section 8. Scientific Evidence Admissibility Standards

Advances in the use and validation of conventional DNA technologies have led courts to admit DNA evidence on the national level under *Daubert,*[63] *Frye*[64] or other standards.[65] This is particularly true for *nuclear (autosomal) DNA (nDNA).*[66] Some courts have also approved the admission of *mtDNA*[67] and *Y-STR*[68] test results under their relevant admissibility standards. However, these latter two technologies may still remain open to challenge. Challenges may also be valid for more novel technologies or those that have not been in use for long, alternative applications of existing technologies, and future developments.

Even when the DNA technology has been accepted, an admissibility hearing may be allowed in some jurisdictions to see whether the testing, analysis and conclusion in a particular case meet the applicable standard.[69] Other jurisdictions require looking at the weight of the evidence and not the threshold admissibility.[70] Counsel should consider whether it is more advantageous to expose any deviations from protocol in an admissibility hearing or wait for trial.

Section 9. Motions in Limine — Statistics Issues

Regardless of threshold admissibility, the DNA evidence offered by the government may be challenged at pretrial, under the following Federal Rules of Evidence (FRE):

- FRE 401, as lacking in relevance.
- FRE 403, as having limited relevance that is substantially outweighed by the risk of unfair prejudice.

The latter issue arises in cases where the crime scene evidence yields only a *partial DNA profile* that matches the defendant or the defendant is one of the DNA contributors. The lower the number of loci where alleles are found in the crime scene evidence and/or the higher the number of callable alleles at the majority of loci, the greater the likelihood of unfair prejudice and the risk of juror confusion.

As a general rule, testimony about partial DNA profiles with low statistical significance has been admitted, with restrictions, during the prosecution's argument to avoid jury misinterpretation.[71] In particular, courts must be responsive to the risk of what is termed the *prosecutor's fallacy:*

> Defendant Palmer also cites to what is more formally called "the fallacy of the transposed conditional, or the prosecutor's fallacy." This fallacy represents incorrect reasoning — i.e., when the jury "will confuse the probability of a random match with the potentially very different probability that the defendant is not the source of the matching samples." ... [T]he Government must be "careful to frame the DNA profiling statistics presented at trial as the probability of a random match, not the probability of the defendant's innocence that is the crux of the prosecutor's fallacy." While this is a very real danger, the courts that have dealt with this potential problem have found that careful oversight by the district court and proper explanation can easily thwart this issue.[72]

When the *random match probability* is high — that is, when the allele combination is found with great frequency across the population — an argument can be made for absolute exclusion of the evidence because of the high likelihood of confusion and the low probative value of the proof. For example, the evidence was properly excluded in a case where "the random match probability with respect to the DNA detected on the [evidence] was approximately 1 in 2 from the African-American population."[73]

Thus, when the prosecution offers partial match evidence, the defense counsel can seek exclusion under Rule 403.

If exclusion is denied, seek to ensure that the expert testimony and the prosecution's argument are carefully restricted so that the evidence is not misrepresented or misused.

Of note is that, in light of the current version of the *SWGDAM Guidelines* — which require the conduct of a statistical analysis in support of any *inclusion* that is determined to be relevant in the context of the case (Guideline 4.1) — the possibility of the prosecution seeking to use a potentially highly prejudicial statistic is likely to occur much more frequently. In the past, the laboratory had the discretion to decide if providing a statistic for a partial profile that contained very little identifying information — but still included a suspect — would be likely to be more prejudicial than helpful. This is no longer the case. Although, in general, the statistics for the DNA typing results that provide the most genetic information and/or the highest discrimination potential are to be reported (Guideline 4.1.1), if a report's conclusion statement for a partial profile result includes the suspect and that partial profile is the only profile generated, the *SWGDAM Guidelines* require that a statistical number be attached — no matter how little data can be derived from the profile.

Section 10. Motions in Limine — Presence of the Defendant's DNA in the Databank

When a "cold hit" or databank check results in the defendant's identification as a suspect, counsel should file a motion *in limine* seeking to preclude testimony regarding this investigative technique. How the defendant came to be a suspect is rarely persuasive. Once jurors learn that the defendant's DNA was in a databank, the inevitable conclusion is that the defendant has a prior criminal record.

This is similar to showing a suspect's *mugshot,* a term that clearly connotes a criminal record. Calling it a photograph can help "sanitize" the idea of a mugshot; however, there is no parallel method for "cleansing" the reference to a DNA databank. The following objections should be made:

- Such proof has no relevance, as the databank hit is not necessary to the proof of guilt;[74] under Rule 404(b), such evidence only establishes the suspect's bad character.[75]
- Under Rule 403, any potential probative value is substantially outweighed by the likelihood of unfair prejudice, to wit, the diminution or removal of the presumption of innocence.[76]

When the defendant has been identified through a databank hit, there may also be concerns about how the probability statistics were calculated. For a discussion on database statistics, please see Chapter 9, Section 6.

Section 11. Motions in Limine — Hearsay, Confrontation and DNA Evidence

Both tactical and legal questions arise when a prosecutor attempts to introduce DNA evidence by submitting lab report results. When the DNA results are not in dispute, it may be to the tactical advantage of the accused for the results to be introduced in a brief, nondramatic reading. However, when results (or conclusions derived from results) are in dispute or when the defense is concerned about identifying and preserving potential appellate issues, a hearsay or Confrontation Clause challenge may be appropriate.

A hearsay challenge to a lab report may be viable under state evidence codes if the report does not qualify as a business record because it is prepared for litigation. If the state business records exception does not include such a limitation, a challenge may be brought under Confrontation Clause principles.

In *Crawford* v. *Washington,*[77] the U.S. Supreme Court reconfigured Confrontation Clause analysis as applied to hearsay in criminal trials. The Supreme Court concluded that the right of confrontation restricted only the admission of testimonial hearsay; the Supreme Court intimated that typical business records would be outside of that classification.[78]

Notwithstanding that dictum, many courts have found forensic lab reports to be testimonial hearsay, and thus inadmissible, if the lab examiner is not present for cross-examination.[79] Treatment of DNA analyses has been mixed, with some courts finding the reports to be testimonial[80] and others concluding that the reports are nontestimonial and thus not subject to Confrontation Clause challenge.[81]

A more recent decision determined that the analyst must introduce a DNA report and findings by live testimony rather than submitting the document(s) or an affidavit (unless the accused waives his or her right of confrontation).

In *Melendez-Diaz* v. *Massachusetts,*[82] the U.S. Supreme Court held that, for Confrontation Clause purposes, a lab analyst is as much a witness as someone who observes the crime. The Supreme Court also ruled that lab reports are not admissible as business records because they are intended for use in court and are "a record for the sole purpose of providing evidence against a defendant."[83]

This raises questions over the constitutionality of having an analyst other than the one who ran the tests serve as the in-court witness. Although the Supreme Court did not address this directly in *Melendez-Diaz,* its language makes this practice problematic for confrontation purposes. As the Supreme Court noted, the accused needs to know "what tests the analysts performed, whether those tests were routine, and whether interpreting their results required the exercise of judgment or the use of skills that the analysts may not have possessed."[84]

By definition, the person who technically reviewed the case file is equally qualified to answer these questions. The Supreme Court explained that cross-examination helps ensure that methods were applied reliably. It emphasized the difference between a witness who can authenticate the document and one who may testify as to its contents:

> [A] clerk ... [may be] permitted to certify to the correctness of a copy of a record kept in his office but [has] no authority to furnish, as evidence for the trial of a lawsuit, his interpretation of what the record contains or shows, or to certify to its substance or effect.[85]

Thus, if the analyst who performed the test (and/ or the technical case file reviewer) is no longer available to testify, confrontation principles may bar a replacement witness from testifying unless that person conducts the testing and analysis again.[86]

In *Melendez-Diaz,* the Supreme Court did approve of "notice-and-demand" statutes — laws or rules of procedure that require the prosecution to give notice of its "intent to use an analyst's report as evidence at trial, after which the defendant is given a period of time in which he may object to the admission of the evidence absent the analyst's appearance live at trial."[87] However, the constitutionality of provisions that place an initial burden on the defendant to subpoena the analyst was left unresolved.[88]

Melendez-Diaz makes clear that, for confrontation purposes, the critical witness is the analyst — not individuals with peripheral involvement such as "chain of custody" witnesses.

The Supreme Court noted that chain of custody may be proved circumstantially and that the chain need not be completely unbroken in order for evidence to be admissible.

Thus, before a DNA report or test result is introduced at trial, defense counsel must determine whether it is desirable to have that proved in court with live testimony. Subsequently, counsel must determine whether the prosecution's method of presenting the results aligns with the defendant's confrontation rights.

Section 12. Admitting Evidence

CODIS Searches of Partial and Mixture Profiles

Not all CODIS hits have 13 loci that match. In some cases, the crime scene evidence may be degraded and have identifiable alleles at some, but not all, of the tested loci. In other cases, the crime scene DNA may yield a full (13-loci) profile, but there may not be an offender or arrestee in the database who fully matches the profile. When this happens with a single-source DNA profile, the typical CODIS search will, with rare exception, come back as stating that no matches were detected. Both circumstances have different potential consequences in a criminal case.

When the crime scene evidence is degraded and a 13-loci profile cannot be generated, the evidence can still be searched against databank profiles developed from other evidence and convicted offender/arrestee profiles. In a databank search, if there is a mismatch between the offender's/arrestee's or previously existing evidence profile at more than one allele, the offender/arrestee or casework profile will not be returned during the search as potentially matching the newly entered partial evidence profile, even when very limited typing results are obtained. However, if there are no mismatches or only one mismatch between the offender's/arrestee's or previously entered casework profile and the newly entered partial evidence profile, the offender/arrestee or casework profile will be returned on the list of matching profile(s).

In these cases where a hit has occurred, the issues are (1) a threshold of relevance and (2) a secondary concern of unfair prejudice. Evidence is relevant if it has any tendency to make a fact of consequence more or less probable. Applying that modest standard, courts have found that even a partial match to DNA evidence is relevant; it is no different than proof that the suspect and the perpetrator had the same color hair or drove the same type of vehicle.[89]

As previously stated, the problem with a partial DNA profile developed from crime scene evidence is the risk of juror confusion and misinterpretation of the statistics — and this is even before the issue of a databank hit is taken into account. A statistical probability of 1 in 2 — which means that one in every two people within the appropriate population group have the same trait — may be misunderstood by jurors as there being a 1-in-2 chance that the DNA came from the suspect. Because of this risk, courts have either placed strictures on prosecution arguments to ensure the clarity of the evidence[90] or have excluded the DNA evidence entirely under a Rule 403 analysis.[91]

When confronted with such evidence, defense counsel would be well advised to consider:

- Ensuring that a comprehensive review has been conducted of the testing results obtained from the crime scene evidence to see whether there might be an identifiable allele at another locus or loci that might exclude the defendant.
- Consulting with an expert to ensure an understanding of the statistics.
- Moving *in limine* to exclude the testimony or restrict prosecutorial argument on its significance to ensure that it is not misstated.
- Presenting a defense expert to ensure that the low statistical significance of the evidence is made clear to the jury.
- Securing a jury instruction that clarifies the limited strength of the proof.
- Making appropriate objection to any prosecutorial argument that misstates the evidence.

Another problem arises when a partial DNA profile is generated and there is no known suspect. The crime lab will upload the partial DNA profile into their local, state and/or national database(s), provided that it meets the criteria for upload. The defense cannot object to such an investigative tool, as there is no basis for objection (and, indeed, no defendant at the time of the search).

As indicated above, searches are governed by the rules of the particular databank and whether it can accept partial profiles. If the search parameters are met, a large number of matching profiles may be generated, even when a relatively high number of loci are available.[92] When such evidence is used to generate a suspect and initiate a prosecution, at least two concerns arise.

First, there is debate over which model to use when calculating the statistical probability: Should it be assessed as the probability of randomly matching that profile in the offender/ arrestee databank or in terms of the rarity of the DNA profile in the population as a whole?[93] Use of a statistic that incorporates the size of the offender databank raises a concern that the jury will learn of the defendant's prior arrests or convictions.

Counsel should explore whether to challenge the statistics used by the prosecution and whether to avoid reference to the defendant having been in the offender or arrestee databank. A motion *in limine* to resolve this must be litigated pretrial. Counsel should seek a resolution that uses the lower statistic and precludes mention of an offender database match.

Just as a partial profile may be uploaded to generate leads, a complete crime scene profile may be checked against the DNA databank and may closely — but not completely — match someone in the pool of arrestees and offenders. Failure to match fully at more than one locus will exclude the individual in the databank search, which precludes this from becoming a problem. However, a modified search that allows for more mismatches can be initiated by the lab such that profiles with a high correspondence of matching alleles at the remaining loci may turn up on a list of "nearly matching" profiles. The indication may be that the perpetrator is a relative of the "near-match" individual. Leads generated through databank checks that then focus police attention on relatives are called "familial DNA searches." The use of familial DNA searches is projected to increase the crime-solving rate from the current 10% to 14%.[94] England and Wales have extensively applied these types of searches.[95] In 2006, the FBI changed its policy to allow familial DNA searches using CODIS.[96] More recently, states have debated whether to permit such testing. California announced a formal adoption of familial DNA searches,[97] whereas Maryland's 2008 legislation authorizing DNA testing of violent crime arrestees bans familial DNA investigations.[98]

Familial DNA searches are of primary interest and receive extensive attention in the media,[99] within the forensic science and law enforcement communities,[100] and in legal research and scholarship.[101]

The results (if any) from a familial DNA search will not be conclusive, but they will provide investigative leads. In this regard, these leads are analogous to a partial license plate number from a vehicle observed at a crime scene. Police may identify numerous cars with the partial plate but, without more information, there is no probable cause to arrest.

From the defense perspective, the legal issues will arise from what the police do next — which family members are targeted, and whether and how police obtain DNA profiles from those individuals. If a family member or relative ends up "matching" the crime scene profile and is arrested, counsel will need to address the search and seizure issues that arose from obtaining the person's sample (see Chapter 7, Section 2).

Unless DNA is obtained by consent, court order or warrant, or from the relative's abandoned property, the results of any comparison will be suppressible as fruits of unlawful searches and seizures. In addition to Fourth Amendment concerns, defense counsel should examine whether any state constitutional provision or law was violated in the conduct of such searches (or in the initial databank search).

Uploading Partial Profiles to CODIS

As previously discussed, the local or casework lab has the option of uploading partial profiles into its local databank, and in some cases the state databank, to see if there are any matches that can be used to generate leads for further investigation.

The limits on this practice will be determined by the rules of the jurisdiction's databank. Current CODIS protocol permits uploading partial DNA profiles into the *National DNA Index System (NDIS)* if identifying alleles are obtained for at least 10 of the 13 core loci. Criteria for uploading into the *State DNA Index System (SDIS)* vary from state to state. *Local DNA Index System (LDIS)* uploading requirements are dependent on local agencies' protocols. When samples are uploaded, they typically remain in the databank permanently. An option separate from uploading is often called a "keyboard search," whereby a one-time search of a partial profile is conducted against all of the resident profiles in the databank at that time. Such an option will depend on the jurisdiction's applicable laws or a court order. (See Chapter 9, Section 13, and Chapter 10 for more information.)

Searching CODIS for Less Than a Perfect 13-Loci Match

As indicated earlier, when the search of a DNA profile fails to produce a match or hit, a search may be initiated for close matches. These searches seek to locate individuals in the databank with a higher than expected number of matching alleles — this number will vary depending on the relationship between the individual in the databank and the source of the searched DNA profile. The search parameters can be set to look for a full sibling, parent, child, or some other defined relationship. The more remote the biological relationship, the broader the search algorithm must be set, which translates into a larger number of close matches in the search report. The search report result will be a list of individuals that may contain a person who is a relative of the actual perpetrator.[102] For example, if a search was set up to look for a full sibling in the national databank, and the resulting search report were to list 18 people who had one or two alleles matching the crime scene profile at 11 of the 13 loci, police might want to investigate relatives of those 18 individuals.

End Notes

[1] 489 U.S. 602, 616 (U.S. 1989).

[2] Id. (citations omitted).

[3] 489 U.S. at 617. See also Maclin, "Is Obtaining an Arrestee's DNA a Valid Special Needs Search Under the Fourth Amendment?" 34 J. L. M ED. & E THICS 165 (Summer 2006); and Kaye, "Who Needs Special Needs? On The Constitutionality of Collecting DNA and Other Biometric Data from Arrestees," 34 J. L. M ED. & E THICS 188 (Summer 2006).

[4] See Va. Code Ann. 19.2-310.2.

[5] Va. Code Ann. 19.2-270.5.

[6] See Ala. Code 36-18-24; Alaska Stat. 44.41.035; Ariz. Rev. Stat. Ann. 13-610; Ark. Code Ann. 12-12-1109; Cal. Penal Code 290.7; Colo. Rev. Stat. 17-2-201; Conn. Gen. Stat. Ann. 54-102g; Del. Code Ann. tit. 29, § 4713; Fla. Stat. Ann. 943.325; Ga. Code Ann. 24-4-60; Haw. Rev. Stat. 706-603; Idaho Code 19-5507; 730 Ill. Comp. Stat. Ann. 5/5-4-3; Ind. Code Ann. 10-1-9-10; Iowa Code Ann. 13.10; Kan. Stat. Ann. 21-2511; Ky. Rev. Stat. Ann. 17.170; La. Rev. Stat. Ann. 15:609; Me. Rev. Stat. Ann. tit. 25, § 1574; Md. Ann. Code art. 88B, § 12A; Mass. Gen. Laws Ann. ch. 22E, § 3; Mich. Stat. Ann. 28.171; Minn. Stat. Ann. 609.117; Miss. Code Ann. 45-33-37; Mo. Ann. Stat. 650.055; Mont. Code Ann. 44-6-102; Nebraska Rev. Stat. 29-4106; Nev. Rev. Stat. 176.0913; N.H. Rev. Stat. Ann. 651-C:2; N.J. Stat. Ann. 53:1-20.20; N.M. Stat. Ann. 29-16-6; N.Y. Exec. Law 995- c; N.C. Gen. Stat. 15A-266.4; N.D. Cent. Code 31-13-03; Ohio Rev. Code Ann. 2901.07; Okla. Stat. Ann. tit. 74, 150.27a(A); Or. Rev. Stat. 137.076; 42 Pa. Cons. Stat. Ann. 4701; R.I. Gen. Laws 12-1.5-8; S.C. Code Ann. 23-3-620(B)(1); S.D. Codified Laws 23-5A-1; Tenn. Code Ann. 40-35-321; Tex. Gov't Code Ann. 411.1471; Utah Code Ann. 53-10-403; Vt. Stat. Ann. tit. 20, 1933; Va. Code Ann. 19.2-310.2; Wash. Rev. Code Ann. 43.43.754; W. Va. Code 15-2B-6; Wis. Stat. Ann. 165.76; Wyo. Stat. Ann. 7-19-403.

[7] 42 U.S.C. § 14135 et seq.

[8] See 42 U.S.C. § 14135a(d). The federal statute provides:

a) Collection of DNA samples.
   1) From individuals in custody. The Director of the Bureau of Prisons shall collect a DNA sample from each individual in the custody of the Bureau of Prisons who is, or has been, convicted of a qualifying Federal offense (as determined under subsection (d)) or a qualifying military offense, as determined under section 1565 of title 10, United States Code.
   2) From individuals on release, parole or probation. The probation office responsible for the supervision under Federal law of an individual on probation, parole or supervised release shall collect a DNA sample ing Federal offense (as determined under subsection (d)) or a qualifying military offense, as determined under section 1565 of title 10, United States Code.
   3) Individuals already in CODIS. For each individual described in paragraph (1) or (2), if the Combined DNA Index System (in this section referred to as "CODIS") of the Federal Bureau of Investigation contains a DNA analysis with respect to that individual, or if a DNA sample has been collected from that individual under section 1565 of title 10, United States Code, the Director of the Bureau of Prisons or the probation office responsible (as applicable) may (but need not) collect a DNA sample from that individual.
   4) Collection procedures.
      (A) The Director of the Bureau of Prisons or the probation office responsible (as applicable) may use or authorize the use of such means as are reasonably necessary to detain, restrain and collect a DNA sample from an individual who refuses to cooperate in the collection of the sample.

(B)The Director of the Bureau of Prisons or the probation office, as appropriate, may enter into agreements with units of State or local government or with private entities to provide for the collection of the samples described in paragraph (1) or (2)

5) Criminal penalty. An individual from whom the collection of a DNA sample is authorized under this subsection who fails to cooperate in the collection of that sample shall be:
(A) Guilty of a class A misdemean or; and
(B) Punished in accordance with
title 18, United States Code.

b) Analysis and use of samples. The Director of the Bureau of Prisons or the probation office responsible (as applicable) shall furnish each DNA sample collected under subsection (a) to the Director of the Federal Bureau of Investigation, who shall carry out a DNA analysis on each such DNA sample and include the results in CODIS.

c) Definitions. In this section: (1) The term "DNA sample" means a tissue, fluid or other bodily sampleof an individual on which a DNA analysis can be carried out. (2) The term "DNA analysis" means analysis of the deoxyribonucleic acid (DNA) identification information in a bodily sample.

d) Qualifying Federal offenses. (1) The offenses that shall be treated for purposes of this section as qualifying Federal offenses are the following offenses under title 18, United States Code, as determined by the Attorney General:

A. Murder (as described in section 1111 of such title), voluntary manslaughter (as described in section 1112 of such title), or other offense relating to homicide (as described in chapter 51 of such title, sections 1113, 1114, 1116, 1118, 1119, 1120 and 1121).
B. An offense relating to sexual abuse (as described in chapter 109A of such title, sections 2241 through 2245), to sexual exploitation or other abuse of children (as described in chapter 110 of such title, sections 2251 through 2252), or to transportation for illegal sexual activity (as described in chapter 117 of such title, sections 2421, 2422, 2423 and 2425).
C. An offense relating to peonage and slavery (as described in chapter 77 of such title [18 USCS §§ 1581 et seq.]).
D. Kidnapping (as defined in section 3559(c)(2)(E) of such title).
E. An offense involving robbery or burglary (as described in chapter 103 of such title, sections 2111 through 2114, 2116, and 2118 through 2119).
F. Any violation of section 1153 involving murder, manslaughter, kidnapping, maiming, a felony offense relating to sexual abuse (as described in chapter 109A [18 USCS §§ 2241 et seq.]), incest, arson, burglary or robbery.
G. Any attempt or conspiracy to commit any of the above offenses.

(2) EIn addition to the offenses described in paragraph (1), the following offenses shall be treated for purposes of this section as qualifying Federal offenses, as determined by the Attorney General:

A. Any offense listed in section 2332b(g)(5)(B) of title 18, United States Code.
B. Any crime of violence (as defined in section 16 of title 18, United States Code).
C. Any attempt or conspiracy to commit any of the above offenses.

(e) Regulations.

1) EIn general. Except as provided in paragraph (2), this section shall be carried out under regulations prescribed by the Attorney General.
2) EProbation officers. The Director of the Administrative Office of the United States Courts shall make available model procedures for the activities of probation officers in carrying out this section.

(f) ECommencement of collection. Collection of DNA samples under subsection (a) shall, subject to the availability of appropriations, commence not later than the date that is 180 days after the date of the enactment of this Act.

[9] 42 U.S.C. § 14135 et seq. 42 U.S.C. § 14135; 468 F. Supp. 2d 261, 262.

[10] See 42 U.S.C. § 14135e, which provides:

(a) In general. Except as provided in subsection (b), any sample collected under, or any result of any analysis carried out under section 2, 3 or 4 [42 U.S.C. § 14135, 14135a, or 14135b] may be used only for a purpose specified in such section.

(b) Permissive uses. A sample or result described in subsection (a) may be disclosed under the circumstances under which disclosure of information included in the Combined DNA Index System is allowed, as specified in subparagraphs (A) through (D) of section 210304(b)(3) of the Violent Crime Control and Law Enforcement Act of 1994 (42 U.S.C. § 14132(b)(3)).

[11]. Proper authorization required. 42 U.S.C. § 14135e(c):

(c) Criminal penalty. A person who knowingly —

1) EDiscloses a sample or result described in subsection (a) in any manner to any person not authorized to receive it; or
2) LObtains, without authorization, a sample or result described in subsection (a), shall be fined not more than $100,000.

[12] United States v. Reynard, 2007 U.S. App. LEXIS 665 (9th Cir. 2007) (DNA collection from convicted felons constitutional, even if convicted prior to date of act — retroactivity challenge rejected); United States v. Hook, 471 F.3d 766 (7th Cir. 2006) (same).The Tenth and Second Circuits have upheld such searches under the "special needs" exception to the warrant requirement. United States v. Kimler, 335 F.3d 1132, 1146 (10th Cir. 2003), cert. denied, 540 U.S. 1083 (mem.) Roe v. Marcotte, 193 F.3d 72 (2d Cir. 1999). Other circuits have relied on the determination that inmates do not have a reasonable expectation of privacy against DNA collections from convicted persons. Ernst v. Roberts, 379 F.3d 373 (6th Cir. 2004) (upholding as reasonable a requirement that federal offenders who were on parole, probation or supervised release submit to compulsory DNA profiling); Ernst v. Roberts, 379 F.3d 373: Rehearing, en banc, granted by, Vacated by: Ernst v. Roberts, 2004 U.S. App. LEXIS 24149 (6th Cir. Nov. 17, 2004): Different results reached on rehearing at: Ernst v. Rising, 427 F.3d 351. Velasquez v. Woods, 329 F.3d 420, 421 (5th Cir. 2003) (per curiam) (same); Jones v. Murray, 962 F.2d 302, 306-07 (4th Cir. 1992) ("While we do not accept even this small level of intrusion for free persons without Fourth Amendment constraint ... the same protections do not hold true for those lawfully confined to the custody of the state. As with fingerprinting, therefore, we find that the Fourth Amendment does not require an additional finding of individualized suspicion before blood can be taken from incarcerated felons for the purpose of identifying them."). See generally, Nicholas v. Goord, 2004 U.S. Dist. LEXIS 11708 (D.N.Y. 2004) (upholding New York statute against Fourth Amendment challenge and summarizing case law); Nicholas v. Goord, 2004 U.S. Dist. LEXIS 11708: Subsequent appellate history contains negative analysis. Recent state court decisions to the same effect include State v. O'Hagen, 914 A.2d 267, 270 (N.J. 2007) ("The Act requires all persons convicted of a crime [or found not guilty by reason of insanity] to give a deoxyribonucleic acid (DNA) sample. We hold that the Act is constitutional under both [the United States and New Jersey] Constitutions."); Commonwealth v. Derk, 895 A.2d 622, 630 (Pa. Super. Ct. 2006).

[13] See In re Calvin S.,150 Cal. App. 4th 443 (Cal. App. 2007) (recognizing juvenile's stronger privacy rights but still finding balance of interests to favor requiring entry of juvenile DNA into databank); United States v. Stewart, 468 F. Supp. 2d 261 (D. Mass. 2007) (finding unconstitutional the DNA Analysis Backlog Elimination Act of 2000 as applied to probationer convicted of Social Security fraud); United States v. Stewart, 468 F. Supp. 2d 261: Subsequent appellate history contains negative analysis. Vermont v. Watkins (Vt. Dist. Ct. App. 24. 2006) (No. 6805-2-04) (invalidating on state constitutional grounds the "suspicionless collection and banking" of DNA samples from all convicted nonviolent felons); Green v. Berge, 354 F.3d 675, 679-81 (7th Cir. 2004) (Easterbrook, J., concurring) (noting, "[f]elons whose terms have expired" form a different category of individuals than supervised releasees for the purposes of a Fourth Amendment inquiry); United States v. Amerson, 483 F.3d 73, 79 (2d Cir. 2007) (discussing that probationers (at issue in Amerson) have a greater expectation of privacy than parolees (at issue in Samson)); United States v. Kincaide, 379 F.3d 813 (9th Cir. 2004) (en banc) (circuit splits 6-5 in upholding an Act requiring probationers, parolees and persons on supervised release to provide DNA for use in a databank, with special concurrence noting that ruling does not apply to persons who have fully served supervision period and paid debt to society); In Re: C.T.L., 722 N.W.2d 484 (Minn. Ct. App. 2006).

[14] Fingerprint at scene may not establish probable cause: California: Birt v. Superior Court (1973) 34 Cal. App. 3d 934 (Cal. App. 1973) (defendant's fingerprints on a lighter found inside a rented van used in a robbery deemed insufficient to support a probable cause finding as the suspect could have left the lighter in the rental van on some occasion long before the robbery); New Hampshire: State v. Maya, 493 A.2d 1139, 1144 (N.H. 1985) (fingerprint at scene establishes probable cause when on item that only the burglar could have left).

[15] Compare United States v. McNeill, 2007 U.S. Dist. LEXIS 56209, 18-21 (D. Pa. 2007) (cigarette butt at crime scene plus numerous additional items of circumstantial evidence proved sufficient for a warrant to obtain defendant's DNA profile).

[16] As of October 2011, 25 states and the federal government require the collection of DNA from arrestees.

[17] 722 N.W.2d 484, 486 (Minn. Ct. App. 2006).

[18] 722 N.W.2d at 491.

[19] 722 N.W.2d at 492.

[20] Anderson v. Commonwealth, 274 Va. 469, 475 (Va. 2007).

[21] 489 U.S. 602, 616 (U.S. 1989).

[22] 384 U.S. 757, 766-767 (1966). Schmerber v. California, 384 U.S. 757: 344 Ill. App. 3d 684, 687; 800 N.E.2d 1227, 1230; 279 Ill. Dec. 644, 647.

[23] 489 U.S. at 619 (citations omitted).

[24] 392 U.S. 1 (1968) (permitting civilian stops and, in some instances, frisks on reasonable suspicion, a standard lower than probable cause).

[25] None of the foregoing implies that a brief detention in the field for the purpose of fingerprinting, where there is only reasonable suspicion not amounting to probable cause, is necessarily impermissible under the Fourth Amendment. Hayes v. Florida, 470 U.S. 811, 816 (U.S. 1985). The Hayes court also explained that removing a person to a police station required probable cause, "at least where not under judicial supervision," a caveat implying that a lower standard might apply when a judge makes the determination.

[26] United States v. Garcia-Ortiz, F. Supp. 2d, 2005 U.S. Dist. LEXIS 38108 (D.P.R. 12/23/05) (2005 WL 3533322); United States v. Garcia-Ortiz, 2005 U.S. Dist. LEXIS 38108: Subsequent appellate history contains possible negative analysis.27 U.S. v. Swanson, 155 F. Supp. 2d 992 (C.D. Ill. 7/11/01); In re Shabazz, 200 F. Supp. 2d 578, 583 (D.S.C. 2002); In re Grand Jury Proceedings Involving Vickers, 38 F. Supp. 2d 159 (D.N.H. 12/4/98); State v. Lee, 964 So. 2d 967 (La. Ct. App. 2007); State v. Rodriguez, 240, 921 P.2d 643, 650 (Ariz. 1996) (based on Arizona Nontestimonial Order [NTO] statute); State v. Rodriguez, 240, 921 P.2d 643.State v. Lee, 964 So. 2d 967 (La. Ct. App. 2007); In re Nontestimonial Identification Order Directed to R.H., 762 A.2d 1239 (Vt. 2000).

[27] Schmerber v. California, 384 U.S. 757 (1966).

[28] Schmerber, 384 U.S. at 757, 769-770.

[29] Hayes v. Florida, 470 U.S. 811, 816 (U.S. 1985) at 813. This rule of exclusion has not been applied in immigration arrest cases if the fingerprints were taken for deportation proceeding purposes and not for an "unanticipated and unforeseen criminal prosecution[.]" United States v. Oscar-Torres, 2007 U.S. App. LEXIS 25988 (4th Cir. 2007). This limited exception to the suppression doctrine should have no applicability in the typical criminal case.

[30] American Bar Association (ABA) Standards for DNA Evidence, Standard 2.2(b)(i)(C), www. abanet.org/crimjust/standards/dnaevidence. html#2.2. The ABA has created Criminal Justice Standards on DNA Evidence. Regarding the collection of DNA from a suspect, Standard 2.2 requires a search warrant or judicial order to collect DNA over a person's objection. The standard further cautions that, "except in exigent circumstances," this should not occur until after a hearing with counsel. The state must establish reasonable suspicion or probable cause, depending on the collection method, and the state must further establish "the sample will assist in determining whether the person committed the crime."

[31] Practically speaking, the state may seek to obtain your client's DNA profile to compare the profile with any DNA that may be found on a certain piece of evidence, such as a gun. Without knowing if there is DNA on the gun, the state cannot possibly assure the court that having the defendant's sample will assist in determining whether he or she possessed the gun. As such, defense counsel should strenuously object to the taking of their client's DNA profile without proof that the state has an item of evidence that has already yielded an unidentified DNA profile.

[32] Schmerber v. California, 384 U.S. at 757, 770.

[33] See, e.g., United States v. McNeill, 2007 U.S. Dist. LEXIS 56209, 18-21 (D. Pa. 2007) (warrant for DNA sample from McNeill approved after his DNA profile was found on a cigarette butt left at the robbery scene; DNA warrant was obtained for confirmatory testing and match).See, e.g., United States v. Wright, 215 F.3d 1020, 1026 (9th Cir. 2000) (blood trail from bank robber who was shot at crime scene is a sufficient predicate for DNA testing of Wright, once evidence was developed identifying him as suspect); People v. Phillips, 336 Ill. App. 3d 1033, 1035-1036 (Ill. App. Ct. 2003) ("One would expect that when oral and vaginal sexual assaults are alleged, hair, semen, and/or blood may be present, thereby establishing a sufficient nexus between the assault and the need for such evidence.").

[34] Jones v. State, 343 So. 2d 921, 923 (Fla. 3rd DCA 1977) (internal quotations omitted). ~

[35] 641 N.E.2d 1328 (Mass. 1994). ~

[36] Id. at 1331 (internal citation omitted). ~

[37] Id. (citing Warden Md. Penitentiary v. Hayden, 387 U.S. 294, 307, 87 S. Ct. 1642, 1650 (1967)). See also Pittman v. United States, 375 A.2d 16, 19 (D.C. 1977) (Tangible objects — like the samples sought from Mr. Smith — must be relevant in some way "either independently or as corroborative of other evidence.").

[38] Id. at 1332. ~

[39] 416 A.2d 1209 (Conn. 1979). ~

[40] Id. at 1211 (emphasis added). ~

[41] See, e.g., State v. Hardaway, 36 P.3d 900, 915 (Mont. 2001) (disallowing warrantless test of blood on the defendant's hand where police knew it was the defendant's and not a crime victim's). ~

[42] Cupp v. Murphy, 412 U.S. 291, 296 (U.S. 1973). ~

[43] United States v. Edwards, 415 U.S. 800, 802 (U.S. 1974). ~

[44] Washington v. State, 922 So. 2d 145, 169 (Ala. Crim. App. 2005); Commonwealth v. Houghton, 2007 Mass. Super. LEXIS 390 (Mass. Super. Ct. 2007). ~

[45] Cf. State v. Madplume, 2005 Mont. Dist. LEXIS 1510 (Mont. Dist. 2005) (finding no exigency where defendant "was in police custody and under arrest in a tribal jail cell with no sink, toilet or water and was under the full control of law enforcement"); 2005 Mont. Dist. LEXIS 1510. ~

[46] Schneckloth v. Bustamonte, 412 U.S. 218 (U.S. 1973). ~

[47] Florida v. Jimeno, 500 U.S. 248, 251 (U.S. 1991) ("The scope of a search is generally defined by its expressed object."). ~

[48] Some courts have held that language limiting consent to a specific use bars all others. See, e.g., State v. Gerace, 437 S.E.2d 862, 863 (Ga. Ct. App. 1993) (consent to alcohol and drug testing in blood did not extend to DNA testing); State v. Binner, 886 P.2d 1056 (Ore. Ct. App. 1994) (consent for alcohol testing of blood and refusal to allow drug testing of blood must be honored).

[49] Cf. United States v. Yang, 478 F.3d 832, 835 (7th Cir. 2007) (relinquishing notebook to police for fingerprint analysis allowed police to read notebook):

Rather, he voluntarily allowed Officer Schneider to take the notebooks in their entirety to the police station and hold them for several days. He placed no limitations on access to the notebooks. He did not separate the notebook covers and keep the written contents to himself. He did not request that the officers perform the fingerprint analysis in his presence. He did not close or secure the contents of the notebooks in anyway so that only the covers could be accessed.

[50] See, e.g., *Wyche* v. *State,* 906 So. 2d 1142, 1143-1149 (Fla. 1st DCA 2005).

[51] See Walker, S., "Police DNA 'Sweeps' Extremely Unproductive: A National Survey of Police DNA 'Sweeps,'" POLICE PROFESSIONALISM INITIATIVE, Department of Criminal Justice, University of Nebraska (2004).

[52] *State* v. *Notti,* 71 P.3d 1233, 1237-1238 (Mont. 2003).

[53] *Pharr* v. *Commonwealth,* 646 S.E.2d 453, 457 (Va. Ct. App. 2007). See also *State* v. *Glynn,* 166 P.3d 1075, 1078 (Kan. Ct. App. 2007) ("We hold there is no constitutional violation or infringement of any rights of privacy when the police use a DNA profile lawfully obtained in one case to investigate and charge the DNA donor in a subsequent and different case or cases.").

[54] For example, in one Michigan case, the police "requested" that all African-American men in a particular neighborhood "volunteer" DNA samples and were warned that refusal would cause grounds for suspicion. Some who refused were, in fact, subject to later search warrants.

[55] After a class action lawsuit, police agreed to destroy the collected DNA. See "DNA Dragnet Police Seek DNA Samples From the Public to Catch the Guilty," available at www.cbsnews.com/ stories/2004/09/10/ 60minutes/main642684.shtml; Peterson, R.S., "Note: DNA Databases: When Fear Goes Too Far," 37 AMER . C RIM. L. R EV. 1219, 1224, 1227 (2000) (citing Willing, R., "Privacy Issue Is the Catch for Police DNA 'Dragnets,'" USA TODAY (Sept.16, 1998); Hansen, M., "DNA Dragnet," ABA J. (May 2004) 38-43). A similar situation occurred in Louisiana, and citizens were allegedly told that failure to cooperate would result in public identification as a suspect. O'Brien, K., "Men Seek Return of DNA From Serial Killer Search: Some Claim Police Bullied Them For Swabs," TIM E S -PICAYUNE (Dec. 28, 2003); Sayre, A., "Tool of DNA Offers Potential for Abuse," THE B ATON R OUGE A DVOCATE (Dec. 22, 2003); "Men Targeted by 'DNA Dragnet' Demand Return, Destruction of Samples," THE NEW S TANDARD (Nov. 9, 2004). The police also searched the DNA of approximately 1,300 other individuals in this case. See "Judge Orders Removal of Wichita Man's DNA Sample from Database," AP N EW SW IRE (March 21, 2005).

[56] *Id.*

[57] *California* v. *Greenwood,* 486 U.S. 35, 40-41 (U.S. 1988) (internal quotation and citation omitted): [H]aving deposited their garbage in an area particularly suited for public inspection and, in a manner of speaking, public consumption, for the express purpose of having strangers take it, respondents could have had no reasonable expectation of privacy in the inculpatory items that they discarded.

[58] See, e.g., *Litchfield* v. *State,* 824 N.E.2d 356, 363 (Ind. 2005); Henderson, "Learning From All Fifty States: How to Apply the Fourth Amendment and Its State Analogs to Protect Third Party Information From Unreasonable Search," 55 CATHOLIC U. L. R EV. 373 (Winter 2006).

[59] *Commonwealth* v. *Cabral,* 866 N.E.2d 429, 433-435 (Mass. App. Ct. 2007).

[60] *State* v. *Athan,* 158 P.3d 27, 33 (Wash. 2007).

[61] Hudson v. State, 205 S.W.3d 600, 604 (Tex. App. 2006).

[62] State v. Reed, 641 S.E.2d 320, 322-323 (N.C. Ct. App. 2007).

[63] Daubert v. Merrell Dow Pharmaceuticals, 509 U.S. 579, 113 S. Ct. 2786, 125 L.Ed.2d 469 (1993).

[64] Frye v. United States, 54 App. D.C. 46, 293 F. 1013, 1014 (D.C. Cir. 1923). Superseded by statute as stated in: Smith v. GE, 2004 U.S. Dist. LEXIS 7011 (D. Mass. Apr. 23, 2004) 2004 U.S. Dist. LEXIS 7011.

[65] See, e.g., Johnson v. State, 264 Ga. 456, 448 S.E.2d 177, 179 (Ga. 1994) (quoting Caldwell v. State, 260 Ga. 278, 286-287(1)(b) (1990)) (admissibility of DNA evidence turns on the trial court's determination of "whether the general scientific principles and techniques involved in [DNA testing] are valid and capable of producing reliable results, [and] also whether [the DNA tester himself] substantially performed the scientific procedures in an acceptable manner."); Spencer v. Commonwealth, 240 Va. 78, 393 S.E.2d 609 (1990) ("The court must make a threshold finding of fact with respect to the reliability of the scientific method offered (i) unless it is of a kind so familiar and accepted as to require no foundation to establish the fundamental reliability of the system; or (ii) unless it is so unreliable that the considerations requiring its exclusion have ripened into rules of law; or (iii) unless its admission is regulated by statute.").

[66] See, e.g., United States v. Morrow, 374 F. Supp. 2d 51, 62 (D.D.C. 2005) (collecting cases under Daubert); People v. Shreck, 22 P.3d 68, 80 (Colo. 2001) (collecting cases under Frye).

[67] See, e.g., United States v. Beverly, 369 F.3d 516, 529 (6th Cir. 2004); Magaletti v. State, 847 So. 2d 523, 528 (Fla. 2nd DCA 2003) (collecting cases); State v. Underwood, 134 N.C. App. 533, 518 S.E.2d 231 (N.C. Ct. App. 1999); State v. Scott, 33 S.W.3d 746 (Tenn. 2000); State v. Council, 335 S.C. 1, 515 S.E.2d 508 (S.C. 1999); People v. Klinger, 185 Misc. 2d 574, 713 N.Y.S. 2d 823 (N.Y. Crim. Ct. 2000); Williams v. State, 342 Md. 724, 679 A.2d 1106 (Md. 1996).

[68]. See United States v. Adams, 189 Fed. App. 120, 124 (3d Cir. 2006); State v. Russell, Wash. App. LEXIS 3041 (Wash. Ct. App. 2007) (holding that Frye hearing is unnecessary, as Y-STR typing is "merely one specific type of STR DNA testing").

[69]. See, e.g., Murray v. State, 838 So. 2d 1073, 1077-1082 (Fla. 2002).

[70]. See, e.g., United States v. Morrow, 374 F. Supp. 2d 42, 50 (D.D.C. 2005).

[71]. United States v. Morrow, 374 F. Supp. 2d 51, 65 (D.D.C. 2005) (collecting cases).

[72]. United States v. Morrow, 374 F. Supp. 2d 51, 66 (D.D.C. 2005).

[73]. United States v. Graves, 465 F. Supp. 2d 450, 459 (D. Pa. 2006): [E]ven with appropriate safeguards, the minimal probative value of the umbrella DNA evidence — in which half of the relevant population cannot be excluded as a contributor to the DNA sample — is substantially outweighed by the danger of unfair prejudice and confusion of the issues. Thus, the sneaker DNA evidence is admissible and the umbrella DNA evidence is not admissible.

[74]. See, e.g., United States v. Sallins, 993 F.2d 344, 346 (3d Cir. 1993) (restricting use of evidence explaining "background" of police investigation where "the need for such evidence is slight").

[75]. United States v. Williams, 113 F.3d 243, 247 n.3 (D.C. Cir. 1997) ("evidence of a prior arrest, without more, would not have been admissible as it would not tend to prove predisposition, but at most general criminal propensity").

[76]. See generally, Old Chief v. United States, 117 S. Ct. 644 (U.S. 1997).

[77]. 541 U.S. 36 (U.S. 2004).

[78]. Citing to authority from the time of the adoption of the Bill of Rights, the Court noted, "Most of the hearsay exceptions covered statements that by their nature were not testimonial — for example, business records or statements in furtherance of a conspiracy." Crawford v. Washington, 541 U.S. 36, 56 (U.S. 2004).

[79] See, e.g., Rollins v. State, 897 A.2d 821, 833 (Md. 2006) (collecting cases).

[80] See, e.g., Roberts v. United States, 916 A.2d 922, 938 (D.C. 2007): [T]he conclusions of FBI laboratory scientists — the serologist, the PCR/STR technician and the examiner — admitted as substantive evidence at trial are "testimonial" under Crawford [and] thus subject to the requirements of cross-examination and declarant unavailability confirmed by that decision.

[81] See, e.g., State v. Crager, 2007 Ohio LEXIS 3355 (Ohio 2007); People v. Geier, 161 P.3d 104, 138 (Cal. 2007).

[82] 129 S. Ct. 2527 (2009).

[83] Id. at 30.

[84] Id. at 25-26.

[85] Id. at 30 (internal quotation omitted).

[86] Also unanswered by this decision is whether the right of confrontation is violated when one analyst testifies to results and to his confirmation or verification by another lab employee. This remains an open issue in confrontation analysis.

87 Id. at 37.

88 This issue may be decided by the U.S. Supreme Court in a review of Briscoe v. Va., 2009 U.S. LEXIS 4947 (U.S., June 29, 2009).

89 See, e.g., United States v. Morrow, 374 F. Supp. 2d 51, 65 (D.D.C. 2005) ("the particular DNA matches identified by Defendants Morrow and Palmer do not show a significant statistical probability that they contributed to those samples; however, they do show that the defendants cannot be excluded as contributors [and thus] the DNA evidence remains probative"); United States v. Graves, 465 F. Supp. 2d 450, 458 (E.D. Pa. 2006) (same; collecting cases).

90 In Morrow (note 89, supra), the court held that "even DNA evidence with relatively low statistical significance may be admitted as probative evidence, provided that certain safeguards are afforded." 374 F. Supp. 2d at 68.

91 In Graves (note 89, supra), the District Court ruled that crime scene DNA on an umbrella was inadmissible because of the risk of undue prejudice and confusion: [T]he minimal probative value of the umbrella DNA evidence — in which half of the relevant population cannot be excluded as a contributor to the DNA sample — is substantially outweighed by the danger of unfair prejudice and confusion of the issues.

92 United States v. Graves, 465 F. Supp. 2d 450, 459 (E.D. Pa. 2006): In the fall of 2005, Arizona ... compared the DNA profiles of each of the 65,493 persons in its database against each other. From this comparison, Arizona DPS reported some remarkable findings: its database had 122 pairs of people who matched at 9 out of the 13 loci [and] 20 that matched at 10 loci ... .Ungvarsky, E. "What Does One in a Trillion Mean?" 20(1) GENE W ATCH 10-14 (January/ February 2007), http://www.wisspd.org/htm/ ATPracGuides/Training/Prog Materials/Conf2011/ CDNAE/11.pdf. See also Chapter 9, Section 7.

93 The debate is aptly summarized in decisional law (see United States v. Jenkins, 887 A.2d 1013, 1025 (D.C. 2005); People v. Nelson, 43 Cal. 4th 1242, 185 P.3d 49 (2008)) and in newspaper accounts ("Debate on Analyzing 'Cold Hit' DNA Matches Swirls in Case Before California Supreme Court," Los ANGELEs TIMEs (May 9, 2008)).

94 "DNA of Criminals' Kin Cited in Solving Cases," W AsHINGToN P osT, A-10 (May 12, 2006) (quoting estimates by Dr. Frederick Bieber).

95 See note: "Less Privacy Please, We're British: Investigating Crime with DNA in the U.K. and the U.S.," 31 HAsTINGs INT'L & C oMP. L. R EV. 487 (Winter 2008).

96 American Prosecutors Research Institute, "Catching Criminals by Investigating Profiles with Allelic Similarities," 10 SILENT W ITNESS, 2 (2006) ("Until recently, the FBI did not release the personal information of partial matches"), www. ndaa.org/publications/ newsletters/silent_witness_volume_10_number_2_2006.html.

97 "Brown Unveils DNA Technique to Crack Unsolved Crimes," OFFICE OF THE ATTORNEY G ENERAL (April 25, 2008), http://ag.ca.gov/ newsalerts/ release.php?id=1548&.

98 Senate Bill 211 (2008), § 2-506(D), http:// mlis.state.md.us/ 2008rs/bills/ sb/sb0211e.pdf.

99 A Not So Perfect Match, '60 Minutes," aired April 1, 2007, www.cbsnews.com/ stories/2007/03/23/ 60minutes/main2600721. shtml?source=search_story.

100 See, e.g., Minutes of the Commonwealth of Virginia Scientific Advisory Committee on Familial DNA (May 8, 2007), www.dfs.virginia. gov/ about/minutes/ saCommittee/ 20070508.pdf. In March 2008, the FBI sponsored a two-day conference on familial DNA issues.

101 See, e.g., Epstein, "'Genetic Surveillance' — The Bogeyman Response to Familial DNA Investigations," SOCIAL S CIENCE R ESEARCH NETWORK (May 5, 2008), http://papers.ssrn.com/sol3/ papers.cfm?abstract_id=1129306.

102 Bieber, Brenner, and Lazer, "Finding Criminals Through DNA of Their Relatives," 312 S CIENCE, 5778 (June 2006), 1315-1316.

8. DNA Basics: Trial Issues

Section 1. Getting Ready for Trial

Finalize Defense Theory

As defense counsel prepares for trial, it is critical to finalize the defense theory with respect to the DNA evidence in the case. At the outset, counsel has two options:

- The DNA profiles match, but there is an innocent explanation for it.
- The client is not the source of the DNA.

If the DNA evidence appears to be rock solid and leaves counsel with no viable angle from which to challenge the evidence itself — for example, if it is not a mixture of body fluids from two or more individuals, there is ample biological material, there are no claims of allelic drop-out, the statistic is in the quadrillions, there is no close relative who might share the profile, and there is no other reasonable explanation — defense counsel may have to concede that the DNA belongs to the client but then challenge the link that the government seeks to establish between the DNA and the crime.

Defense counsel might challenge the time or manner in which the DNA was deposited. Perhaps the defendant had a legitimate reason to be at the crime scene before the crime happened, and his DNA could be there for a noninculpatory reason. In this type of case, counsel could concede the validity of DNA science (because it reliably supports a fact consistent with defense theory) but then contrast it with other forensic evidence that may have a poorly established empirical foundation.

Alternatively, perhaps the location of the DNA or the cellular type from which it was extracted — for example, skin cells rather than blood — is inconsistent with the government theory of events or the manner by which the DNA is alleged to have arrived on the scene. Perhaps the defendant was never at the crime scene, and his DNA was transferred from someone or something to the crime scene or to a piece of evidence after it was collected from the scene.

Given evidence in support, counsel might also choose to argue that the evidence was contaminated at the crime lab when the defendant's *reference sample* was analyzed on the same day as the crime scene evidence. Perhaps the defendant's DNA was planted. These latter lines of argument — *contamination* or a plant — should be approached with caution.

On the other hand, if the case involves a DNA mixture to which the client allegedly contributed, or if it involves a *degraded DNA sample,* the defense theory may be that the DNA analyst made a series of subjective judgment calls throughout the analysis. This is the phenomenon known as observer bias, contextual bias, confirmation bias, ascertainment bias or *expectation bias,* in which an individual unconsciously interprets ambiguous evidence to fit a preconceived notion (in this case, that law enforcement arrested the right person).

In a *"cold hit"* case, the defense theory may be that perhaps the statistic used to represent the probability of a random match dramatically overestimates the significance of finding a match when it was obtained by searching a large database, or ignores the fact that the match may have been coincidental. In the case of Y-STR or mtDNA, the defense theory may be that perhaps the population database used as a basis for the statistic is not a proper sampling of the relevant population and thus cannot serve as a reliable basis for the statistic that purports to describe the frequency with which the profile in question is expected to occur in the population.

Regardless of the theory that counsel selects to challenge the DNA evidence, it should be "beta tested," using both DNA specialists and non-DNA specialists to ensure that the theory comes through clearly and understandably despite the complexity of the evidence. For additional strategy information, see Chapter 10, Section 1.

Section 2. Trial Advocacy

Although all criminal trials require careful adherence to the principles of trial advocacy, special care is needed in cases with forensic DNA evidence. Jurors may not understand the limitations of DNA evidence regarding the risk of an unreliable match and what DNA does not prove in a particular case.

Because of these concerns, counsel must develop a consistent theme and theory — in consultation with an expert in forensic DNA science and trials — that will inform the pretrial motions practice, jury selection, opening statements, witness examination, closing arguments and jury instruction requests. Given that many jurors today are more receptive to receiving information visually rather than by testimony alone, counsel must also assess whether a PowerPoint® or other presentation will enhance the jury's comprehension of the defense theory. As part of this overarching strategy, counsel will have to decide how to address the DNA evidence — to acknowledge it briefly, embrace it as part of the defense theory, or attack it.

Section 3. DNA and the Jury

Some lawyers and experts may forget how to influence the people they are trying to persuade. Moreover, some judges may ignore the people whose responsibility it is to actually decide the case — the jurors. One question that must be addressed is whether jurors really understand what lawyers and experts are trying to communicate.[1]

The issue of juror comprehension of forensic DNA evidence transcends the adversary roles in the courtroom. Sometimes, DNA will favor the defendant; at other times, it will be the prosecution's key tool in seeking a conviction. Regardless of which party relies on the DNA evidence, the tools for juror comprehension remain the same.

A judge has the authority to ensure that evidence is presented in a comprehensible manner. This derives primarily from Rules 611 and 614 of the Federal Rules of Evidence (FRE), and their counterparts at the state level. In forensic DNA cases, processes that have been suggested or applied include the following:

- Juror note-taking.
- Multipurpose notebooks for jurors (with glossaries, witness names and photos, pre-admitted exhibits, etc.).
- Written preliminary instructions, with copies to all jurors.
- Pretrial tutorials for jurors and judges.
- Funding for representatives of indigent persons to obtain a teaching expert for complex scientific evidence.
- Copies of an expert's PowerPoint® slides to jurors for review during testimony.
- Exhibit management and indexing.
- Introduction of court experts.
- Sequential expert testimony: In a case with both prosecution and defense expert witnesses, a judge may have the defense experts testify immediately after the

prosecution experts to allow the jury to digest all expert opinion and reasoning at one time.

- A decision tree to help jurors comprehend the steps necessary to reach a conclusion based on scientific evidence.
- Language or images that convey probabilities and other complex issues.[2]
- Juror questions submitted to witnesses (in particular, scientific witnesses).
- Juror discussions of evidence during trial breaks.
- Interim commentaries or summaries by attorneys in lengthy trials.
- Judicial intervention when expert testimony is incomprehensible.
- Reopening or reclosing upon impasse (if jurors need help on a particular issue).
- Copies of written final instructions to all jurors.
- Jury instructions in nontechnical language to facilitate assessment of expert testimony.
- Completion of responses to juror questions and requests during deliberations, including alternatives to responding, "Rely on the instructions previously given."

This list does not endorse these practices or suggest that they are fair and appropriate in particular cases. Rather, the list highlights the types of presentation innovations that have been tested and implemented because of the advent of forensic DNA evidence. In any case, proposed techniques must be assessed to determine whether they diminish the prosecution's burden of proof or unduly emphasize the forensic evidence. Techniques must also comply with jurisdictional law and due process requirements.

Section 4. Jury Selection

The term *jury selection* is actually a misnomer. Lawyers do not select jurors directly; instead, they reject those who appear antagonistic to their side. Jury selection, when conducted properly, may serve multiple purposes:

- Educating prospective jurors.
- Humanizing the defendant.
- Identifying jurors who might favor one side (or, at least, might be neutral).
- Highlighting those antagonistic to one side and then developing grounds for a challenge for cause.
- Developing potential issues for appeal.

Whether each goal can be achieved may depend on whether the judge conducts the process entirely or permits attorney questioning, and how sophisticated is the attorney's understanding of the law and practice of *voir dire.*

Counsel must be familiar with the governing principles of *voir dire* — in particular, the constitutional mandate that jurors be impartial,[3] and the ban on race[4] and gender[5] as factors in juror rejection or selection. Beyond that, counsel must develop the skills required for *voir dire.*

The first of these required skills is in the shaping of questions. Telling jurors what the law is, and then asking them about their ability to follow the law, is a useless practice. Prospective jurors are likely to give the response, "Of course I can and will follow the law." Open-ended questions — asking the juror what he or she thinks — are much more likely to reveal prejudices, biases or fixed opinions. Questions such as "What do you think about DNA in criminal cases?" or "Do you think police labs ever make errors when they collect and analyze DNA?" are much more likely to get an honest and accurate response than "Jurors must be open to all possibilities, such as lab error. You can do that, can't you?"

Closed-ended (leading) questions are best used when developing a challenge for cause. At this point, subtly and with control, the lawyer should try to direct the prospective juror in a way that he or she expresses strong allegiance to a partiality or other impairment.

Important questions to ask jurors in a DNA-based prosecution may include:

- What do you know about DNA?
- Do you have a science background?
- Do you have a strong belief that if a person's DNA is allegedly found at a crime scene, he or she must be guilty? Why or why not?
- Do you watch television shows about police investigation, such as *CSI*-type programs? If so, what do they make you think about DNA and crime solving?
- Are you aware that labs can make mistakes in handling and testing DNA?

Questions about juror trust in law enforcement officers and law enforcement laboratories are also critical, as jurors who unquestionably accept police testimony will not be open to challenges regarding evidence collection or lab practices. Some questions that might help explore these issues include the following:

- Tell me the biggest mistake you (or someone you know) ever made at work.
- Did anyone find out about the mistake?
- How did you handle it?
- Do you wish you had handled it differently?
- If two people testified about an event, one a police officer and one a civilian, and they had different versions, would you be more likely to believe the police officer?
- Some people believe that if there is DNA evidence, it does not matter what the other evidence does or does not show. How do you feel about that?

Some of these questions may also be appropriate in a case where DNA evidence would be expected but was not found. In such cases, juror expectations may support an argument that guilt has not been proven beyond a reasonable doubt.

Ultimately, the defense theory and counsel's understanding of how many questions the judge will tolerate must inform the selection of questions. Experienced lawyers know that if a judge allots limited time to the jury selection process, more latitude will be given when lawyers focus on relevant questions quickly.

In a complex or high-profile case, judges may be amenable to the use of juror questionnaires.

Questionnaires are useful screening tools before in-court questioning takes place. Judges can pose the types of questions set forth above, and a quick review of the written answers can determine whether and what additional questioning is worthwhile.

Section 5. Opening Statement

Defense counsel will almost always need to address the DNA evidence in the opening statement. There are three explanations for DNA evidence: (1) the DNA profiles match and originated from the same person, (2) there was contamination or error, or (3) the DNA profiles match by chance. The opening statement is an opportunity to explain to the jury that the DNA evidence:

- Is not relevant (e.g., the defendant lives at the location, or there was an innocent transfer of DNA from an item to the evidence).
- Is not important (e.g., there was consent in a rape case).
- Could be interpreted differently (e.g., a DNA mixture profile in a gang rape case where neither the defendant nor 50 percent of the general population could be excluded as potential contributors).

Using technical terms such as *loci* or *electropherogram* in an opening statement is not recommended. If the DNA issue is easy for the average person to comprehend, use simple language that a high school student could understand — even if counsel can use the technical terms with ease. If the DNA issue is complicated, find a way to make it easy by using pictures, simple charts, analogies or examples. This is not the time to be embarrassed to call and ask other attorneys for examples of simple ways to explain DNA concepts. Just as colleagues have stories to illustrate circumstantial evidence and reasonable doubt, more and more attorneys are developing stories to illustrate DNA concepts. This is particularly true in public defender offices in major cities. Resources may also be available through national defense attorney associations, such as NACDL and NLADA. An experienced expert witness will be able to help explain the DNA evidence to the jury in simple terms.

Section 6. Witness Preparation

If the defense is considering calling an expert witness at trial, counsel must thoroughly prepare beforehand. Witness preparation begins with the informed selection of a qualified expert. Throughout the investigative stages of the case, defense counsel must regularly consult with the expert to determine what issues will need to be brought before the jury. In addition, the expert may be able to assist counsel in determining how best to present the evidence, particularly when exhibits are used.

For a detailed explanation of the witness preparation process in a forensic DNA case, see Chapter 4, Section 1.

Section 7. Objections During the State's Direct Examination of a DNA Expert

A well-prepared prosecution expert will know not to stray from the facts or go beyond his or her expertise. When confronted with such a witness, defense counsel may conclude that objections will be counterproductive — they may alienate jurors and prolong the witness's testimony with what may be terribly damaging evidence.

Thus, a strategy of "Don't object" or "Get the witness off the stand quickly" may be best.

Nonetheless, counsel must be alert to areas where objections are appropriate and potentially beneficial to the defense. These are areas in which:

- The expert has testified beyond his or her qualifications (e.g., when someone with no training in statistics offers probability data).
- There is no foundation for a particular claim, or when the answer assumes facts not in evidence.
- The expert's explanation is misleading.

The last category is one in which defense counsel may prefer not to object if, in fact, the defense has a qualified expert who can explain to jurors how the prosecution witness erred. Defense counsel must be familiar with the applicable evidentiary rules governing expert testimony, particularly the rules that limit what an expert may rely on, and what reliable evidence may be repeated in the courtroom.[6]

Similarly, counsel must consider whether confrontation guarantees a successful result — currently a hotly debated topic. Hearsay rules limit testimony to one expert witness for each result or analysis that is prepared by lab personnel other than the expert witness.[7]

Section 8. Taking Juror Questions During Testimony (If Allowed)

In many (but not all) jurisdictions, trial courts are given the discretion, or are mandated by court rules, to allow jurors to ask questions of the witnesses during the trial.[8] Because of the "CSI effect," many jurors are particularly interested in forensic science issues and expert witnesses. Several studies have shown increased juror satisfaction and attentiveness in jurisdictions where jurors are allowed to ask questions.[9]

In these jurisdictions, jurors can submit written questions to all trial witnesses following the completion of counsel's questioning of each witness. Trial judges must instruct jurors of their right to ask questions and the procedures to be followed.

Typical court rules provide that, after the trial judge asks whether the jurors have any questions for the witness, the jurors' written questions — unsigned — are handed to a member of the courtroom staff. After the judge reviews the questions, defense and prosecution counsel approach the bench, review the questions, and make any objections outside the hearing of the jury. The court will determine the acceptability of each question asked and, if deemed appropriate, typically will ask the witness the juror's question. Both sides are then permitted to ask follow-up questions, limited to the answer given by the witness.

If the court sustains an objection to the question, the question will not be asked. Jurors are instructed that their questions will be treated the same as questions asked by counsel, subject to objections, and that the jury should not attach any significance to counsel not asking a particular question.

Section 9. Effective Cross-Examination of a DNA Expert

There is no prepackaged script for cross-examining a DNA expert. The substance and tone of the cross-examination will depend on the facts, the defense theory of events, the expert witness and other case-specific considerations. At the outset, defense counsel should determine exactly what they want to communicate to the jury about the DNA evidence. How can the defense make the DNA expert a *defense* expert and make the DNA evidence support the defense theory of events, while contradicting the prosecution's theory (or at least showing that the DNA provides no evidence in support of the prosecution's theory)?

Broadly speaking, there are two directions that defense counsel can go in cross-examining a government DNA expert. One direction is adversarial, by which counsel seeks to undermine the expert's testimony and ultimate conclusion. The other direction is nonadversarial; counsel does not dispute the analyst's central premise but, rather, tries to turn the expert's testimony into evidence that actually supports the defense theory.

The Adversarial Cross-Examination

FOR CASES INVOLVING DNA TRANSFER:

Q. Are you familiar with the concept of DNA transfer?
A. Yes.

Q. Is it true that DNA can transfer from a person to an object?
A. Yes.

Q For example, if a person bleeds onto the floor, his DNA transfers from inside his body onto the floor?
A. Yes.

Q. This is called primary transfer?
A. Yes.

Q. If someone stepped into the blood on the floor, we'd expect some of the blood to transfer from the floor onto the shoes?
A. Yes.

Q.That's an example of secondary transfer?
A. Yes.

Q. If the person then walked across the room, some of that blood would transfer from the bottom of his shoe onto the floor with each step?
A. Yes.

Q. That's called tertiary transfer?
A. Yes.

Q. Because of DNA transfer, you can get blood into a room in which the bleeding person was never present?
A. Yes.

Q. This is true for other DNA sources as well?
A. Yes.

Q. For example, skin cells can transfer from my hands to the counsel table?
A. Yes.

Q If you came by the table and put your hands on it, some of my DNA could transfer onto your hands?
A. EYes.

Q. If you then went to the door and opened it, you could leave some of my DNA on the door handle?
A. Yes.

FOR CASES INVOLVING DNA TRANSFER OF A PRIMARY WEARER OR MAJOR PROFILE:

Q.Isn't it true that people sometimes transfer DNA onto clothes when they wear them?
A. Yes.

Q You would expect to find my DNA on the collar of my shirt, for example; is that correct?
A. Yes.

Q This is called wearer DNA?
A. Yes.

Q. If two people wore the same shirt, you could find two DNA profiles on the shirt; is that correct?
A. Yes.

Q Isn't it also true that some people leave more DNA behind than others?
A. Yes.

Q Is that because people shed their skin cells that contain DNA at different rates?
A. Yes.

Q. The major profile refers to the contributor who has more DNA than the other person in a mixed sample?
A. Yes.

Q Isn't it true that the major DNA profile doesn't tell you who wore the item of clothing last?
A. Yes, that's true.

Q Isn't it true that the major DNA profile doesn't tell you who wore the item of clothing most often?
A.That's correct, it doesn't.

Q. So, what your DNA testing tells us is that a person with that profile cannot be excluded as having had contact with that item at some point in time; is that correct?
A. Yes.

Q. And it doesn't tell us when that happened, where it happened or how it happened; isn't that true?
A. Yes.

FOR CASES INVOLVING THE INNOCENT DEPOSIT OF DNA:

Q.The DNA test results can't tell when the DNA was left; is that correct?
A. Correct.

Q. The DNA test results can't tell you the time it was left; is that correct?
A. Correct.

Q. The DNA test results can't tell you the date it was left; is that correct?
A. LiCorrect.

Q. The DNA test results can't tell you whether it was deposited consensually; is that correct?
A. Correct.

If defense theory favors opposing the government's claims regarding the DNA evidence — that is, if the defense theory is that the defendant's DNA is not where the prosecution says it is — then cross-examination of the government witnesses may be the primary strategy for undermining the evidence and showing the jury why they should discount it. The goal of the adversarial cross-examination should not be to spar with or outwit the expert but, instead, to systematically highlight the shortcomings of the procedures that led to the DNA report asserting that, for example, the defendant's DNA profile cannot be excluded as admissible evidence.

Time, Place, Transfer and Contamination

In most cases — except some sexual assault cases — analysts cannot say exactly when or under what circumstances DNA came into contact with a piece of evidence. DNA at a crime scene may have been left there days, weeks or months before the crime, or after the crime was committed. A person or object may have transferred the DNA there. If the defendant's DNA is present at the crime scene or on an object recovered from the crime scene, this does not mean with any certainty that he or she was present at the crime scene at any time. The government analyst should concede these points easily; they may be worth exploring on cross-examination if the defense theory suggests contamination, transfer or the innocent presence of the defendant at the scene at a different point in time.

Many labs do not attempt to distinguish between vaginal and skin cells. A scientist may be able to obtain a DNA profile but not be able to testify that the source definitely was vaginal fluid or skin. Contrast this with the *confirmatory tests* for semen. Most of the time, scientists can confirm that the DNA profile came from sperm cells.

The defendant's DNA may have come into contact with an item of evidence through contamination. As a preliminary matter, counsel should look carefully at chain-of-custody logs for every step of the process — from the crime scene to the laboratory to the analyst's workstation, and any other movement or handling in between (including any time evidence was removed from storage and then returned). If the defendant's known DNA sample was handled on the same day as, and in particular before, an item of evidence — which the laboratory's protocols may prohibit — there may be reason to think that the defendant's DNA was transferred to the evidence through mishandling. If the defense theory is that the DNA was contaminated, counsel should proceed with caution and be prepared to elicit evidence in support, through either the DNA analyst or others who came in contact with the evidence during the chain of custody.

Under either a transfer or contamination theory, counsel will want to find a compelling way to illustrate to jurors how little DNA is required for it to register on the analyst's instrument. Depending on which DNA testing kit is used, one nanogram or less is considered to be an optimal amount of DNA for testing. Defense attorney Bob Blasier famously illustrated this concept: Hold up a packet of sugar and note that it contains approximately 1 gram of sugar. Assume the packet contains 1,000 individual granules. Confirm with the analyst that given this premise, to obtain a nanogram of sugar they would need to divide a single crystal by 1,000, and then divide one of those pieces by 1,000, and then again. Finally, the expert will agree, you are at 1 nano-gram, or less than the eye can see — and that amount, or less, is all that is needed for a person's DNA profile to appear in a test result. That amount of DNA might be transferred by a small number of skin cells that came in contact with a person or object, which later came in contact with another object. If counsel picks up a pen in the courtroom, the expert will probably agree that there is a fair chance that counsel's DNA is now on that pen by way of shed skin cells.

In short, miniscule amounts of DNA can be transferred easily. This line of cross-examination can be effective under a theory of simple transfer or contamination. Factors such as evidence packaging, handling and chain of custody are critical to developing the facts necessary to support such a theory.

Statistics

Unless there is an opposing legal ruling, to be in compliance with the current *SWGDAM Guidelines,* a DNA analyst should be presenting a statistic to characterize the significance of a match involving DNA found on an item of evidence and someone involved with the case. In non-mixture cases — and often *in distinguishable DNA mixture* cases — that statistic will take the form of a random match probability, as discussed in Chapter 6, Section 7. In an indistinguishable mixture case, the statistic will likely take the form of a *combined probability of inclusion (CPI), a combined probability of exclusion (CPE),* also called random man not excluded (RMNE), or a *likelihood ratio (LR).*

One section of the adversarial cross-examination should focus on the analyst's statistical claims. It is important to dispel what is known as the *prosecutor's fallacy,* which is a common misinterpretation of the random match probability statistic. If the probability that a randomly selected individual would match the DNA profile found at the scene is 1 in 1 trillion, that does not mean that there is a 1-in-1-trillion chance that the DNA came from someone other than the defendant. It means that if a person is picked at random out of the general population, the probability that he or she will match the detected profile is 1 in 1 trillion.

The question of "What is the probability that the evidence DNA profile came from the defendant is not one that DNA testing that is supported with a traditional random match statistical calculation can address directly; however, some laboratories now use source attribution statements in their DNA reports. A source attribution statement is used to definitively state that, to a reasonable degree of scientific certainty, this DNA profile originated from this person, or their identical twin. Source attribution testimony must actually rely on demonstration via a statistical calculation (typically in the case notes) that the obtained DNA profile meets or exceeds the value set by the lab in order to make such a statement. The defense's expert should be able to assist counsel in locating the statistical data and publications on which the lab relied to reach its conclusion and design appropriate challenge questions.

Other Genetic Locations (Loci)

Another potential area for cross-examination is that the inculpatory claims are being made on the basis of a finite number of genetic locations — most typically, the *13 core CODIS loci,* the 15 STR loci in the Identifiler® kit, or the 15 STR loci in the PowerPlex® 16 kit — of the literally billions of genetic locations comprising the complete human DNA chain.

The entire human DNA genome for each person is unique (even identical twins). In forensic DNA testing, however, only a finite number of autosomal STR genetic markers are examined (as noted above, typically 13–15 areas). Research strongly supports that it is necessary to examine 10 or more of these areas of DNA in order to be able to distinguish between people, even those who are related (with the exception of identical twins, who will have the same autosomal STR DNA profiles). The prosecution's expert should concede that other genetic loci developed for forensic identification were not used in the current case. If the defendant was excluded at just one of those other locations, the expert would have to agree that the DNA must have originated from another person — however, the analyst did not test those other locations. Defense counsel should be aware that forensic scientists within a laboratory use whichever commercial test kit their laboratory protocols specify. Although there are a number of kits available that test for additional genetic markers beyond the 13 core CODIS loci, not all labs use the same kits.

Although the analyst's laboratory may not be in the practice of testing those other DNA locations, other laboratories do if they use a kit that includes those additional loci. The argument would therefore be that the analyst did not send the DNA evidence to another lab to see whether the defendant was excluded at one of those other locations; instead, the analyst simply offered a conclusion based on his or her lab's limited technology.

To support this line of cross-examination, counsel should learn exactly which locations were tested, which locations are routinely tested at the lab (determined by the commercial test kit used), and what other locations are tested at other labs (including the labs' names and how long they have been testing those other locations). Similarly, counsel should know the names of both the test kit used in the case and the kits used in labs that cover two or more locations.

Other Evidence Not Tested

A critical component of cross-examination may be to highlight evidence that was not tested but might have provided significant information. Evidence may not have been tested because:

- Items were not collected.
- Items were collected but not submitted to the lab.
- A decision was made not to test a submitted item because of case circumstances, direction from the submitting agency or limited resources.
- A decision was made to test only a portion of an item (for example, only some of several stains on a bed sheet).

Also, the analyst may not have tested the DNA of other potential suspects and thus did not compare the profiles of additional suspects to the DNA found at the crime scene or qualifying run mixtures through the databank to see if an alternative suspect(s) might have been identified as a potential contributor(s). Such inquiries may be particularly fruitful when the case involves a DNA mixture with unaccounted-for alleles.

Mixtures and Degraded or Low-Level DNA

Mixtures and degraded or low-level DNA samples also provide fertile ground for the adversarial cross-examination. Mixture interpretation can be subjective, extremely complicated and susceptible to contextual bias. Interpretation of unresolvable mixtures, meaning ones where a major and/or a minor contributor(s) profile(s) cannot be discerned, is even more complicated — as is the interpretation of complex mixtures, which typically contain DNA from three or more individuals. There have been cases where subjective interpretations of mixtures have led to wrongful convictions, such as in the case of Josiah Sutton in Texas.[10] Consultation with an expert is critical when interpreting and understanding mixtures.

The relative peak heights of alleles might be an appropriate area for cross-examination. For example, if four alleles are found at a *locus,* it may indicate a two-person mixture. When two of these alleles are found in much greater concentration (major contributor) than the other two (minor contributor), it can be presumed that the two major alleles are from one source and the two minor alleles are from another source. If the laboratory reports that one major and one minor allele go together to include the defendant, this can be an area ripe for challenge. As noted in the section on discovery (Chapter 3, Section 2), counsel should have a copy of the lab's protocols as well as the protocols from other labs — including the FBI — to screen for discrepancies between the protocols governing this lab's practices and others in the field. Pay particular attention to areas where the analyst's discretion factors into the analytic outcome.

Another fertile area for cross-examination with mixtures, as well as low-level DNA, deals with what is observed below the *analytical threshold.* Each laboratory will determine the analytical threshold for their instruments, based upon their *validation studies.* The analytical threshold defines the minimum height requirement at and above which detected peaks can be reliably be distinguished from background noise on the electropherograms. Peaks above the threshold are typically not due to noise and are either true alleles or artifacts. Conversely, any peaks below the threshold cannot reliably be distinguished from background noise. DNA analysts should acknowledge that, generally speaking, alleles corresponding to genetic material can be observed below the analytical threshold set by their lab and that these peaks would not be listed in their report.

The current version of the *SWGDAM Guidelines* (released January 14, 2010) specifically prohibit this practice, as stated in Guideline 3.1.1.2: "...the analytical threshold should be established based on signal-to-noise considerations (i.e., distinguishing potential allelic peaks from background). The analytical threshold should not be established for purposes of avoiding artifact labeling, as such may result in the potential loss of allelic data."

It is essential to determine whether there is any additional information (especially below the lab's reportable/analytical threshold) that did not appear in the final report but may be contained in the supporting data or electropherogram printouts.

Even data analyzed by the lab using an analytical threshold supported by empirical data can and should be reviewed by the defense expert, particularly when the testing involves a mixture, partial, or low-level DNA sample. For example, data detected below the threshold may suggest the presence of a third party. DNA present at low levels may also contradict the conclusions made by the analyst based on the above-threshold DNA. Particularly in low-level DNA samples, there may be information below the threshold that could potentially exclude the defendant. The defense expert should assist counsel in ascertaining whether there is additional information that could help the defense's case.

The lab that conducted the DNA analysis may have an exception to application of their set analytical threshold in place — either you or your expert should look for this in their protocol. The purpose of this exception is to allow the analyst, in conjunction with their technical reviewer, to use analytical data below the threshold to exclude a suspect when that additional data supports such an exclusion. It is important to note that this practice also requires an exception to the standard operating procedure of interpreting the data (identifying all peaks that are suitable for comparisons) prior to all comparisons. While that would still be done, this exception allows the lab to take allelic information below threshold into account after they have compared the suspect's profile to the reportable alleles. If there is a concern about the lab not having reported alleles deemed by the defense expert to support a conclusion that would exclude your client (rather than include him/her or report the results as inconclusive/uninterpretable), it is appropriate to ask the analyst if they are aware of cases/ information in their protocol that allows the lab to engage in this exception and exclude based on data below threshold. If the lab is not applying this exception, this may serve to support the defense expert's contention.

Artifacts

Another area for the adversarial cross-examination is any instance in which the government expert characterizes a blip on the electropherogram as stutter, a dye blob, a spike or an artifact of any kind — or, in the absence of a blip/peak, as with allelic drop-out. In some cases, an alternate interpretation of these phenomena — that is, that the blip/peak represents an allele from genetic material rather than a technical artifact, or that the absence of a blip/peak means there really is no DNA present — may lead to an interpretation that excludes the defendant.

These phenomena have been observed to exist (see Chapter 6, Section 3). Every determination in a case involves a certain element of subjectivity, and contextual bias can be particularly problematic when distinguishing between legitimate stutter and actual alleles from genetic material that is located in a stutter position.

Counsel should pay particular attention to the laboratory's stutter protocols; any deviation from those protocols should raise a red flag — however it is important to be aware that an exception to the general guidelines listed in the lab's protocol will always be a possibility. The current version of the *SWGDAM Guidelines* takes care to point this out, as stated in Guideline 3.5.8.3, which addresses interpretation of potential stutter peaks in a mixed sample. Guideline 3.5.8.3 states: "If a peak is at or below this expectation [with being of allelic origin], it is generally assigned as a stutter peak. However, it should also be considered as a possible allelic peak, particularly if the peak height of the potential stutter peaks(s) is consistent with (or greater than) the heights observed for any allelic peaks that are conclusively attributed (i.e., peaks in non-stutter positions) to the minor contributor(s)." Accordingly, when there is a mixture sample in which minor contributors are determined to be present, a given peak might fall within both the technical range for stutter and the relative range for a heterozygous counterpart to another peak identified as an allele corresponding to genetic material. This situation provides an opportunity for a different in interpretation of the findings.

A Chain of Assumptions

When preparing to cross-examine the government DNA analyst, it is critical to work with the defense expert to identify, first and foremost, that the electronic data counsel receives match the report in accordance with the laboratory's protocols.

The expert should evaluate every judgment call, every below-threshold peak and "blip" on the electropherogram that was discounted as stutter, a dye blob, a spike, noise, or other artifact as a possible ground for cross-examination. In addition, counsel/their expert should review any instance where the absence of DNA was determined to be allelic drop-out. Counsel should also explore every alternative explanation for every determination made by the analyst.

This series of judgments can be characterized as a "chain of assumptions" on which the analyst's testimony hinges. The jury must be firmly convinced that every single assumption is correct. If any one of those assumptions turns out to be incorrect, this can be used to infer that the conclusion drawn by the analyst may not be correct. Counsel may be able to persuade the jury that a chain of assumptions — each of which was arguably reached on the basis of an assumption of guilt — is a violation of the defendant's presumption of innocence and cannot meet the standard for a criminal conviction.

The Nonadversarial or "Teaching" Cross-Examination

Alternately, the defense may concede (for strategic reasons) that the DNA evidence is what the government analyst says it is — namely, the evidence profile matches the defendant and was extracted from a particular type of body fluid (see Chapter 2, Section 2). In short, the analyst is exactly right.

It may be that the defendant had intercourse with the murder victim some time before the murder, which the defense does not dispute. It may be that the defendant was in the home where the homicide took place, and the DNA supports this — which is consistent with defense theory because the defendant visited the victim the day before. Whatever the facts, there are scenarios in which defense counsel may not want to dispute claims made by the DNA analyst.

In such cases, counsel may opt to use cross-examination as an opportunity to teach the jury about how "good science" is done to contrast it with the other forensic work in the case.

The teaching cross-examination can highlight any scientific information in a forensic analysis. A range of academic and scientific disciplines supports the foundations of forensic DNA typing, from cellular biology to population genetics. DNA examiners have highly detailed protocols for what constitutes a match and always rely on validated population databases to generate statistics estimating the probability that two samples have the same profile. Some DNA analysts will not make absolute claims of identity.

Regardless, all DNA analysts rely on advanced technology to generate charts of observed alleles, which are associated with specific frequency values which serve as the basis of statistical or identity claims. The databases and associated statistics have been validated. Furthermore, all of the underlying data that serve as the basis of the analyst's claims are provided to the defense.

In short, forensic DNA typing analysis is a rigorous scientific process, which can be developed in detail on cross-examination and later contrasted with the practices of other forensic disciplines in the case.

Section 10. Special Considerations for Trying mtDNA Cases

Basic mtDNA Typing Premise — Not an Identification

mtDNA is different from nuclear (autosomal) DNA — the most commonly encountered forensic variety — in a number of ways. A careful challenge to mtDNA evidence should hone in on these differences while following the general guidelines above. Below is a synopsis of the core areas of forensic mtDNA that can be developed through admissibility challenges under *Frye, Daubert* or other evidentiary standards governing expert testimony and scientific evidence, or as a basis for challenging the expert directly through cross-examination. Refer to Chapter 2, Section 3, for information regarding the biology of mtDNA.

Any attempt to characterize a purported mtDNA "match" as evidence of conclusive identity is scientifically unsupportable on the basis of its method of inheritance alone. Scientists are in agreement that "[m]itochondrial DNA typing does not provide definitive identification."[11] This should be brought to the attention of a judge or jury considering such evidence.

Contamination Concerns in mtDNA Typing

Because mtDNA analysis almost always involves working with very low amounts of template DNA, from the outset of the *DNA sequencing* typing process there is considered to be a far higher risk of *contamination* than with nuclear DNA typing. In addition, this work with small amounts of DNA dictates the need for more cycles of PCR amplification to generate typing results for each sample, plus the detection method is very sensitive. The risk of contamination in mtDNA analysis has been deemed substantial and, indeed, even more precautions must be undertaken in a lab that conducts mtDNA testing to avoid the introduction of contamination than in a lab that solely conducts nuclear DNA testing.

The *SWGDAM Guidelines for Mitochondrial DNA (mtDNA) Nucleotide Sequence Interpretation* (released April of 2003) [see http://www.fbi.gov/ about-us/lab/forensic-science-communications/ fsc/april2003/swgdammitodna.htm] states that low levels of exogenous DNA contamination and/or background is commonly observed. With respect to contamination, the *SWGDAM mtDNA Guidelines* require that mtDNA labs:

- Take precautions to minimize contamination.
- Monitor contamination.
- Have a method to define and quantify contamination.
- Determine their maximum allowable threshold for contamination through internal validation studies.
- Have standard operating procedures in place to deal with contamination.
- Establish evaluation criteria for controls, including but not limited to a positive control, a negative control, and a reagent blank control, each of which has been processed through sequencing along with the sample.

If more than one person's DNA is extracted and amplified, the sequencing results may reflect this mixture. In extreme cases, the contaminating DNA can greatly exceed the donor's DNA and thereby yield a false positive result.[12] To address this issue, the FBI has established a "contamination ratio" of 10:1 — meaning that the FBI considers one part contamination per 10 parts mtDNA sample to be suitable for interpretation.[13] Defense counsel should question the underlying data on which the FBI, or another laboratory, relies in permitting the use of their contamination ratio. Counsel should also be aware that the FBI supports four regional mitochondrial DNA testing labs, located in Arizona, Connecticut, Minnesota and New Jersey, to perform mtDNA testing, which means these labs are likely to have similar protocols.

In 2001, a commercial laboratory reported the results of a two-year study of thousands of trials to determine the effect of *heteroplasmy,* contamination and other factors on lab mtDNA test results.[14] The researchers found the presence of contaminants in 2.4 percent of cases.[15] The source of the contamination was not lab staff; the researchers determined that the contaminants came from a source outside the laboratory. In at least two cases, the contamination affected the interpretation of results.[16]

The Role of Heteroplasmy in Forensic Typing of MtDNA

Forensic analysts account for the biological anomalies of mtDNA by applying a more flexible standard for declaring a match, reported as "cannot exclude," between two profiles. Whereas a match between nuclear DNA profiles is straightforward in non-mixture cases, it is less so with mtDNA. At the outset, a DNA analyst sequences the HVI and HVII regions of both the evidence sample and the suspect's sample. If the two profiles fail to match at all base positions, most laboratories — in accordance with their written protocol — will not exclude the suspect as the source of the evidence sample. Instead, most crime labs "will only definitively exclude a suspect if there are two or more base-pair differences between the samples with no evidence of heteroplasmy, on the theory that one difference may be the result of heteroplasmy."[17]

The defense's challenge to a purported mtDNA match might begin by highlighting the fluidity of the definition crime laboratories have adopted to accommodate the biological realities of mtDNA. Crime labs hold that a mtDNA match is always a match and, because of heteroplasmy, what appears to be a nonmatch at one or two base positions may not exclude an individual. From a legal perspective, a protocol that not only allows but requires interpretation of apparently exculpatory evidence in a manner that renders the same evidence inculpatory may raise concerns about observer bias.

Although some crime labs will only declare an automatic exclusion in cases where one or more differences are observed between the evidence sample and the suspect sample, they will declare an inclusion (or "failure to exclude") under a variety of scenarios. For example, if an analyst concludes that the profiles being compared are identical at each of the bases in the HVI and HVII regions, the suspect will be deemed "included (or not excluded) as a possible contributor of the evidence sample."[18] If either the suspect or evidence sample displays heteroplasmy, the analyst will not necessarily exclude the suspect as a possible source when two or less differences are noted between the sequences based upon the evaluation of the number, position, and nucleotide composition of polymorphic sites.[19] Moreover, if the two sequences differ by a single base pair but neither profile appears *heteroplasmic,* the analyst will not necessarily exclude the suspect; instead, the analyst may characterize the results as inconclusive based upon the evaluation of the number, position, and nucleotide composition of polymorphic sites, notwithstanding the seemingly different profiles.[20] The exact criteria for distinguishing a "failure to exclude" from an actual exclusion vary from laboratory to laboratory, with little standardization.[21]

Basic mtDNA Database Issues

After determining that there is a failure to exclude, or "determining that the mtDNA profile of a reference sample and an evidence sample cannot be excluded as potentially originating from the same source" using one of the above suppositions, the analyst will then compare the sample to a database(s) of profiles to estimate its significance.[22-23] The population databases used to determine a random match frequency estimate for mtDNA results differ from those used in nuclear autosomal STR DNA cases to generate a random match probability. With mtDNA testing results, the counting method is the most common approach used. The counting method involves counting the number of times that a particular mtDNA sequence is seen in a database. The larger the number of unrelated individuals in the database, the better the statistics will be for a random match frequency estimate. It has been argued that using a database of mtDNA profiles representative of the relevant population is essential.

An admissibility challenge to mtDNA evidence can focus principally on the apparent shortcomings of the databases used to derive the statistical expression that serves as a prerequisite to admission of the evidence. The FBI maintains the primary mtDNA database used for forensic analysis in the United States (the *SWGDAM* mtDNA Population Database). In 2004, a team of scientists assessed the reliability of the FBI mtDNA database. Using the African-American mtDNA sub-database to serve as the basis of a "thorough inspection,"[24] they found that the SWGDAM database contained "a number of major deficiencies."[25] Steps have been taken since that time to address the identified concerns.

mtDNA Database Quality Control Issues

On the basis of the 2004 review, scientists have been critical of the SWGDAM mtDNA database; at that time, errors relating to the accuracy of the profiles that populate the database were observed.

The lack of sufficient quality control standards governing the initial typing of the profiles resulted in a number of errors. Molecular biologists Y.G. Yao and colleagues identified "five major and common types of errors, namely, base shifts, reference bias, phantom mutations, base mis-scoring and artificial recombination" within forensic mtDNA databases.[26] They described the need for quality control standards as "urgent" and recommended "[e]xtreme caution ... at all stages of data collection and proofreading processes."[27] As indicated above, steps have been taken to address these and other deficiencies noted in the mtDNA databases during the intervening period. In addition, the International Society for Forensic Genetics (ISFG) EMPOP mtDNA database is available for use (see http://empop. org/), if preferred and/or to compare the random match frequency estimates obtained. The mtDNA haplotypes in the EMPOP database are stored in difference-coded format, relative to the *revised Cambridge Reference Sequence (rCRS)* and aligned using the phylogenetic approach described by Bandelt and Parsons in 2008. [See Bandelt, H.J., and W. Parsons (2008), Consistent treatment of length variants in the human mtDNA control region: A reappraisal. *Int'l J. Legal Med.* 122(1), 11-21.] Per Bandelt and Parsons:

In forensic science, as well as in molecular anthropology and medical genetics, human mitochondrial DNA (mtDNA) variation is being recorded by aligning mtDNA sequences to the revised Cambridge reference sequence (rCRS). This task is straightforward for the vast majority of nucleotide positions but appears to be difficult for some short sequence stretches, namely, in regions displaying length variation. Earlier guidelines for imposing a unique alignment relied on binary alignment to a standard sequence (the rCRS) and used additional priority rules for resolving ambiguities. It turns out, however, that these rules have not been applied rigorously and led to inconsistent nomenclature. There is no way to adapt the priority rules in a reasonable way because binary alignment to a standard sequence is bound to produce artificial alignments that may place sequences separated by a single mutation at mismatch distance larger than 1. To remedy the situation, we propose a phylogenetic approach for multiple alignment and resulting notation.

In their 2004 analysis of the African-American subpopulation database, Bandelt and colleagues "detected as many as five artificial combinations of totally unrelated mtDNA segments stemming from different samples, which suggest fatal sample mix-up in the lab or during data transcription."[28] Even following their report and a series of revisions by the FBI, "several obvious clerical errors still remain in the revised database."[29] According to Bandelt and colleagues, the remaining errors "could only be corrected through thorough resequencing of the original samples."[30] The FBI has not published the results of any such resequencing.

The Counting Method

One of the database's deficiencies is that it may not be accurately representative of the sub-populations it claims to represent. As of October 2008, the entire SWGDAM mtDNA database contained 5,071 profiles divided across 14 racial subpopulations. The populations and number of profiles at that time are listed in Figure 25.

Some population/geographic groups are represented by fewer than 100 profiles — in one case, by as few as eight for a portion of a population group. The profiles were gathered from a collection of blood banks, paternity-testing labs, scientific research groups and FBI agents.[31] The racial classifications are based on self-reporting, not genetic ancestry. The samples are not geographically defined. The SWGDAM mtDNA database is an example of a "convenience sample" obtained from only a handful of locations; no effort was made to randomize the selection.[32]

The number of profiles in the EMPOP database in mid-April of 2011 was more than 8,000. To find out how many mtDNA haplotypes are in this collaborative database at any point in time, see http://empop.org/modules/ overview/. The number of haplotypes in the database for each geographic affiliation can be seen by selecting each region, using the "geographic affiliation filter."

Because mtDNA is maternally inherited and not recombinant, mtDNA profiles are not randomly distributed across the population. The distribution of a given mtDNA sequence is a function of women's migration and reproduction rates. An individual and all of his or her siblings — as well as their mother, grandmother, great-grandmother and maternally related third cousins, and so on — are expected (absent mutations) to share identical mtDNA profiles. Over generations, profiles stay intact or mutate to a very similar sequence.

Race	**Number of Profiles**
African-Americans	1,148
Apaches	180
Caucasians	1,814
Chinese/Taiwanese	356
Egyptians	48
Guamanians	87
Hispanics	759
Indians	19
Japanese	163
Koreans	182
Navajos	146
Pakistanis	8
Sierra Leone	109
Thai	52
TOTAL	5,071

Source: SWGDAM Guidelines for Mitochondrial DNA (mtDNA) Nucleotide Sequence Interpretation (October 2008), Federal Bureau of Investigation.

Figure 25. mtDNA Database.

In addition, the high mutation rates characteristic of the HVI and HVII regions create unique variants. More recently created variants in the human population may not have had time to spread from their location of origin. This creates geographical areas where certain *haplotypes* (the collective genotype of closely linked loci on an area of DNA) or *haplogroups* (groups of similar haplotypes) are prevalent and other areas where they are wholly or largely nonexistent.

To calculate the statistical frequency within the mtDNA database, a forensic analyst counts the number of times a particular profile occurs in the database(s) being used. This is known as the *counting method*.[33] Given that vast majority of mtDNA profiles have most likely not been generated, the "count" of a given profile is often zero out of the number of profiles already observed in the population database.

When a profile is observed at least once, the conventional statistical calculation involves dividing the number of observations by the size of the database.[34] For example, if the profile was observed once in the African-American database (n = 1,148), the frequency would be reported as 1/1,148 or 0.0008711. The analyst would then place a 95-percent confidence interval around that number as a margin of error in estimating the frequency in the larger human population.[35] The laboratory would report the upper-bound frequency: For an observed frequency of 0.0008711, the upper confidence limit is 0.004839, or 0.48 percent. Thus, the lab would report that the frequency of occurrence in unrelated individuals of the observed haplotype in the African-American population is 0.48% — this could also be reported as 99.52% of all African-Americans are excluded as having the observed haplotype.[36] The lab may report the observed counts within subpopulations as well as from the overall database. Regardless, there should always be a qualifier that explains that individuals within the same matrilineage will have the same mtDNA haplotype (barring mutation).

mtDNA Database Profiles May Not Be Representative of Some Subpopulations

Dozens of *phylogeographic*[37] studies have been performed to identify the geographic distribution of mtDNA haplotypes in regions all over the world. These studies are fairly limited in the United States.

These studies show that mtDNA is not randomly distributed throughout the world; different haplogroups and haplotypes are concentrated within certain geographical populations.[38] Scientists rarely encounter new nuclear DNA haplotypes when studying new population subgroups, but the opposite is true for mtDNA. Although some haplogroups of mtDNA sequences are widely distributed throughout the population,[39] many exist only within certain geographic clusters.[40] Nonrandom distributions also exist *within* geographic locations because of subtle linguistic, religious or economic/caste distinctions.[41] This is not a surprising observation.

These distributions are not limited to rare or ancient populations — today, different geographic regions demonstrate different mtDNA patterns.[42] For example, a particular cluster of mtDNA sequences called Haplogroup J is widely distributed in western and central Europe, but is rare in Iberia.[43] A sub-haplogroup of that cluster has been observed primarily in Britain, with one other occurrence from an ancestor in Italy.[44] A mutation that has an 8-percent frequency within the Canary Islands has never been found outside the Islands.[45] One study identified a considerable number of matches between Mozambique and American sequences from African haplogroups, including some sequences that had never been observed outside Mozambique and others observed only in American populations.[46] MtDNA population genetic linkage in North America — discussed in detail in the next two sections — is also well documented in scientific research.[47]

Whether the heterogeneous geographic distribution of mtDNA lineages reflects genetic clustering, inadequate sampling or some combination of the two, the sampling of mtDNA profiles should take into account geographic heterogeneity and stratification in order to create representative databases for use in forensic typing. This concern applies to the various ethnic groups in the United States — whether African-American, Hispanic, Asian or otherwise (although less so for Caucasians).

Detailed information about the effect of migration patterns on African-American,[48-64] Hispanic,[65-66] Native-American,[67-69] Asian,[70-74] and Caucasian[75-76] populations is readily available.

The Role of Ethnicity in Trying a mtDNA Case

The defendant's ethnic background may play a critical role in informing the appropriate defense strategy in a mtDNA case — given the population substructuring described above and its influence on the distribution and frequency of a given mtDNA profile in both the region in which the defendant resides and the region(s) from which the population databases were collected.

At minimum, defense counsel should learn about the historical and contemporary migration patterns from the defendant's ancestral origin to both the geographic location where he or she resides and the location(s) from which the database samples were drawn. If, for example, a defendant's ancestral origins are not represented in the mtDNA database, but he is a member of a significant immigrant community of common origin in his place of residence, a strong argument can be made that the lack of a match to his mtDNA profile in a database collected from a random set of individuals from elsewhere in the country or world is devoid of meaning. It offers no information about the frequency of that profile in the relevant geographic area.

mtDNA Defense Experts

If an admissibility challenge fails, the defense may want to consider retaining its own expert.

The same strategic considerations generally apply: If the government calls a forensic scientist and the necessary points cannot be developed sufficiently on cross-examination, the defense may want to call a scientist with expertise in the underlying science and/or population genetics to challenge the government expert's claims.

In a mtDNA case, the defense may also want to call an expert on the migration patterns of the defendant's ancestors to show that his or her haplogroup is not properly represented by the database the government relied on for its statistical representation. Counsel should be aware that the SWGDAM database is not the only mtDNA database available, as noted above.

Alternate databases can be located on the Internet, including, but not limited to, www.mitomap.org and www.empop.org.

mtDNA Treatment in the Courts

A number of challenges have been made to the admissibility of mtDNA evidence as an inculpatory tool, but they have been generally unsuccessful thus far. Many courts have acknowledged that "mtDNA analysis is more applicable for exclusionary, rather than identification, purposes,"[77] but have nonetheless admitted the evidence on grounds that the discriminatory power and other limitations of mtDNA evidence are questions of weight rather than admissibility.[78]

Appellate courts in at least 11 states[79] as well as one federal district court80 have found the results of mtDNA testing admissible under various evidentiary standards.

Fresh admissibility challenges are critical, however, in light of the studies that called the reliability of mtDNA forensic databases — and, thus, the inculpatory claims hinging on them — into question.

Section 11. Special Considerations for Trying Y-STR Cases

Y-chromosome DNA — from which Y-STR forensic markers are derived — differs from traditional nuclear DNA on a number of counts. Just as with mtDNA, a well-planned challenge to Y-STR DNA evidence should hone in on those differences while following the general guidelines for DNA cases outlined above.

Below is a synopsis of the core substantive areas of forensic Y-STR DNA that can be developed through admissibility challenges under *Frye, Daubert* or other evidentiary standards governing expert testimony and scientific evidence, or as a basis for challenging the expert directly through cross-examination. Refer to Chapter 2, Section 3, for basic biological information about Y-STRs.

Any attempt to characterize a purported Y-STR "match" as evidence of conclusive identity is scientifically unsupportable on the basis of its method of inheritance alone. Any lab report conclusion of an inclusion, or match, must be qualified in a manner that clearly points out that other males in the same lineage will have the same Y-STR haplotype (barring mutation) as the one generated from the item of evidence.

The Limited Discriminatory Power of Y-STRs

Y-STR profiles are exactly the same among related males within patrilineal lines. Although Y-STR DNA has proven to be a powerful exclusionary tool, its ability to inculpate is less powerful. Far short of identifying a particular person as the source of a Y-STR profile, observing that a suspect profile is consistent with an evidence sample does no more than reduce the population of possible contributors to the defendant plus "all patrilineal related male relatives and an unknown number of unrelated males as being the donor of the evidence sample."[81] Consequently, "the observation of a match with Y-STRs does not carry the power of discrimination and weight into court as an autosomal STR match."[82]

Frequency Estimates and Y-STR Population Databases

As with mtDNA, Y-STR population databases may also provide fertile ground for defense challenges. As with any DNA match evidence, the ability to assign significance to the match hinges on the reliability of the databases used to conduct the statistical calculations.

Refer to http:// www.cstl.nist.gov/ strbase/y_strs.htm for Y-STR haplotype databases commonly used in the forensic community. The current number of haplotype profiles available for searching using the Consolidated U.S. Y-STR Database (http://www. usystrdatabase.org/) can be obtained by selecting the "Database Descriptive Statistics" tab at the top of the Web page. The resulting page for release 2.4 of the database (as of January 2, 2011) lists the "Total Number of Haplotypes (N)" as 18,547 (see http://www.usystrdatabase. org// pdf/DatabaseDescriptive Statistics.pdf).

Y-STR Profiles Cluster Regionally

Not surprisingly, Y-STR profiles cluster geographically, following migration and settlement patterns. These clusters are discernible among major population groups (for example, Caucasian, African-American and Hispanic). Studies have shown that there are statistically significant differences in the frequency of Y-STR profiles within some discrete ethnic groups, depending on which geographic locations are sampled.[83] Some Y-STR profiles are very common in certain geographic locales and much less common in others. Consult with an expert to determine whether ethnic clustering issues may exist in the particular case.

As a result of the substantial geographic sub-structuring of Y-STR DNA profiles, some scientists have expressed "particular concern [over] the sampling of multiple populations and their assembly into global databases."[84] When Buckle-ton et al. surveyed available scientific literature in 2005, they found "no report in the literature yet of how to interpret Y-chromosome haplotypes accounting for population subdivision."[85] They reported that "further investigation into how to compensate for population subdivision at the Y chromosome is warranted urgently" and said that "it is imperative that every effort should be made to use appropriate local databases" when attempting to estimate the frequency of a given Y-STR profile.86 Guideline 5.6 of the SWGDAM Y-chromosome Short Tandem Repeat (Y-STR) Interpretation Guidelines (released January of 2009) [http://www.fbi.gov/about-us/forensic-science-communications/fsc/jan2009/standards/2009_01_standards01.htm] states, "It is recognized that population substructure exists for Y-STR haplotypes. Studies with current population databases have shown that the FST values are very small for most populations. Thus the use of the counting method that incorporates the upper-bound estimate of the count proportion offers an appropriate and conservative statistical approach to evaluating the probative value of a match."

The Role of Ethnicity in Trying a Y-STR Case

The defendant's ethnic background may play a critical role in informing the defense strategy in a Y-STR DNA case. At minimum, defense counsel must be aware of the defendant's ethnic background because he may be a member of a significant immigrant community of common patrilineal origin. If the defendant's ethnic background is not represented by the database(s) searched, defense counsel should consider hiring an expert to help determine the significance — or lack thereof — of the statistics provided.

For example, consider a defendant from San Miguel (in eastern El Salvador), who immigrated to Washington, D.C., along with many of his countrymen in the last 20 or 30 years.[87] The Salvadoran population of the D.C. region is a new community, having emerged within a single generation; in fact, this phenomenon of recent immigration helped the District earn the title "immigrant gateway."[88] With limited opportunity for intermingling with established local communities, the Salvadoran population in the district may be genetically insulated.

The defense can argue that the District's Salvadoran community is unique, due not only to the recency of its emergence but also to its unique genetic ancestry. The Salvadoran source population is unlike the more common sources of Hispanic immigrants to the United States, such as those from Mexico or Puerto Rico. In El Salvador, the native Indian population remained largely intact despite Spanish conquests.[89] In Mexico, by contrast, the majority of inhabitants have been classified as "mestizos," who are genetically traceable to a mixture of European and African ancestry. Puerto Ricans are heavily of European and West African descent.[90] Thus, Salvadoran immigrants, as a group, might be expected to be genetically distinct from other Hispanics residing in the United States.

There may also be some genetic variation within the Salvadoran population. For example, Salvadorans now residing in the D.C. metropolitan area have a markedly different ancestry than their countrymen who immigrated to other major U.S. destinations such as Los Angeles and other parts of California. Those who immigrated to the District of Columbia came predominantly from eastern El Salvador, from rural communities insulated from the urbanized centers of the West like San Salvador, where presumably the majority of genetic mixing would occur.[91] Salvadoran immigrants who relocated to California, on the other hand, came largely from the major metropolitan areas in western El Salvador.[92] In this example, the defendant and his countrymen now calling the District of Columbia home remain genetically akin to the narrow subset of Salvadoran natives occupying that particular, insulated region in eastern El Salvador.

Further amplifying the insulation and uniqueness of the Salvadoran genetic fabric is that immigrating families tend to follow family members who have gone before them.[93] The effect is even more dramatic when the gateway is new, when the newcomer population has not yet had the opportunity to mingle genetically with more established populations in the region.

Arguments along these lines can be further substantiated with census data and other research on migration and settlement patterns — and then contrasted with information on the source populations of the Y-STR database that was used for the statistical significance requirement for the matched evidence haplotype. With respect to the Hispanic branch of the Y-STR database, it can be argued that neither eastern Salvadorans nor the Hispanic population of Washington, D.C., are represented.

Defense Experts

If the government calls a forensic scientist and the necessary points regarding Y-STR database limitations cannot be developed sufficiently on cross-examination, the defense may want to call a scientist with expertise in the underlying science to challenge the government expert's claims. In a Y-STR case, the defense may also want to call an expert on the migration patterns of the defendant's ancestors to show that the database the government used for its statistical representation does not properly represent the defendant's Y-STR profile.

Y-STR Treatment in the Courts

A number of challenges have been made to the admissibility of Y-STR evidence as an inculpatory tool, but they have been generally unsuccessful thus far. Y-STR "inclusion" evidence has been admitted in several jurisdictions.[94] Not all of these cases have been fully litigated. Fresh admissibility challenges are worthy of consideration, depending on the facts of the specific case.

Section 12. Voir Dire of the Prosecution's DNA Expert

Several questions emerge when preparing a *voir dire* of a DNA expert. The first question is whether to conduct a *voir dire* in the first place. *Voir dire* has three basic goals:

- To exclude the witness's expert testimony.
- To limit the witness's expert testimony.
- To highlight for jurors why they should give little or no weight to the opinion, even though the witness is permitted to provide an expert opinion.

Depending on defense strategy, counsel may opt to conduct a detailed *voir dire* prior to the expert's direct testimony on the merits of the case. Counsel may also opt to weave *voir dire* questions based on answers provided into the "bias" segment of cross-examination. Generally, *voir dire* is conducted in front of the jury; however, especially in cases where the defense is seeking exclusion or limitation of the expert opinion, counsel can ask the court to allow the *voir dire* to occur outside the jury's presence, either before or after the jury is sworn in.

Regardless of strategy, counsel should review the analyst's curriculum vitae and perform an extensive Internet search for all publications and occurrences of the analyst's name. To the extent possible, investigate the contents of the curriculum vitae, such as trainings, certifications and professional associations. If a claimed credential can be earned by simply submitting a fee, counsel should know how much the fee is and, depending on the circumstances, consider becoming a member. This line of *voir dire* can help affect the testimony to come, creating a degree of skepticism and potentially undermining the expert's authority in the minds of jurors.

Depending on the expert, counsel may also want to challenge his or her educational background. For example, if the prosecution's expert does not hold a master's degree or Ph.D. in a hard science, counsel may make some headway with questions that illuminate that the analyst is primarily a technician with minimal scientific training or little to no training in molecular biology, statistics and/or population genetics. Such questions can suggest to the jury that the witness has minimal comprehension of the underlying science or little ability to form judgment regarding its accuracy. Counsel may consider delving into the specifics of what the analyst studied in school, including undergraduate and graduate work, if applicable. Highlight the irrelevancy and inadequacy of the training to support the defense's position — that the examiner is not trained to make the complex judgments he or she is in court to express.

Counsel will also want to ask the expert to acknowledge the authoritativeness of certain publications — such as Dr. John Butler's *Forensic DNA Typing*,[95] Dr. John Buckleton's *Forensic DNA Interpretation*,[96] one or both of the National Research Council reports,[97] or articles from authoritative journals that contain specific points that bolster the defense theory. These publications can serve as the foundation for some of the areas counsel will develop during cross-examination and potentially with subsequent witnesses, including, if applicable, the defense's DNA expert.

Section 13. Stipulations — Qualifications and/or Results

In certain cases, counsel may wish to stipulate to either the qualifications of the scientist or the results of the DNA tests. These strategy decisions must be thought out well in advance.

Reasons for stipulating to a scientist's qualifications may include a desire to downplay the scientist's training and expertise. Equally valid, however, would be a desire to save time and avoid the recitation of education and training. For example, consider a criminal sexual conduct case where the defense is arguing that the sex was consensual. Because defense counsel is not going to dispute the DNA evidence *per se*, there is really no need for the state to bolster its evidence by establishing that the scientist is well trained. Another consideration is not to draw the ire of the court or bore the jury with unnecessary discussion of the expert's qualifications.

The same example can be used to illustrate when counsel may wish to stipulate to the results. In light of the fact that consent is being asserted, both the prosecution and the defense may wish to spare the jury a lengthy explanation of how *PCR* and *capillary electrophoresis* work. The same may be true in a self-defense case or when an insanity claim is asserted.

Caution should be used when considering stipulation. It is the exception, not the rule, in a DNA case.

Section 14. Questioning Law Enforcement on Evidence Collection and Chain-of-Custody Issues

Ideally, every evidence collection would be videotaped. Practically speaking, this is not going to happen. As basic preparation, discovery should provide defense counsel with a timeline of when things were gathered and when they arrived at the laboratory. The lab notes should tell counsel the condition in which evidence arrived at the lab — for example, how was the container sealed; if there was moisture inside a heat-sealed plastic bag; or a brown paper bag said to contain one seat cover also included two socks, ChapStick®, a comb, fingernail clippers and condom in sealed wrapper, in addition to the seat cover.

Using the police reports, counsel will likely learn which officers were at the scene at the same time. Question the officers about any training they have had regarding DNA evidence collection. An excellent outline of police officer responsibilities can be found in the "Officers' Responsibilities" section of *What Every Law Enforcement Officer Should Know About DNA Evidence: Investigators and Evidence Technicians.*[98]

Cross-examination should focus on developing the defense theory. When counsel knows the answers, issues of contamination or transfer should be addressed by asking leading questions on the following topics:

- Having contact with the defendant before collecting evidence.
- Playing a role in the defendant's arrest.

- Having contact with the victim/witness/person of interest (whose DNA has appeared in the case) before collecting evidence.
- How the evidence was collected:
  - Were gloves worn?
  - Were gloves changed?
  - How often were gloves changed?
  - Was a list of collected evidence written down?
  - Was the pen used to record the list of collected evidence cleaned to remove any potential DNA on it before it was used at the scene?
  - Is it your custom to write each thing as it is collected or to do a laundry list at the end?
  - As a general habit, do you put your pen in your mouth or use it to scratch yourself?
  - Did you cough or sneeze during the evidence collection process?
  - Were you wearing a face mask over your nose and mouth?
  - What did you do with the collected evidence?
  - Where was it stored or held while all of the evidence was collected?
  - Was each piece of evidence sealed immediately after collection or were all bags sealed at the end?
  - How long did it take to transport the evidence?
  - What were the storage conditions in the evidence room?
  - How did the evidence get to the lab?
  - Was the evidence in the trunk, glove box or body of the car?
  - Was the evidence in the front seat or back seat?
  - What was done to minimize the risk of contaminating DNA from getting on the outside of the evidence containers?
  - Were all of the evidence envelopes and bags put into one box for transport?

Cautionary note: It is a good idea to tour the evidence room if given the opportunity. An unsecured evidence room — or a secured evidence room with a door that can be propped open during nice weather — could have an impact on the integrity of the evidence.

Section 15. Defense Expert Testimony Issues

In most cases, an expert should be retained for consultation. It is rare, however, that a defense expert is called to testify.

Some considerations for whether defense counsel should call an expert to testify include:

- Can counsel develop evidence effectively and persuasively through cross-examination of the government expert?
- Will the defense expert's testimony be more helpful than harmful?
- Can the defense expert communicate the information in a way that keeps the jurors' attention?

- Will having a defense expert neutralize or minimize the state's DNA evidence and allow the jurors to focus on other evidence in the case?

To support a public defender request for funding, a lengthy record should be made, establishing that an expert is necessary to adequately defend the client. Counsel should emphasize supporting case law and statutory law, where applicable.[99]

If the defense decides to call an expert to testify, the selection of that witness is critical. Factors to consider include:

- Education.
- Experience.
- Prior testimony.
- Objectivity (worked for both sides).
- Demeanor.
- Ability to communicate complex issues in understandable language.
- Considerations for the defense expert's testimony include:
- Limit the expert to the important points.
- Make it interesting.
- Keep it relevant.
- Recall that jurors (like the rest of us) have varying learning styles. Try to incorporate visuals in addition to verbal presentation to keep their attention and assist in their understanding.
- Anticipate cross-examination.

Section 16. Defense Case Stay on Theme

The decision to call an expert witness in the defense case is significant. The defense may call an expert who provides an explanation of the DNA evidence contrary to the prosecution's explanation. The jury may find that the defense expert is more credible or, at a minimum, conclude that the DNA evidence is not clear-cut and thus decide the case on the basis of other evidence. Although this approach has its rewards, it also carries risks, primarily in the implicit suggestion that the jury can decide between the two experts and not look only to the prosecution to carry out its burden to prove the case beyond a reasonable doubt.

Defense counsel may elect not to present its own expert testimony. The defense may prefer to have the jury focus exclusively on the quality of the prosecution's DNA evidence and whether, in light of the cross-examination and other evidence in the case, the DNA evidence supports the prosecution.

The decision whether to offer the testimony of a defense DNA expert derives from the question, "Does the testimony advance the theory of the defense?" For example, if the defense theory is that there was consent, there is little reason to call an expert to testify that the state laboratory failed to follow its own protocols in conducting the DNA analysis. Conversely, if the defense theory is that evidence was contaminated, counsel may benefit from calling an expert to testify to quality assurance norms. The expert may be able to comment on ways in which the state failed to safeguard the crime scene DNA evidence from evidence seized from the defendant or even from the defendant's reference sample. Also, in cases where the defense has developed DNA evidence inconsistent with the defendant's guilt or consistent with the profile of a third-party perpetrator, such results can have substantial impact on the jury.

When considering whether to call an expert, prudent counsel will think carefully about how the prosecutor will cross-examine the expert. Do the positives of the testimony outweigh the negatives? Defense counsel should consider whether the DNA evidence points to an important fact (for example, the DNA profile on the hat identified as worn by the assailant) or an unimportant matter (for example, an old beer can found on the street, down the block from the shooting). Once the expert is on the stand, the prosecution might seek to turn the expert into a witness for the prosecution, highlighting material favorable to its case. Therefore, defense counsel should never take lightly the decision to call an expert.

Section 17. Defense Counsel's Closing Argument in a DNA Case

Before trial commences, counsel already should have developed the defense, incorporating the defense theory about the DNA evidence. Make sure that the closing argument fits the defense and works with the evidence that has been presented.

Keep closing arguments simple and do not overstate the defense's case. Do not get overly technical. Counsel's job is to demystify DNA. If the defense is attacking DNA, show that DNA can, in part, be subjective. A simple presentation will help covey that message. A simple, less technical presentation also will help empower the jury to consider the DNA evidence critically. If jurors believe they can understand and critically evaluate the evidence, counsel can persuade them to look beyond the analyst's conclusion of a DNA match.

Visual aids can be powerful with the jury. Counsel should use visuals that are clear and make the point they want jurors to understand. One chart that has been used successfully shows each allele call that involved a subjective interpretation by the analyst. In a case with a partial profile, a mixture, or low-level DNA sample, there may be a discrete number of times that an analyst made an interpretation adverse to the defendant, and a different interpretation at any point could reasonably be interpreted to exculpate him or her. Visually presenting this information to the jury can be very compelling. Keep visual displays simple and keep the connection with jurors.

Finally, anticipate and respond to the prosecution where appropriate. In particular, in jurisdictions where the prosecution gets the last word, make sure the jury understands this and pre-empt any arguments the prosecution might make in response to the defense's closing argument.

Section 18. The Prosecution's Closing Argument in a DNA Case

Defense counsel must be alert to possible factual and legal errors that may arise in a prosecutor's closing argument, including:

Misuse of the DNA evidence: For example, the failure to exclude a defendant from a DNA mixture from clothing left at the crime scene does not prove that the defendant was at the scene or that he or she was the last person to wear the clothing. Counsel should object to the argument of facts that are not supported by, or cannot be inferred from, the trial evidence.

The prosecutor's fallacy: This has been explained by one court as "incorrect reasoning," that is, when the jury confuses the probability of a random match with the potentially very different probability that the defendant is not the source of the matching samples.[100] If the random match probability is 1 in 1 million, this does not mean that there is a 1-in-1-million chance that the DNA came from someone other than the defendant. This can be addressed in a pre-closing motion *in limine* or by objection during the argument.

Burden shifting: Very often, the defense will not present an expert to challenge the prosecution's DNA evidence. In closing, a prosecutor might argue that the evidence is "unrebutted," "the defense could have brought in an expert to say that something was wrong with this analysis," or "the defense could have done its own DNA testing but did not." Such comments may be viewed as shifting the burden of proof, as the defense has no obligation to present such evidence.[101]

Comments on silence: Decisional law is strong in precluding use of a defendant's failure to testify and explain the evidence, as such comments trespass on the individual's privilege barring compelled self-incrimination.[102]

It must be noted that some comments otherwise forbidden will be allowed if defense counsel has "opened the door" or invited such a response. Thus, the defense must carefully design its closing argument to overcome an objection on these grounds.

End Notes

[1] Some articles and studies have examined the issue of juror comprehension of scientific testimony. See Dann, B.M., V.P. Hans and D.H. Kaye, "Can Jury Trial Innovations Improve Juror Understanding of DNA Evidence?" 255 NIJ J. 2 (2006); Reinstein, Myers, and Griller, "Complex Scientific Evidence and the Jury," 8 J UDICATURE 150 (1999).

[2] Studies on juror comprehension of DNA include:

Dann, B.M., V.P. Hans and D. Kaye, Testing the Effects of Selected Jury Trial Innovations on Juror Comprehension of Contested mtDNA Evidence: Final Technical Report, National Institute of Justice, Office of Justice Programs, U.S. Department of Justice, Dec. 30, 2004.

Goodman, J., E. Greene and E.F. Loftus. "What Confuses Jurors in Complex Cases," TR IAL (November) 65-74 (1985).

Faigman, D.L., and A.J. Baglioni, "Bayes' Theorem in the Trial Process: Instructing Jurors on the Value of Statistical Evidence," 12(1) LAw & H UM. B EHAV. 1-17 (1988).

Kaye, D.H., and J.J. Koehler. "Can Jurors Understand Probabilistic Evidence?" J. ROYAL S TAT. S OC'Y, Series A, 154, part 1, 75-81 (1991).

Koehler, J.J. "When Are People Persuaded by DNA Match Statistics?" 25 LAw & H UM. B EHAV. 493-513 (2001).

Koehler, J.J. "Error and Exaggeration in the Presentation of DNA Evidence at Trial," 34 J URIMETICS 34, 21-39 (1993).

Thompson, W.C. "Are Juries Competent to Evaluate Statistical Evidence?" 52 LAw & CONTEMP. P ROBS. 9-41 (1989).

[3] Irvin v. Dowd, 366 U.S. 717 (U.S. 1961).

[4] Batson v. Kentucky, 476 U.S. 79 (U.S. 1986); overruled in part as stated in Coleman v. Deloach, 1998 U.S. Dist. LEXIS 9615 (S.D. Ala. May 7, 1998).

[5] J.E.B. v. Ala. ex rel. T.B., 511 U.S. 127, 129 (U.S. 1994).

[6] For example, Rule 703, Federal Rules of Evidence, allows experts to rely on evidence that may be inadmissible at trial if such evidence is of a type normally relied on by experts in the pertinent field. That same rule, however, forbids the expert from disclosing to the jury what the inadmissible evidence is.

[7] Although the U.S. Supreme Court ruled in Melendez-Diaz in 2009 (124 S. Ct. 2527 (2009)) that an analytical report cannot be entered into evidence on its own without the supporting testimony of the analyst who performed the work, there may still be instances in which a court will allow another analyst or supervisor to testify in lieu of the analyst who did the work.

[8] See, for example, Rule 18.6(e) of the Arizona Rules of Criminal Procedure.

[9] National Center for State Courts, Jury Trial Innovations (Munsterman, G.T., P. HannafordAgor and M. Whitehead, eds., 2d ed. 2006); American Bar Association, Principles for Juries and Jury Trials 91-124 (2005); Heuer, L., and S. Penrod, "Juror Notetaking and Question Asking During Trials," 18 LAW & H uM. B EHAV. 142 (1994); Mott, N.L., "The Current Debate on Juror Questions: 'To Ask or Not to Ask, That is the Question,'" Symposium: The Jury at a Crossroad: The American Experience, 78 CHICAGO-KENT L. REV. 1099-1125 (2003); Heuer, L., and S. Penrod, "Increasing Juror Participation in Trials Through Note Taking and Question Asking," 79 J uDICATuRE 256 (1996); Penrod, S., and L. Heuer, "Tweaking Commonsense: Assessing Aids to Jury Decision Making," 3 PSYCHOL. P uB. P OL'Y & L. 259 (1997); AOC State of New Jersey Jury Subcommittee, Report on Pilot Project Allowing Jury Questions (unpublished AOC report); Diamond, S., M. Rose and B. Murphy, "Jurors' Unanswered Questions," (Spring) COuRT R EV. 20-29 (2004), http:// aja.ncsc.dni.us/courtrv/cr-41-1/CR41-1Diamond. pdf, and a law review version: Diamond, S.S., M.R. Rose, B. Murphy and S. Smith, "Juror Questions During Trial: A Window Into Juror Thinking," 59 VANDERBILT L. R EV. 1927 (2006); Mize, G.E., P. Hannaford-Agor and N.L. Waters, National Center for State Courts, The State-ofthe-States Survey of Jury Improvement Efforts: A Compendium Report, pp. 34-7 (2007), available at www.ncsconline.org/D_Research/cjs/pdf/ SOSCompendiumFinal.pdf.

[10] www.innocenceproject.org/Content/268.php.

[11] Parsons, T.J., and M.D. Coble, "Increasing the Forensic Discrimination of Mitochondrial DNA Testing Through Analysis of the Entire Mitochondrial DNA Genome," 42(3) CROATIAN M ED. J. 304, 304 (2001).

[12] Wilson, M.R., M. Stoneking, M.M. Holland, J.A. Dazinno and B. Budowle, "Guidelines for the Use of Mitochondrial DNA Sequencing in Forensic Science," 20 CRIME LAB DIG. 68–77 (1993).

[13] See Fisher, C.L., et al., Mitochondrial DNA: Today and Tomorrow, presented at E LEVENTH ANNuAL INT'L S YMPOSIuM ON HuMAN IDENTIFICATION, at 1 (2000).

[14] See Melton, T., and K. Nelson, "Forensic Mitochondrial DNA Analysis: Two Years of Commercial Casework Experience in the United States," 42 CROATIAN M ED. J. 298 (2001).

[15] Id. at 300.

[16] Id.

[17] Kaestle, F.A., et al., "Database Limitations on the Evidentiary Value of Forensic Mitochondrial DNA Evidence," 43 AM. CRIM. L. R EV. 53, 62 (2006).

[18] FBI Laboratory DNA Unit II, Mitochondrial DNA Sequencing Protocol (2004) [hereinafter FBI MtDNA Protocol (2004)], at § 11.3.3.

[19] Id.

[20] Id.

[21] Id. (citing Statement of Dr. M. Thomas P. Gilbert, submitted in United States v. Chase, D.C. Super. Ct. Crim. No. F-7330-99 (July 9, 2004) (reviewing protocols for all major mtDNA testing laboratories and observing that "forensic laboratories come to no consensus as to how to interpret heteroplasmic sequences. ... [T]he interpretation guidelines vary when determining what would be labeled as 'inconclusive' and what would be labeled as an 'exclusion.'").

[22] Technically, only the differences between the sample and the reference (CRS/Anderson) sequence are compared with the database profiles. Isenberg, A.R., and J.M. Moore, "Mitochondrial DNA Analysis at the FBI Laboratory," 1 FORENSIC S CI. C Omm. 1 (1999), available at www. fbi.gov/hq/lab/fsc/backissu/july1999/ dnalist.htm.

[23] See, e.g., United States v. Porter, 618 A.2d 629 (D.C. 1992) (requiring expression of statistical significance of a DNA "match" as a prerequisite to admission); People v. Axell, 235 Cal. App. 3d 836 (2001) (same).

[24] Bandelt, H.J., et al., "Problems in FBI MtDNA Database," 305 S CIENCE 1402, 1403 (2004).

[25] Id. SWGDAM is the Scientific Working Group on DNA Analysis Methods.

[26] Yao, Y.G., C.M. Bravi and H.J. Bandelt, "A Call for MtDNA Data Quality Control in Forensic Science," 141 FORENSIC S CI. I NT'L 1, 1 (2004).

[27] Id. at 4.

[28] Bandelt, supra note 24.

[29] Id.

[30] Id.

[31] Budowle, B., et al., "Mitochondrial DNA Regions HVI and HVII Population Data," 103 FORENSIC S CI. I NT'L 23, 25 (1999).

[32] Id.

[33] FBI MtDNA Protocol (2004), supra note 18, at § 11.1.

[34] Laboratories use a slightly different statistical calculation when the sequence is not observed in the database. See Holland, M.M., and T.J. Parsons, "Mitochondrial DNA Sequence Analysis: Validation and Use for Forensic Casework," 11 FORENSIC S CI. R EV. 31-32 (1999).

[35] A 95-percent confidence interval means that, if a series of such margins of error were constructed in estimating the frequency of the sequence in the population, approximately 95 percent of them should include the true frequency of the sequence in the population. Alternatively stated, there is approximately a 5-percent chance that the margin of error does not contain the true frequency of the sequence in the population. See Witte, R.S., Statistics at 215 (2d ed., 1985). As the sample size grows, the confidence interval will become narrower, indicating 95-percent confidence in a smaller range of possible values for the frequency (Id. at 216).

[36] Kaestle et al. (2006), supra note 17, at 64-65.

[37] Phylogeography "is a field of study concerned with the principles and processes governing the geographic distributions of genealogical lineages, especially those within and among closely related species [and] deals with historical, phylogenetic components of the spatial distributions of gene lineages. In other words, time and space are the jointly considered axes of phylogeography onto which (ideally) are mapped particular gene genealogies of interest." Avise, J.C , Phylogeography: The History and Formation of Species (Harvard University Press, 2000) at 3.

[38] See, e.g., Mishmar, D., et al., "Natural Selection Shaped Regional MtDNA Variation in Humans," 100 PROC. N AT'L ACAD. S CI. 171 (Jan. 7, 2003) ("extensive global population studies have shown that there are striking differences in the nature of the mtDNAs found in different geographic regions").

[39] See Richards, M., et al., "In Search of Geographical Patterns in European Mitochondrial DNA," 71 Am. J. H um. G ENET. 1168, 1170 (2002).

[40] Although phylogenetic analysis — reconstructing genetic relationships within a population — has been conducted on many of the SWGDAM racial sub-databases, such studies show, at most, only that a particular database accurately reflects most of the haplogroups that exist in the relevant population, for example, that the Caucasian database contains all major haplogroups in the Caucasian population. Such studies do not, however, take into account the geographical distribution of the sequences within the population and thus cannot be cited as evidence that a database accurately reflects the frequency of a profile in a particular geographic area. Only phylogeographic studies — those that focus on the spectrum and area-specificity of major haplogroups and the haplotypes within them — can accurately determine true frequencies. See Rando, J.C., et al., "Phylogeographic Patterns of MtDNA Reflecting the Colonization of the Canary Islands," 63 ANNALS H um. G ENET. 413, 424 (1999) [hereinafter Rando et al. (1999)].

[41] See, e.g., Bamshad, M., et al., "Genetic Evidence on the Origins of Indian Caste Populations," 11 GENomE R ES. 994 (2001) (discussing economic and caste distinction); Dutta, R., et al., "Patterns of Genetic Diversity at the Nine Forensically Approved STR Loci in the Indian Populations," 74 Hum. B ioL. 33 (2002) (same); Merriweather, D.A., et al., "Mitochondrial DNA is an Indicator of Austronesian Influence in Island Melanesia," 110 Am. J. P HYS. A NTHRoPoL. 243 (1999) (linguistic distinctions); Rudan, P., et al., "Anthropological Research of Hvar Islanders, Croatia — From Parish Registries to DNA Studies in 33 Years," 28 CoLLEGium ANTHRoPoLoGiCum 321 (2004) (religious); Zhivotvsky, L.A., et al., "The Forensic DNA Implications of Genetic Differentiation Between Endogamous Communities," 119 FoRENSiC S Ci. i NT'L 269 (2001) (no obvious subdivision).

[42] See, e.g., Balding, D., Weight-of-Evidence for Forensic DNA Profiles 105-06 (2005) [hereinafter Balding (2005)] ("[M]aternally-related individuals might be expected to be tightly clustered, possibly on a fine geographical scale. Reports of Fst estimates for mtDNA drawn from cosmopolitan European populations typically cite low values, reflecting the fact that this population is reasonably well-mixed, as well as the effects of high mtDNA mutation rates. However, researchers rarely are able to focus on the fine geographic scale that may be relevant in forensic work, and there are some large Fst estimates at this scale.") (emphasis added); Brandstatter, A., et al., "Mitochondrial DNA Control Region Sequences From Nairobi (Kenya): Inferring Phylogenetic Parameters for the Establishment of a Forensic Database," 118 iNT'L J. L EGAL m ED. 294 (2004) (describing new forensic database containing sequences from Nairobi and finding that there were significant differences in mtDNA compositions of this new database and the African-American SWGDAM database as well as of published sequences from Sierra Leone, Mozambique and United States); Forster et al., "Continental and

Subcontinental Distributions of mtDNA Control Region Types," 116 iNT'L J. L EGAL m ED. 99-108 (2002); Kaestle, F.A., and K.A. Hors-burgh, "Ancient DNA in Anthropology: Methods, Applications, and Ethics," 119(S35) Am. J. P HYS. ANTHRoPoL. 92, 95 (2002) ("[M]itochondrial markers are also often geographically specific, and in some cases are limited in distribution to a single tribe (private polymorphisms)."); Kittles, R., and S.O.Y. Keita, "Interpreting African Genetic Diversity," 16 AFRiCAN ARCHEoL. R EV. 87-91 (1999); Pereira, L., et al., "Prehistoric and Historic Traces in the mtDNA of Mozambique: Insights Into the Bantu Expansions and the Slave Trade," 65 Am. J. Hum. GENET. 439-458 (2001) [hereinafter Pereira et al. (2001)]; [Rando et al. (1999), supra note 40, at 413, 424; Salas, A., et al., "The African Diaspora: Mitochondrial DNA and the Atlantic Slave Trade," 74 Am. J. H um. G ENET. 454-65 (2004) [hereinafter Salas et al. (2004)]; Yao, Y.G., et al., "Phylogeographic Differentiation of Mitochondrial DNA in Han Chinese," 70(3) Am. J. H um. G ENET. 635, 649 (2002).

[43] Richards, M.B., et al., "Phylogeography of Mitochondrial DNA in Western Europe," 62(3) ANNALS H um. G ENET. 241, 255 (1998) (discussing J Haplogroup).

[44] Id. at 254 (discussing J1b1 Haplogroup).

[45] Rando et al. (1999), supra note 40, at 420, 424.

[46] Pereira et al. (2001), supra note 42, at 439, 451-452. ("There remain a large number of sequences from African haplogroups sampled in the Americas and Europe for which no match can be found in the current African database. This may be due in part to the fact that the main regions from where slaves were taken, such as Angola and the Slave Coast, remain uncharacterized.") (citation omitted). See also Lorenz, J., et al., African-American Lineage Markers: Determining the Geographic Source of mtDNA and Y Chromosomes, presented at 73rd annual meeting of the American Association of Physical Anthropologists, Tampa, FL, Apr. 15, 2004, available at www.physanth.org (discussing study suggesting that there is a large proportion of unexamined, undocumented mtDNA variability among individuals indigenous to sub-Saharan Africa).

[47] See, e.g., Eshleman, J., R.S. Malhi and D.G. Smith, "Mitochondrial DNA Studies of Native Americans: Conceptions and Misconceptions of the Population Prehistoric of the Americas," 12 EVoLuTioN. A NTHRoPoL. 7-18 (2003) (noting that, whereas Haplogroup X is found in low frequency in Europe and Western Asia, the Native American variant is significantly different, possessing mutation that distinguishes it from Old World versions); Jorde, L.B., and S.P. Wooding, "Genetic Variation, Classification, and 'Race,'" 36 nature g enet. S28, S29 (Nov. 2004) ("[I]ndividuals tend to cluster according to their ancestry or geographic origin."); Malhi, R.S., et al., "The Structure of Diversity Within New World Mitochondrial DNA Haplogroups: Implications for the Prehistory of North America," 70(4) AM. J. Hum. genet. 905 (2002) (Native Americans have haplogroups whose frequencies vary greatly among Canada, United States and Mexico); Parra, E.J., R.A. Kittles, et al., "Ancestral Proportions and Admixture Dynamics in Geographically Defined African Americans Living in South Carolina," 114 am. J. p HyS. a ntHropol. 118 (2001) [hereinafter Parra and Kittles (2001)]; Parra, E.J., A. Marcini, et al., "Estimating African-American Admixture Proportions by Use of Populations-Specific Alleles," 63 AM. J. H um. g enet. 1839 (1998) [hereinafter Parra and Marcini (1998)]; Tishkoff, S.A., and K.K. Kidd, "Implications of Biogeography of Human Populations for 'Race' and 'Medicine,'" 36 nature g enet. S21, S26 (November 2004) (frequency of mtDNA haplogroups are unevenly distributed within and among geographic regions and "knowledge of ethnicity (not just broad geographic ancestry) and statistical tests of substructure are important [to the] proper design of case control association studies"). Cf. Melton, T., et al., "Diversity and Heterogeneity in Mitochondrial DNA of North American Populations," 46 J. f orenSic S ci. 46 (2001) (while arguing that the North American population is homogeneous, this identifies, without exploring, a population of Hispanics in Pennsylvania who differed significantly from any other population in the study).

[48] Cann, R.L., M. Stoneking and A.C. Wilson, "Mitochondrial DNA and Human Evolution," 325 nature 31 (1987). See also Curtin, P.D., The Atlantic Slave Trade: A Census (U. Wisc. Press, 1969; Lovejoy, 2d ed., 1994) [hereinafter Curtin (1969)]. Curtin's calculations were later refined by David Northrup: Northrup, D., The Atlantic Slave Trade (1994). See also Watson, E., et al., "MtDNA Sequence Diversity in Africa," 59 am. J. Hum. g enet. 437 (1996).

[49] See, e.g., Melton, T., et al., "Extent of Heterogeneity in Mitochondrial DNA of sub-Saharan African Populations," 42 forenSic S ci. i nt'l 582, 588-89 (1997) (finding numerous haplotypes with SSO frequencies of greater than 10 percent in particular African population and "substantial subpopulation heterogeneity" in "continental African populations"). The authors conclude that "control region sequencing would be a good alternative for forensic identifications in African or African-derived populations where there is uncertainty about whether subpopulations are present, at least until further populations are studied" (Id. at 589).

[50] See generally, Salas et al. (2004), supra note 42, at 455-56.

[51] Parra and Kittles (2001), supra note 47, at 19.

[52] Morgan, P.D., Slave Counterpoint: Black Culture in the Eighteenth Century Chesapeake and Lowcountry (1998) at 33-44.

[53] Id. at 34-36.

[54] Jackson, F.L., "Concerns and Priorities in Genetic Studies: Insights from Recent African-American Biohistory," 27 S eton Hall l. r ev. 951, 961-62 (1997); Parra and Marcini (1998), supra note 47, at 1839 (listing countries of Africa by economic region). This very same resistance makes African-Americans whose ancestors come from the Gold Coast more susceptible to sickle cell trait and sickle cell disease. Muniz, A., et al., "Sickle-Cell-Anemia and Beta-Gene Cluster Haplotypes in Cuba," 49 am. J. H ematol. 163 (1995); Pante-De Sousa, G., et al., "Betaglobin Haplotypes Analysis in Afro-Brazilians from the Amazon Region: Evidence for a Significant Gene Flow from Atlantic West Africa," 26 annalS H um. B iol. 365 (1999).

[55] Curtin (1969), supra note 48, at 83.

[56] See generally, Grossman, J.R., Land of Hope: Chicago, Black Southerners, and the Great Migration (1991).

[57] Id. at 28-30.

[58] Id. at 112-13 (migration from Mississippi delta to Chicago); Lemann, N., The Promised Land: The Great Black Migration and How It Changed America (1991) (migration from the Carolinas and Virginia up the East Coast) at 109-120.

[59] See www.census.gov/geo/www/mapGallery/ images/black.jpg (pictorial depiction of geographical distribution of African-Americans in United States).

[60] See Parra and Marcini (1998), supra note 47, at 1845-47; Parra and Kittles (2001), supra note 47, at 19; Salas et al. (2004), supra note 42, 454-65.

[61] See Parra and Marcini (1998), supra note 47, at 1845-47; Chakraborty, R., "Gene Admixture in Human Populations: Models and Predictions," 29 Y.B. P HYS. A NTHRoPoL. 1-43 (1986); McLean, Jr., D.C., et al., "Three Novel mtDNA Restriction Site Polymorphisms Allow Exploration of Population Affinities of African Americans," 75 HUM. B ioL. 147 (2003)

[62] See www.census.gov/geo/www/mapGallery/images/americanindian. jpg (visual depiction of heavy Native American clustering in the western part of the United States); Ogunwole, S.U., The American Indian and Alaska Native Population: 2000, at 4-6 (U.S. Census Bureau, February 2002) (noting that 43 percent of American Indians lived in the West, 31 percent lived in the South, 17 percent lived in the Midwest, and 9 percent lived in the Northeast).

[63] Parra and Marcini (1998), supra note 47, at 1845. The admixture study reports two results from Philadelphia, based on two independent sample sets taken from patients in two separate hypertension studies. These sample sets exhibited significant differences in their percentage of admixture (Id.). Thus, even within a single city, different groups of African-Americans display significantly different mtDNA profiles.

[64] Id. at 1845-47.

[65] See Bonilla, C., M.D. Shriver, et al., "Admixture in the Hispanics of the San Luis Valley, Colorado and Its Implications for Complex Trait Gene Mapping," 68 ANNALS HUM. G ENET. 139, 140 (2004) (the term Hispanic applies to individuals from several continents with "diverse cultural features and genetic backgrounds").

[66] See id. (reporting differences in admixture among Puerto Rican, Cuban and Mexican groups as well as within smaller region of San Luis Valley); Irwin, J., et al., "Development and Expansion of High-Quality Control Region Databases to Improve Forensic MtDNA Evidence Interpretation," 1(2) FoRENSic S ci. i NT'L: G ENETicS 154 (2007) (showing significant regional differences between "Hispanic" populations).

[67] Malhi, R.S., et al., "Native American mtDNA Prehistory in the American Southwest," 120 AM. J. P HYS. A NTHRoPoL. 108, 113 (2003) [hereinafter Malhi et al. (2003)].

[68] Id. In addition, the Navajo and Apache tribes are not representative of the variation present in haplotypes/haplogroups among all North American Native Americans. Tribal groups in the United States share few haplotypes. See Malhi et al. (2002), supra note 47, at 914, Table 2 (estimating sharing at approximately 29 percent).

[69] Malhi et al. (2003), supra note 67, at 121-22.

[70] The primary published analysis of this database concerns only the Chinese samples, and although the analysis suggests that the frequencies of the haplogroups in the data set are similar to those in another Han Chinese dataset of 263 individuals, the authors' data reveal significant differences in almost all cases. Allard and Wilson et al., "Control Region Sequences for East Asian Individuals in the Scientific Working Groups on DNA Analysis Methods Forensic mtDNA Data Set," 6 LEGAL M ED. L11, L18 Fig. 2 (2004). Other studies also show significant genetic variation among and within Asian populations. See, e.g., Kivisild and Tolk et al., "The Emerging Limbs and Twigs of the East Asian mtDNA Tree," 19 M oL. B ioL. E VoL. 1737 (2002) (other Asian populations not represented in the SWGDAM East Asian database have significantly different

frequencies of mtDNA haplogroups than those in the database); Melton, T., and M. Stoneking, "Extent of Heterogeneity in Mitochondrial DNA of Ethnic Asian Populations," 41 J. F oRENSic S ci. 591-602 (1996) (same); Yao et al. (2002), supra note 42, at nn. 76-78 and accompanying text (combining all Han Chinese would be inappropriate).

[71] See Reeves, T.J., and C.E. Bennett. We the People: Asians in the United States, Pub. No. CENSR-17, Census Bureau, U.S. Department of Commerce, at 1 and Table 1 (2004), available at www.census.gov/prod/2004pubs/censr-17.pdf (listing major Asian groups in the United States, many of which are not included in SWGDAM Asian databases).

[72] Id. at 4 and Fig. 1.

[73] Yao et al. (2002), supra note 42, at 635.

[74] See id. at 649 ("The comparison of the regional Han mtDNA samples revealed an obvious geographic differentiation in the Han Chinese, as shown by the haplogroups-frequency profiles. ... Hence, the grouping of different Han populations into just "Southern Han" and "Northern Han" or the use of one or two Han regional populations to stand for all Han Chinese ... does not appropriately reflect the genetic structure of the Han.") (citations omitted).

[75] See Branicki, W., K. Kalista, et al., "Distribution of mtDNA Haplogroups in a Population Sample from Poland," 50 J. F ORENSIC S CI. 732, 733 (2005) (H Haplogroup observed in 37.8 percent of samples in population from southern Poland); Dubut, V., and L. Chollet, "MtDNA Polymorphisms in Five French Groups: Importance of Regional Sampling," 12 EuR. J. H um. G ENET. 293-300 (2004) (within France alone, frequency of H varies between 35 percent and 50 percent in two separate communities in Brittany); Gonzalez, A.M., and A. Brehm, "Mitochondrial DNA Affinities at the Atlantic Fringe of Europe," 120 J. P HYS. A NTHROPOL. 391-404 (26.3 percent in Norway, 34 percent in England, 36.4 percent in Northern Germany, 38.5 percent in France, and 42.2 percent in Galicia); Malyarchuk and Grzybowski, "Mitochondrial DNA Variability in Bosnians and Slovenians," 67 ANNALS H um. G ENET. 412-25 (2003) (frequency of H Haplogroup is 24 percent in Finland, 26.8 percent in Scotland, and 45 percent in Poland).

[76] See also Pereira, L., et al., "Evaluating the Forensic Informativeness of mtDNA Haplogroup H Sub-Typing on a Eurasian Scale," 159(1) FOR E N-SIC S CI. I NT'L 43, 50 (2006) (use of SNPs to more closely examine haplogroups demonstrates significant interrelatedness below the haplogroup level and suggests that "phylogenetic dissection of mtDNA haplogroups is revealing gradients previously hidden on the Eurasian scale").

[77] Vaughn v. State, 646 S.E.2d 212, 214 (Ga. 2007).

[78] See, e.g., id. at 215 ("The conflicting expert opinions on the [mtDNA] test results go to the weight rather than the admissibility of the testimony"); People v. Ko, 757 N.Y.S.2d 561, 563 (App. Div. 2003) ("mitochondrial DNA analysis has been found reliable by the relevant scientific community; issues regarding contamination go to the weight to be given such evidence"); People v. Ko, 757 N.Y.S.2d 561: State v. Pappas, 776 A.2d 1091 (Conn. 2001) (holding that "issues regarding contamination are important and may bear on the weight of mtDNA evidence in a particular case, but that those issues do not undermine the admissibility of the results of the mtDNA sequencing process") (internal citation omitted).

[79] Wagner v. State, 864 A.2d 1037 (Md. App. 2005) (finding mtDNA "inclusion" evidence properly admitted); State v. Council, 515 S.E.2d 508, 518 (1999) (finding the underlying science of mtDNA reliable and "inclusion" evidence was properly admitted); State v. Council, 515 S.E.2d 508: State v. Underwood, 518 S.E.2d 231, 240 (N.C. 1999) (holding that mtDNA testing is sufficiently reliable to warrant its admissibility into evidence); State v. Scott, 33 S.W.3d 746, 756 (Tenn. 2000) (holding that mtDNA was properly admitted without an admissibility hearing); Adams v. State, 794 So.2d 1049, 1064 (Miss. App. 2001) (holding that science of mtDNA sequencing is adequately proven); State v. Pappas, 776 A.2d 1091, 1110 (Conn. 2001) (finding no error in admitting mtDNA evidence); People v. Holtzer, 660 N.W.2d 405, 411 (Mich. 2003) (holding that use of mtDNA for identification of the defendant is admissible under the test for novel scientific evidence); Magaletti v. State, 847 So.2d 523, 528 (Fla. Dist. Ct. App. 2003) (holding that use of mtDNA analysis to prove identity satisfied Frye); People v. Ko, 757 N.Y.S.2d 561, 563 (2003) (upholding the trial court's admission of mtDNA evidence). Admission of mtDNA evidence also has been upheld in a number of unpublished appellate decisions. People v. Ko, 757 N.Y.S.2d 561: See State v. Smith, 100 Wash. App. 1064, 2000 WL 688180 (Wash. Ct. App. 2000); State v. Ware, 1999 WL 233592 (Tenn. Crim. App. 1999); Sheckells v. Texas, 2001 WL 1178828 (Tex. Ct. App. 2001).

[80] United States v. Coleman, 202 F. Supp. 2d 962, 970-71 (E.D. Mo. 2002) (denying the defendant's motion to exclude mtDNA and finding it reliable, helpful to the jury, and not unduly prejudicial).

[81] Butler, J.M., Forensic DNA Typing (2d ed., 2005) at 214.

[82] Id. at 213-14.

[83] See e.g., Bonilla, C., et al., "Admixture in the Hispanics of the San Luis 81 Valley, Colorado, and Its Implications for Complex Trait Genemapping," 68 ann. H um. g enet. 139 (2004) (reporting wide variation in genetic profiles of various ethnic groups are falling under the cultural rubric of "Hispanic"); Hedman, M., et al., "Analysis of 16 Y STR Loci in the Finnish Population Reveals a Local Reduction in the Diversity of Male Lineages," 142 forenSic Sci. int'l 37 (2004) (a particular 16-loci Y-STR profile is shared by 13 percent of the Finnish population); Roewer, L., et al., "Online Reference Database of European Y-Chromosomal Short Tandem Repeat (STR) Haplotypes," 118 forenSic S ci. c omm. 106 (2001) (the most frequent minimal haplotype is observed in 3 percent of the continental European population); Weale, M.E., et al., "Armenian Y Chromosome Haplotypes Reveal Strong Regional Structure Within an Single Ethno-National Group," 109 Hum. g enet. 659 (2001) (finding significant regional stratification of Y-STR DNA profiles and observing that the London Armenian subsample was insufficient to describe genetic variation); Zarrabeitia, M.T., et al., "Significance of Micro-Geographical Population Structure in Forensic Cases," 117 int'l J. l egal m ed. 302 (2003) (studying Y-chromosome profiles in Cantabria region of Spain and finding that the substantial overstatement of evidential strength frequently results from the use of population databases collected on too broad a geographical scale); Zerjal, T., et al., "The Genetic Legacy of the Mongols," 72 am. J. Hum. g enet. 717 (2003) (8 percent of 2,100 males from Central Asia region closely matching males from an area of Genghis Khan's former Mongol Empire had unique Y-chromosome lineage).

[84] Buckleton, J., C.M. Triggs and S.J. Walsh, Forensic DNA Evidence Interpretation (2005) at 324.

[85] Id.

[86] Id.

[87] See, e.g., Cordova, C.B., The Salvadoran Americans 69 (2005) ("Since 1979, the influx of Salvadoran immigrants to the United States has risen at a high rate."); Latinas in the United States: A Historical Encyclopedia 135 (Ruiz, V.L., and V.S. Korrol, eds., 2006) ("Central American immigration increased exponentially [in the 1980s], quintupling the Salvadoran population [in the United States].").

[88] See Price, M., et al., "The World Settles In: Washington, DC, As an Immigrant Gateway," 26 urBan g eog. 61, 63 (2005) ("Unlike the more established urban immigrant destinations, the District of Columbia is not built upon a rich history of immigration and has only recently become an immigrant destination. Thus there are few historically ethnic immigrant neighborhoods or enclaves."); see also, The Rise of New Immigrant Gateways (Brookings Institute, February 2004).

[89] See El Salvador: A Country Study (Haggerty, R.A., ed., 1990) at 67 ("observers have estimated that much of the Salvadoran population in the 1980s could be said to possess an Indian racial background").

[90] See, e.g., Bonilla, C., et al., "Ancestral Proportions and Their Association with Skin Pigmentation and Bone Mineral Density in Puerto Rican Women from New York City," 115 Hum. g enet. 57 (2004); Buentello-Malo, L., et al., "Genetic Structure of Seven Mexican Indigenous Populations Based on Five Polymarker Loci," 15 am. J. Hum. B iol. 23 (2003).

[91] See Cordova (2005), supra note 87, at 78 ("This population [including that which immigrated to Washington, D.C.] is mainly rural, or coming from provincial Salvadoran cities and towns."); Id. ("large numbers of persons from the eastern part of El Salvador relocated in metropolitan centers in the East Coast" of the United States, including Washington, DC).

[92] Id. ("Large numbers of urban dwellers and those with more education have relocated in the Los Angeles and San Francisco metropolitan areas. In San Francisco, for example, many people are from San Salvador, Sonsonate and other major [western] provincial cities.")

[93] Id. at 78 ("These new immigrants arrived in the United States as a result of already established ethnic and family networks.").

[94] See Curtis v. State, 205 S.W.3d 656 (Tex. App. 2006) (finding Y-STR "inclusion" evidence sufficiently reliable under Daubert); State v. Unsworth, No. L-03-1189, No. L-04-1165 (Ohio App. Sept. 2, 2005) (admitting Y-STR evidence under Daubert); State v. Unsworth, State v. Sanders, State v. Russell, State v. Avila, State v. Temple-State v. Sanders, No. CR-2000 2900 (Ariz. Super. Ct. Dec. 16, 2003) (admitting Y-STR evidence but limiting statistical characterization to number of occurrences of profile in database); State v. Russell, No. 05-1-02485-2 (Wa. Super. Ct. Jan. 2006) (finding Y-STR admissible under Frye, with no need for a new admissibility hearing); State v. Avila, No. 02CF1862 (Ca. Super. Ct. Feb. 17, 2005) (finding Y-STR from Y-PLEX kit and statistics based on ReliaGene database are admissible under Frye); State v. Temple, No. 02040491 (Minn. Dist. Ct. Apr. 14, 2005) (finding Y-STR admissible under Frye); State v. Polizzi, 924 So.2d 303 (La. App. 2006) (admitting Y-STR without a challenge); Shabazz v. State, 592 S.E.2d 876 (Ga. App. 2004) (same).

[95] Butler, supra note 81, at 270.

[96] Buckleton, J., C.M. Triggs and S.J. Walsh, eds., Forensic DNA Evidence Interpretation, Boca Raton, FL: CRC Press, 2005.

[97] NRC I and NRC II: National Research Council, DNA Technology in Forensic Science, Washington, DC: National Academy Press, 1992; National Research Council, The Evaluation of Forensic DNA Evidence, Washington, DC: National Academy Press, 1996.

[98] See also, What Every Law Enforcement Officer Should Know About DNA Evidence: First Responding Officers, http://dna.gov/training/ letraining.

[99] Saks and Kohler, "The Coming Paradigm Shift in Forensic Identification Science," 309 S CIENCE 892 (August 2005).

[100] United States v. Morrow, 374 F. Supp. 2d 51, 66 (D.D.C. 2005); National Research Council, The Evaluation of Forensic DNA Evidence, Washington, DC: National Academy Press, 1996, p. 133, http://books.nap.edu/openbook.php? record_id=5141&page=133.

[101] Decisional law in this area is mixed. See, e.g., State v. Seager, 2001 Iowa App. LEXIS 671 (Iowa Ct. App. 2001) (collecting cases); compare Hayes v. State, 660 So. 2d 257 (Fla. 1995) and United States v. Mason, 59 M.J. 416 (U.S.C.A.F. 2004) (error for prosecutor who argued defense had opportunity to test DNA evidence) with Teoume-Lessane v. United States, 931 A.2d 478 (D.C. 2007) (not improper burden-shifting for prosecutor to ask witness about defense's ability to test DNA evidence). A particular argument, in language and tone, may be subject to challenge on this ground, especially where the comment may be read by the jury as highlighting the defendant's failure to testify. See, e.g., United States v. Triplett, 195 F.3d 990, 995 (8th Cir. 1999).

[102] See Griffin v. California, 380 U.S. 609, 614 (1965) ("we ... hold that the Fifth Amendment, in its direct application to the Federal Government ... forbids ... comment by the prosecution on the accused's silence").

9. Delayed Prosecutions, Cold Case Hits and CODIS

Section 1. Statute of Limitations Defenses

Statute of limitations legislation serves a number of purposes:

[T]he applicable statute of limitations ... is ... the primary guarantee against bringing overly stale criminal charges. Such statutes represent legislative assessments of relative interests of the [s]tate and the defendant in administering and receiving justice; they are made for the repose of society and the protection of those who may [during the limitation] ... have lost their means of defence.[1]

From the defendant's vantage point, there is particular "concern that the passage of time has eroded memories or made witnesses or other evidence unavailable."[2]

The following principles of law are not in dispute:

- Once the period for commencing prosecution has expired, it cannot be retroactively extended by new legislation.[3]
- This is true even in cases where DNA evidence conclusively establishes identity.[4]
- Conversely, when a legislature extends the statute of limitations for a particular criminal act before it expires, the extended period applies and no statute of limitations defense applies.[5]

The advent and success of using DNA to prove identity — particularly in sex crimes with biological material — have led to legislation changing the time period in which specified crimes may be prosecuted. In some instances, the time period has been lengthened or eliminated entirely.

An extended period has been granted in cases in which DNA evidence exists and has been preserved.[6]

Section 2. John Doe Warrants

Typically, the period of limitations is tolled when a charging document with some information about the perpetrator's identity has been properly filed. "John Doe" warrants — warrants without a known name but with some identifying information — have begun to be used, particularly in DNA cases.

The first issue is whether John Doe DNA warrants satisfy the "particularity" requirement of the Fourth Amendment or parallel provisions of state constitutions. Generic descriptions of suspects generally do not meet this standard.[7] However, courts that have considered the issue to date have found that John Doe warrants with a numeric DNA profile as the identifier meet the Fourth Amendment standard.[8]

A separate argument contends that a warrant should give notice to the perpetrator so that he or she can gather evidence and prepare to meet the charges. Clearly, a DNA-profile warrant does not give notice to the average citizen. However, the one court to consider this claim to date has rejected it.[9] This type of claim would apply only in states where the statute of limitations has been extended but not eliminated; there would be no claim of entitlement to such notice in states where the legislature has abolished a particular period for commencing prosecution.

Section 3. Due Process

Regardless of whether the limitations period has been extended or abolished, delayed prosecution may raise due process concerns if the right to present a defense has been severely compromised. The U.S. Supreme Court has explained that the Fifth Amendment requires the dismissal of an indictment — even if it is brought within the statute of limitations — if the defendant can prove that the government's delay was a deliberate device to gain an advantage over him and that it caused him actual prejudice in presenting his defense.[10]

The difficulty in applying this test is twofold. First, it requires proof of the prosecution's ill motive in delaying, unless state law is more solicitous.[11] Second, the prejudice must be substantial.[12] Nonetheless, it is an issue that warrants examination in any case where there is a significant gap between commission of the offense and commencement of actual prosecution.

Section 4. The Databank Hit Case

Overview of the CODIS DNA Databanks

In 1990, the FBI Laboratory began a pilot project called *CODIS,* creating proprietary software that enabled and continues to enable federal, state and local laboratories to electronically upload, exchange and compare DNA profiles.

The Federal DNA Identification Act was enacted as part of the Violent Crime Control and Law Enforcement Act of 1995 (Public Law No. 103- 322). This law authorized the FBI to establish a national DNA index for law enforcement. Since then, federal and state governments have invested significant resources toward developing and maintaining a national databank system. NDIS became fully operational in October 1998.

CODIS users predominantly access two indexes: the forensic index and the offender index.[13] The forensic index contains DNA profiles from crime scene evidence. The offender index contains DNA profiles of individuals who have been convicted of offenses defined by state or federal law. The FBI maintains the CODIS databank.

CODIS has three levels:

- NDIS — the National DNA Index System — is the highest level in the CODIS hierarchy. It enables participating labs to upload, exchange and compare qualifying DNA profiles on the national level. Profiles deemed "allowable" by NDIS are then searched against profiles from all other SDIS participating labs accepted at the national level. As of August 2010, NDIS had more than 8.7 million offender profiles and more than 330,000 casework profiles.[14]
- SDIS — the State DNA Index System — allows laboratories within each state to exchange DNA profiles. Each state has a single statewide databank — SDIS. The FBI serves as the SDIS lab for the District of Columbia. The U.S. Army Crime Lab is also an SDIS lab. Each SDIS Administrator acts as the gatekeeper for determining the acceptability, based on that state's guidelines, of profiles submitted by each of the state's LDIS labs. Profiles accepted by the SDIS Administrator can be searched against those entered by other LDIS labs in the same state. Profiles accepted by an SDIS lab will also be searched against the convicted offender and arrestee (when applicable) profiles entered by the SDIS lab. SDIS custo-dians can share their data with the national CODIS community by forwarding it for consideration for inclusion in NDIS.
- LDIS — the Local DNA Index System — is the databank where regional, county and municipal labs within a state enter their profiles. Bench-level DNA examiners, or the lab's designee, use CODIS software to enter DNA evidence profiles into LDIS, where they are searched against other profiles that have been entered previously by their lab. Local labs can then forward their profiles to the state level for consideration for upload. Local labs must go through their SDIS lab to get profiles into the national level of CODIS.

The three-tiered system allows state and local agencies to operate their individual databases within the confines of state laws, which vary by jurisdiction. The exchange of information within this secure system is controlled by and strictly limited to law enforcement.

CODIS allows for the entry of qualifying DNA profiles into indexes based on specimen categories. The most commonly used specimen categories are as follows:

- Convicted offender: DNA profiles of people convicted of a crime.
- Forensic: DNA profiles developed from crime scene evidence.

- Arrestee: DNA profiles of arrested persons (if state law permits the collection of arrestee samples).
- Missing persons: DNA profiles from missing persons — either known or deduced to be known profiles from missing persons.
- Unidentified humans: DNA profiles from recovered unidentified human remains (UHR) as well as from humans who are unable or unwilling to identify themselves.
- Biological relatives of missing persons: DNA profiles voluntarily contributed by relatives of missing persons.

Other databank indexes exist (such as those that contain RFLP profiles), and the ability to enter mtDNA and Y-STR data has been added for certain specimen indexes. Federal and state laws govern access, disclosure, compatibility, expunction and penalties for unauthorized disclosure of information contained within CODIS.[15]

The DNA Identification Act of 1994, which established NDIS, also created the DNA Advisory Board (DAB) to develop standards for quality assurance. The board's work culminated with the promulgation of the first set of standards document for the forensic DNA casework analysis community, which became effective nationally on October 1, 1998, issued by the FBI director. These standards superseded the existing TWGDAM Guidelines that had previously been used as the guiding document by forensic DNA labs. A second set of standards for convicted offender databasing laboratories, which became effective on April 1, 1999, was issued by the DAB before the group disbanded on March 9, 2000. Currently, the responsibility for maintaining the Quality Assurance Standards (QAS) documents falls to the director of the FBI. Recommendations for updates are provided by the Scientific Working Group on DNA Analysis Methods (SWGDAM).[16]

To participate in NDIS, states must sign a Memorandum of Understanding verifying that the submitting laboratory is in compliance with the FBI's quality assurance standards.

Forensic DNA databanks were originally limited to samples only from adults convicted of felony sex offenses and a few other violent crimes. Databanks have now been expanded to include many other offenses as well as other classes of offenders.

All 50 states, the District of Columbia, and all federal jurisdictions now require certain classes of convicted offenders to provide a biological sample for entry into a DNA database. Each jurisdiction's statute determines whether a person convicted of an offense will be required to submit a biological sample for inclusion in a DNA database. (For more information, see http:// forensic.dna.gov/module9/1/.) The trend is clearly moving toward including larger categories of people, including those with misdemeanor convictions, juveniles and arrestees.[17]

CODIS contains limited information, such as a specimen identifier, the sponsoring laboratory's identifier, the initials or name of DNA personnel associated with the analysis, and the actual DNA profile. Depending on lab protocol, the specimen identifier of profiles submitted to the forensic (casework) index may identify the type of bodily fluid, whether the source is known, and/or whether the entered profile was deduced from results of mixed-sample DNA typing. CODIS does not store criminal history information or the names of convicted offenders/arrestees.

When CODIS software recognizes the same DNA profile in the forensic and offender indexes, it identifies the two profiles as a match. These matches are commonly referred to as "hits." Qualified personnel from both involved labs then analyze the reported match to either validate or refute it. This critical review of all matches is standard operating procedure and is used to ensure that a match produced by a search of the databank "makes sense." With a hit generated by a search of CODIS that involves a 13-loci match between an offender profile and a single-source evidence profile, the review process is fairly straightforward. Once both labs have agreed that the profiles do indeed match, the convicted offender lab will then research which offender corresponds to the specimen identifier in its system and will pull the corresponding sample and rerun it to confirm that the archived sample bearing the offender's name generates the same profile as the one entered into CODIS for that individual. This quality check is to ensure that sample results were not inadvertently switched during analysis or data entry. Once the profile has been confirmed in this manner, the convicted offender lab will subsequently provide basic information regarding the offender, such as name, available Department of Corrections information, and recorded date of birth, race and sex to the casework lab. The casework sample lab will then issue a hit report to the investigating agency to notify it of the databank match. This report typically requests submission of a newly obtained buccal sample from the identified offender to the casework lab as another quality check to ensure that a DNA profile obtained from the offender does indeed match the profile generated for the evidence profile. This report should also specify that the hit information is only intended to provide potential investigative leads that must be pursued by the investigating agency. If subsequent investigation supports that the CODIS match is meaningful, this can be used as the basis for probable cause to obtain the requested biological sample from the offender.

There are times, however, when the DNA profile generated from crime scene evidence that has been entered into CODIS is a mixture of more than one person's DNA. In those cases, the analysts will still critically compare the profiles to see if the offender's profile is included as part of the mixed DNA casework profile. It is not uncommon in these circumstances for the analysts to dismiss a match proposed by CODIS as not "making sense" on the basis of analytical data. When this occurs, unless agency policy states otherwise, no hit report will be issued by the casework lab; however, all information regarding the comparison and disposition of the hit will be maintained in the corresponding case file. When the labs determine, on the basis of their review, that the analytical data support the hit, a similar process to the one noted above is followed by the offender lab to research, confirm and share offender information with the casework lab.

When a DNA profile in the forensic index matches another profile in the forensic index, crime scenes can be linked together. CODIS hits involving two casework profiles will still go through a verification process. If both labs are in agreement that the profiles match, both casework labs will typically provide hit reports to the corresponding investigating agencies. These hits enable investigators to identify repeat offenders, coordinate investigations and share leads, even across multiple jurisdictions.

Introduction: The Hypothetical Databank Hit Case

A woman alleges that she was raped, but she cannot make an identification and the police do not have a suspect. Semen found in her vagina is typed for a DNA profile, and the profile is developed and entered into the state's DNA databank. It is compared with the profiles in the convicted offender databank, and there is a match with the defendant. The police use this hit as probable cause to ultimately take the client's DNA sample, which is then tested and compared with the evidence sample. This may result in prosecution but could result in a delayed prosecution when the following occurs:

- The testing by the lab is conducted well after the alleged incident occurs.
- The testing is conducted and the databank match occurs, but the suspect is not available to provide a sample for direct comparison with the evidence profile until later.
- The databank hit occurs in reasonably close proximity to the alleged incident; however, it takes a while for the government to build a case for prosecution.
- The databank hit occurs in reasonably close proximity to the alleged incident; however, other pending prosecutions against the defendant delay ability of the prosecution to initiate the case at hand.

This section covers some of the concerns and opportunities for defense attorneys dealing with these increasingly common "cold hit" cases.

How to Approach a CODIS or Cold Hit Case: The Basics

A *"cold hit"* case is generally like a "normal" DNA case, except that the government may have little to go on, other than the cold hit. Defense counsel can still use the typical defense theories that do not involve challenging the DNA evidence, such as consent, or fabrication or planting of evidence. Of course, sometimes such a defense is not the best option. In those cases, counsel may want to challenge the DNA match evidence under a general challenge to the reliability of the match, a specific third-party perpetrator theory (involving an unknown person or a relative), or a *contamination* or lab mix-up theory. Reviewing discovery information is critical. T

he defense should review the same discovery information it would review in other DNA cases — evidence reports, physical evidence recovery, sex assault kit collection and hospital records, and police reports. Crime lab records can be somewhat different in a cold hit case. Counsel must review lab reports on the chain of custody and analysis of the evidence sample, the client's known sample, and any other samples that were compared.

Counsel should also ensure that they obtain copies of all match reports generated as a result of previous searches of the DNA databank for the profile in the case. The defense's basic investigation — for example, of the client's alibi, the complaining witness, or evidence of a third-party perpetrator — should proceed as in any other case. In addition, counsel should consider the client's relatives, especially siblings, and any potential unknown relatives as third-party perpetrators. Interview all lab personnel involved in the case and thoroughly review the match report and other lab documents. A cold hit case can seem intimidating. It is important to remember all of the other types of evidence that a cold hit case tends to lack. Generally, a person is identified on the basis of a cold hit precisely because the police lack a suspect. Often, the prosecution has no eyewitness identification, and any post-cold-hit identifications are suspect — assuming the witness was unable to give a detailed and accurate description before the defendant was matched through a databank hit. The prosecution may lack other types of forensic information, such as fingerprints.

The client may not be a "usual suspect," such as a significant other or close associate. Defense counsel probably will not have to deal with a confession and can criticize the police investigation for not discovering anything of value. The questioned cold hit will stand alone, uncorroborated. For more information on cold case resources, visit: http://ncstl.org/ education/Cold%20 Case%20Toolkit.

Section 5. Review the Match Report Carefully

There are five basic types of DNA databank hits:

- The client's DNA is already in a databank. When the lab enters the forensic casework profile into the databank, the profile matches the client's profile.
- A casework profile attributed to the client was previously entered into the databank (from a separate case) and is found to match the newly entered casework profile for which there is no suspect.
- Initial upload of the casework or offender profile does not result in a databank hit. However, at some later point, the upload of either an offender profile or another casework profile results in a hit.
- There is a one-time keyboard search for the client's DNA profile in the databank that results in a hit.
- A databank search results in a hit matching the suspected perpetrator or the individual convicted of the crime — from which the profile was generated when the offender's profile is uploaded into CODIS — referred to as a *benchwork match.*

If the match is to a suspect profile generated and entered by the same LDIS lab, be sure to compare when the client's DNA was originally entered into the databank and when the evidence profile was entered. Was the evidence profile generated before the client's profile was generated for *any* case? Or was it generated before the client's profile was generated for the *present* case? The development of the evidence profile before the client's profile minimizes the risk that the evidence was mistyped or cross-contaminated.

As in other DNA cases, check which loci match; be especially careful when the evidence profile is a mixture. Check the match report to see how many loci match between the client's known sample and the evidence sample. The client can be tested at more than 13 locations, yet evidence samples can contain only 13 — or sometimes fewer — matching loci.

Figure out which databanks were searched: Was it an LDIS, an SDIS, or an NDIS hit? Look at how many samples other than the client's were searched, especially if the match did not result from a search of NDIS. Compare that number with how many samples would have been searched had the government searched against their state or other state databanks and/or the national databank. This can be done by obtaining the size of relevant databanks through a discovery request or Internet searches. Conversely, because local databanks may have less stringent requirements for profile inclusion or can house suspect profiles that cannot be uploaded to NDIS, they may contain profiles not included in NDIS. Was the most local — and, arguably, most relevant — databank searched?

It is also critical to determine how the government came to possess the defense client's DNA. Through the match report or additional discovery, find out when the client's DNA profile was first entered into the databank. Samples are usually entered at the SDIS level because a person was arrested or convicted. Sometimes, samples are included in LDIS and/or SDIS because suspect profiles are allowable, a person "voluntarily" supplied his or her DNA to law enforcement (as in a DNA dragnet), or law enforcement surreptitiously tested an item (like a cigarette butt) for a person's DNA, which was then entered into the local or state databank. Alternatively, in a rare circumstance, the client (a) may have been a prior victim of a crime whose known sample required DNA testing, (b) was the victim of a crime who contributed DNA to a mixed profile that could not be deconvoluted, (c) provided an elimination sample to exclude that person as a possible contributor in another case, or (d) their profile was entered in a local lab databank that allows entry of victim profiles.

Probable Cause and Unreasonable Searches and Seizures

Do not assume that a CODIS hit creates probable cause to arrest the client or take his or her DNA. Look for what the report, issued by the lab providing notification of the databank hit, states or does not mention. If the report states only that there is "consistency" between the profiles or that the client cannot be excluded, then defense counsel can challenge the claim of probable cause.[18] For example, in a Chicago case, police were told of a DNA hit, but the suspect was in prison at the time of the crime. The police later learned that the hit was only a partial match; however, the lab did not mention this in its paperwork.[19] A police source recognized that this type of error could lead to probable cause challenges.[20]

It is important to keep in mind, however, that DNA databank hits must always be confirmed by collection and analysis of a new known sample from the named individual and are only intended to provide potential investigative leads that must be pursued by the investigating agency. Statements similar to the following are typically provided in laboratory hit notification reports:

> This information is provided only as an investigative lead, and any possible connection or involvement of this individual to the case must be determined through further investigation. In order to complete the direct DNA profile comparison, a buccal (cheek) sample from [the offender whose case profile produced a hit] must be submitted to the Laboratory.

Consistency between profiles does not mean that the defendant is the only possible source of the evidence sample, or even a probable source. Without an appropriate measure of the match's statistical significance, the court may not be in a position to find probable cause. The evidentiary value of a purported hit is directly linked to the number of matching loci and whether or not the search evidence profile was a mixture. For example, a match at five loci with three alleles missing may not be particularly strong evidence. Contrast this with a 13-loci match from a single-source sample that produced a hit with a convicted offender sample — in this case, it would be difficult to dispute probable cause.

Remember the context of discovered DNA. A cold hit does not necessarily mean there is probable cause to believe a crime was committed and/or the named individual committed a crime. For example, a Florida man whose DNA profile was in a databank because of a DNA dragnet was matched to an earlier unsolved and unrelated rape. The police arrested the man and publicly announced that they were able to catch the rapist because they had retained DNA samples from a previous case. The next day, the victim, who had not been consulted by the police before the arrest, came forward to proclaim the arrestee's innocence. She had consensual sex with the arrestee shortly before the unknown assailant raped her.[21]

If the police make an arrest only on the basis of a cold hit match, defense counsel should challenge the causal link between a simple match of DNA and the conclusion that the client committed the crime. Hold the government to meeting its burden of proof of probable cause that the client committed the crime, not just that his or her DNA was found at the scene. For example, a defendant's DNA on a cigarette butt found outside a house where a crime occurred does not — without more evidence — establish probable cause for arrest.

Fourth Amendment Challenges to the Client's DNA in a Databank

Defense counsel should file a Fourth Amendment motion concerning the initial or continued inclusion of the client's DNA profile in the databank. The Fourth Amendment can be used to challenge the legality of law enforcement taking the client's DNA, retaining the DNA profile, and comparing it with an evidence sample profile.[22] See Chapter 7, Sections 2 through 8, for more information.

The strongest challenge to databank inclusion is if the government has failed to follow the jurisdiction's controlling statute for DNA. Determine whether the defendant's sample was taken and entered into the databank in compliance with the statute. For example, the databank statute may allow for entry of DNA into the databank upon conviction; however, the client's DNA may have been retained and entered into the databank after he or she was acquitted. Even if the government followed the statute, the defense may be able to challenge the constitutionality of the client's inclusion in the databank.[23]

Even in cases where the client's sample was lawfully obtained and entered into the DNA databank, the court should exclude cold hit evidence if it is unreasonable for the client's DNA profile to remain in the databank.[24] Courts have nearly unanimously ruled for the government on these motions;[25] however, as databanks have expanded, some courts have ruled in favor of the defendant's right not to have his or her DNA taken, searched or retained in a databank.[26] The clear trend across jurisdictions is to include wider classes of individuals in databanks. Arguably, as databanks include new classes, Fourth Amendment reasonableness challenges become stronger.

Every state has a statute authorizing the collection of DNA from offenders and inclusion of the samples in DNA databanks. However, the scope of individuals subject to entry in the databank varies by state. The earliest statutes included only people convicted of a small number of violent felonies and sex offenses. Some jurisdictions still have these limitations. The trend, however, has been for states to include increasingly broad groups of people in the databank, such as all people convicted of felonies or certain misdemeanors.[27] More than half of the states include juveniles as well as adults.[28] Half of the states, plus the federal government, require the collection of DNA from arrestees.[29]

Depending on the statute, an acquitted person may have his or her information removed from the databank — sometimes presumptively and sometimes only after petitioning the court. In jurisdictions where a court petition is required, counsel should advise clients of their right to petition to have their information removed from the database. In many cases, arrestee information may remain in the databank, even after dismissal of the charges or acquittal.

To date, courts have been nearly unanimous in upholding the validity of DNA databanks; however, the courts have dealt mostly with narrow statutes affecting people with reduced privacy interests, such as prisoners. Courts have commonly held that databanks are constitutional as applied to convicted felons who are incarcerated, particularly those convicted of violent crimes. In upholding the statutes, courts have taken into account that there is no particularized suspicion in most cases for running an individual's DNA profile through a databank, either through a "totality of the circumstances test" or a "special needs" test.[30]

In *Samson* v. *California,* the U.S. Supreme Court — using a totality of the circumstances test — upheld a California law permitting suspicionless and warrantless samples to be taken from parolees.[31] The majority in *Samson* found that the "[s]tate's dual interest in integrating probationers back into the community and combating recidivism" outweighed the defendant's privacy interests, particularly because of the "severely diminished expectations of privacy" a parolee faced. The Supreme Court noted that "parolees are more akin to prisoners than probationers" in their expectation of privacy.[32]

For courts, the essence of the Fourth Amendment challenge in a databank case is balancing the government's interests against the defendant's privacy interests.

Expectations of Privacy

On the question of privacy rights, there is a continuum from prisoners (who have no reasonable expectation of privacy)[33] to parolees, then probationers. Arrestees — especially ones who are ultimately acquitted or have the charges against them dismissed — should have the full privacy rights afforded by the Fourth Amendment. There are also people who voluntarily gave their DNA to law enforcement, and people whose DNA was surreptitiously collected by law enforcement and entered into databanks.

Even though the government may have a right to take DNA samples from incarcerated individuals, defense counsel may argue that the government must remove a person's profile from the databank once he or she is released (or upon completion of any period of parole, supervised release or probation). Otherwise, the person is still subject to databank searches, even though he or she no longer has a reduced expectation of privacy. For example, when upholding a DNA databank statute, the Ninth Circuit qualified that the case did not involve "a petitioner who has fully paid his or her debt to society, who has completely served his or her term, and who has left the penal system. ... Once those previously on supervised release have wholly cleared their debt to society, the question may be raised, 'Should the CODIS entry be erased?'"[34]

If the statute allows taking DNA from nonviolent offenders — such as statutes mandating DNA collection from all convicted felons — then there is a stronger challenge if the defendant was entered into the databank for a nonviolent offense because the prosecution may not be able to connect the DNA collection with deterrence. The government interest in reducing recidivism through each convict's fear of being caught in future crimes is lessened in the case of people convicted of crimes where DNA evidence is not relevant. Although some courts have upheld DNA databanks requiring samples from broad classes of offenders, two courts have found statutes that allow broad DNA collection to be unconstitutional.[35]

Rights of Juvenile and Adult Arrestees

Defense attorneys should consider challenging DNA databank statutes as applied to juveniles, even if they have been upheld as constitutional for adult offenders. Convicted juveniles may have increased privacy interests compared with adults convicted of the same crimes. A "juvenile's relationship to the state and the state's public policy favoring confidentiality of juvenile proceedings are factors that should be considered in balancing the interests" between a person's Fourth Amendment rights and the state's interests.[36]

To date, one state court has declared the practice of seizing DNA from all arrestees unconstitutional.[37] This issue is likely to receive more attention. Federal legislation enacted in 2006 allows DNA testing of all persons arrested for federal criminal felonies,[38] and the trend among the states is toward arrestee databanks.

When DNA of a Person Who Was Not Convicted or Arrested Is Entered into a Databank

During the investigation of a crime, law enforcement officers might take DNA samples from discarded objects without the knowledge or consent of the person whose DNA is being taken. This information might then be compared with an evidence profile. Refer to Chapter 7, Section 7, for additional information.

Law enforcement has increasingly used DNA dragnets to collect genetic information that is sometimes compared with evidence in the DNA databanks.39 In a DNA dragnet, law enforcement officers investigating an offense ask the eligible population of a community for "voluntary" DNA samples. After the investigation, police might then submit the profiles of these innocent people for entry into DNA databanks. See Chapter 7, Section 6, for additional information.

Understanding the Evidentiary and Statistical Significance of a CODIS Match

To be meaningful, DNA evidence must be accompanied by an accurate statistical estimate of its probative value.[40] In a cold hit case, when the profile generated is from a single source, the government analyst will generate a *random match probability (RMP)* statistic. Although use of RMP is a well-established way to express a DNA match's statistical significance, some scientists are divided on the appropriate methodology in cold hit cases. Defense counsel must understand competing methodologies.

Prosecuting authorities and crime lab personnel may prefer to use RMP in cold hit cases, arguing that the databank search process has no effect on the RMP.[41] However, another view holds that random match probability is not the appropriate method to determine the chance that the evidence profile might coincidentally match the suspect in a databank. This view holds that the search process must be taken into account in expressing the statistical significance of a DNA match that originated as a cold hit.[42]

Statistically speaking, using this approach, the more profiles that are searched, the more likely a coincidental match will be found. To use an analogy, suppose the perpetrator's birthday is known. If a suspect case is developed on the basis of other evidence, such as eyewitness identification, and then a DNA sample is taken, the RMP is an appropriate measure of the chance of a *coincidental match* — in this case, 1 in 365. If the suspect's birthday matches that of the perpetrator, it is useful additional evidence. On the other hand, imagine the suspect was identified by pulling 200 names out of a phone-book and investigating each person's birthday (analogous to a form of databank search) — and one person matched. The RMP is still 1 in 365, but it cannot be discounted that the police would expect to find a person with a matching birthday after searching 200 people.

In 1992, the National Research Council, the principal operating arm of the National Academies of Science, established the Committee on DNA Forensic Science (NRC I) to address various aspects of the science of DNA forensics, including the methodology for computing the statistical significance of a cold hit.[43] NRC I recommended the confirmatory loci approach, which involves first finding a match at the minimum number of loci needed to make a unique match within the databank (or a higher number if the lab wants to consistently search a set number of loci). Once a match is identified, the lab would then compare the remaining loci from the evidence sample with the same loci from the sample identified by the databank search.

For example, if the laboratory can test a total of 13 loci, and it used six loci to make the search that resulted in the cold hit, it would then compare the remaining seven loci. If the two samples match at the additional "confirmatory" loci, then the lab would calculate an RMP based solely on the confirmatory loci. The RMP associated with the confirmatory loci would accurately express the statistical significance of the match between the suspect sample and the evidentiary sample at those loci. In other words, the second confirmatory step of this approach removes any effect the databank search may have on the calculation of the likelihood of a coincidental match.

The NRC I approach will often present the highest statistical likelihood — and thus the most favorable statistic for the defendant — because the loci used in the search that resulted in the cold hit are not included in the statistical figure presented to the jury. This approach was recommended at a time when fewer loci were available for testing, and some criticized it for wasting information. However, 15 autosomal STR loci are currently able to be routinely typed in forensic labs using multiplex DNA typing kits that have been validated for forensic use and are on the market.[44] In addition, 26 other *autosomal STR locations* have been identified as suitable for forensic application (though not yet validated for use in forensic casework),45 and the National Institute of Standards and Technology (NIST) is actively developing these and other genetic locations that have been identified "so as to permit use of the product rule when combining data between CODIS STR and [the new] loci."[46]

In 1996, the National Research Council on DNA Forensic Science convened another expert committee (NRC II), which, among other things, proposed an alternate methodology for computing a cold hit's statistical significance. As in the preceding committee, NRC II avowed that the statistic "should take into account the search process."[47]

NRC II did not recommend use of RMP in a cold hit case. As its report points out, if you toss 20 coins, the chance of getting all heads is about 1 in 1 million (you might think the coins were weighted if you got all heads). However, if you repeat the experiment 1 million times, the fact that one of the experiments yields all heads would not seem strange. Although it reiterated that the NRC I method was "a sound procedure," NRC II was concerned that the NRC I method did not use all loci in deriving the statistic.[48]

Instead of RMP, NRC II recommended a methodology called the *databank match probability (DMP)*; it is also known as *Np,* from the mathematical formula that is used. Under the method's simplified form, one would multiply the probability representing the profile's frequency (*p*, equivalent to its RMP) by the number of items searched (*N*, the size of the databank) to derive the match's statistical significance.[49] Applying the rough formula to the birthday example yields: 1/365 × 200 = 200/365, or 1 in 1.8 — about a 55- percent chance of getting a match, having made 200 comparisons.

Accordingly, the DMP is calculated by simply multiplying the RMP by the size of the databank. For example, if you had an RMP of 1 in 10 billion and a databank size of 1 million individuals, the DMP statistic would result in a 1-in-10,000 chance of finding that profile in the database: (1 ÷ 10 billion) × 1 million. NRC II's DMP method is usually more discriminating than the NRC I confirmatory loci approach.

In 2000, DAB, an FBI commission responsible for promulgating standard practices relating to the use of DNA evidence at that time, released a report on cold hit statistics and proposed its own solution. DAB approved of NRC II's recommendation of relying on the DMP "for the evaluation of DNA evidence from a databank search" but also noted that the RMP is "of particular interest" in a cold hit case.[50] The DAB recommended that both RMP and DMP be reported to the jury. Using the birthday example, this would entail reporting both "1 in 365" and "1 in 1.8," explaining that the 1-in-365 statistic represents the probability that a random person would match the evidence profile, and the 1-in-1.8 statistic represents the probability that someone in the databank would match by coincidence. Per this DAB approach, if both the RMP and DMP are reported to the jury, the jury will learn that the defendant was previously in the database.

A fourth approach to deriving a cold hit's statistical significance was developed by genetic statisticians, Drs. David Balding and Peter Donnelly, of Imperial College of London and Oxford University, respectively.[51] These scientists use a *Bayesian approach* and make what proponents consider to be a critical distinction at the outset — the difference between the forward-looking probability of finding a coincidental match when searching a databank of a given size and the probability that a particular match is coincidental, having already conducted the search.[52] In general, U.S. courts have not accepted Bayesian statistics, and the same has been true for the Balding and Donnelly approach. Still, it is notable — especially for those in a *Frye* jurisdiction — that yet another competing statistical methodology exists for calculating the significance of a cold hit.

Despite the different approaches outlined above, the prosecution frequently includes the RMP statistic. The government relies on several arguments:

- Random match probability and other methods answer different questions.
- RMP is always relevant, even if another statistic, such as DMP, is also relevant.
- Retesting the client's sample after the initial cold hit allows the use of RMP.[53]

The counter to these prosecution arguments is that, first, the RMP does answer a different question: It determines the odds that an individual randomly selected from the population would match the evidence sample. The question is not about the odds of finding a coincidental match by conducting a cold hit. In every case, the central question is whether the statistic properly explains the significance of the match — that is, whether it demonstrates the likelihood of a coincidental match in a particular case.

The government's related argument — that the RMP is relevant because it gives information about the rarity of the profile — has gained traction with some courts.[54] However, even though some contend that the expected frequency or rarity of a particular profile is a relevant statistic, that information is not directly dispositive of the central question before the jury, which is the likelihood of mistaken identification or false inclusion by coincidence.[55]

Finally, retesting DNA evidence at the exact same genetic locations is a useful quality check to ensure that the profile in the databank is, in fact, the defendant's, but it does not alter the coincidence assessment.56 If a databank search results in a coincidental match, retesting the same genetic markers simply repeats the coincidence. The factor that brings a particular case into the class of a "cold hit" for statistical purposes is the manner in which the suspect was first identified. Whether the prosecution provides the jury with the RMP or some other derivative statistic that accounts for the search process, defense counsel should explore the fundamental assumptions that underpin the RMP in every cold hit case.

All this being said, it is important to be aware that as of 2011, the relevant scientific community has generally accepted the use of the RMP calculations for cases involving a databank hit. Furthermore, counsel must keep in mind that a statistical approach that includes databank size in the calculation raises a serious concern that the jury will learn of their client's prior arrests or convictions.

Cautionary Note: Counsel is strongly encouraged to consult with a statistician in a cold hit case.

Section 6. Arizona Databank Matches and Use of Random Match Probability in Discovery Litigation

As previously discussed, the RMP is determined by following a population genetics model that rests upon several assumptions. The model assumes the following:

- The estimated allele frequencies are accurate.
- The autosomal STR loci used to develop the DNA profile are independent of one another. Forensic scientists multiply the allelic frequency estimates of each forensic locus by the others' estimates by using the product rule.
- The population from which the RMP is derived has been demonstrated to be in Hardy-Weinberg equilibrium.[57]

This theoretical model governs the admission of DNA match evidence in courtrooms throughout the United States. However, a recent study, showing a number of matches at nine or more loci in the Arizona databank, has raised preliminary questions about the meaning of RMP, particularly with respect to cold hit cases. Defense counsel should consider filing a motion to compel the state's databank authority or the FBI to run a search similar to the one discussed later on the state databank (or NDIS, in the case of the FBI).

In 2001, the Arizona Department of Public Safety Crime Laboratory (Arizona DPS) searched its convicted offender databank and observed a nine-STR-loci match between two apparently unrelated individuals (one a Caucasian, the other an African-American).[58] This was the first report of a coincidental match of more than six loci in the United States, so the forensic science community treated it as unusual and noteworthy, meriting heightened attention.[59]

As recently as spring 2005, well-known geneticists who frequently testify for the prosecution treated the 2001 Arizona DPS report as an outlier. They testified, under oath, that matches at 9, 10 or more loci are extremely rare. Scientists associated with crime laboratories testified that a 9- or 10-loci match in two individuals is exceedingly unusual and that the only known 10-loci match involved an incestuous relationship.[60]

In fall 2005, Arizona DPS compared the DNA profiles of each of the 65,493 people in its convicted offender databank with each other. Arizona DPS reported some remarkable findings: Its databank contained 122 pairs of people matching at nine of the 13 loci; 20 pairs that matched at 10 loci; one pair that matched at 11 loci; and one pair that matched at 12 loci. The last two matches were confirmed to be siblings.[61] If the RMP was calculated for each pair, the statistics would be exceedingly rare.[62]

Conversely, such matches are expected.[63] Recent studies illustrate the possibility that siblings may have nearly perfect matches across the 13 STR loci, an issue of particular importance when suspects are charged on the basis of a cold hit.[64]

When comparing each sample in a databank, the number of actual comparisons climbs to astronomical numbers. The Arizona databank examples may be perfectly explained by the presence of related people in the databank (studies of this began in 2008).[65] However, the Arizona matches may indicate some additional problems with the meaning of random match probability, as there is some evidence that the number of matches cannot simply be accounted for by the presence of close relatives.[66] In addition, the Arizona matches may provide jurors with a different perspective regarding a match's significance.

Defense counsel should try to learn the number of coincidental matches in the searched databank. This will support the defense's case in at least three ways:

1) It provides additional support for the argument that the RMP is not relevant in a databank case. The coincidental match illustrates that the RMP may overestimate the odds of a coincidental match when many comparisons, instead of just one, are made.
2) Observing a high number — any number smaller than the RMP, what is expected, according to government claims — helps minimize the significance of the DNA evidence.
3) Using empirical data available through a CODIS search, counsel may be able to give the jury a number that more accurately reflects the chance that the match is coincidental.

Presented simply and straightforwardly, these matches can assist the defense in providing a significant, relevant statistic as to the probability of a coincidental match.

Empirical inquiry into the databanks can also help support the proposition that DNA profiling errors occur. For example, just as there should be fewer 13-loci matches than 12-loci matches across the 13 loci, there should be fewer 25- out-of-26 allele matches than 24-out-of-26 allele matches. If the converse turns out to be true, that might suggest the presence of single-allele typing errors in the entry of databank profiles, or transcription errors.

Some judges have ordered searches of their state convicted offender databanks to determine whether there were pair-wise matches at nine or more loci.[67] The Illinois databank contained 220,000 convicted offender samples; 903 pairs matched at nine or more loci. The Maryland convicted offender databank contained fewer than 30,000 profiles; 32 pairs matched at nine or more loci, and three pairs matched at all 13 loci. Although a news article reported that, as of July 2008, Maryland officials had not researched whether the matches were duplicates, identical twins, brothers or unrelated people,[68] the Maryland State CODIS Administrator had already confirmed, by entering the profile of the second twin — in each of these three pairs of samples from identical twins — that the samples were truly from identical twins. Fingerprint comparisons were used to confirm that each twin was a different individual and that the samples were not duplicate submissions from the same person; the profile from each twin was left in the databank because the offender samples were collected from different individuals compelled to provide DNA under Maryland state law).

In court proceedings, the FBI has routinely refused to examine the national CODIS databank for pair-wise comparisons, as was done in the Maryland, Arizona and Illinois SDIS databanks.

Requesting Access to DNA Databanks or Asking the Government to Run a Comparison Match

The prosecution may oppose a defense counsel request to run a comparison search based on an argument that would impose a time burden.

The prosecution has also argued that labs are not authorized to run this kind of search on the basis of one interpretation of the authorizing statute (42 U.S.C. §14132). States have argued that, under the statute, only databank records relating to DNA analysis for that particular case may be provided to a criminal defendant. The statute actually provides:

> (3) ... pursuant to rules that allow disclosure of stored DNA samples and DNA analyses only (B) in judicial proceedings, if otherwise admissible pursuant to applicable statutes or rules; (C) for criminal defense purposes, to a defendant, who shall have access to samples and analyses performed in connection with the case in which such defendant is charged; or (D) if personally identifiable information is removed, for a population statistics database, for identification research and protocol development purposes, or for quality control purposes.

The government crime lab may object to the release of personal identification information. In response, the defense can argue that NDIS does not contain personal information such as names. It can also be argued that if the government were to turn over the data to the defense's expert, he or she could remove any identifying information. Even if a personal identifier number were included, the defense would have no way of connecting that number to an actual person without access to the FBI's operational identifiers. Most important, the defense does not need identifying information; it needs only a report of the number of matches at nine or more loci. The prosecution or crime lab may also argue against disclosure because the lab's Memorandum of Understanding with the FBI, which allows its use of CODIS, states that (1) the lab must take reasonable precautions to prevent unauthorized persons from accessing the CODIS software, and (2) labs will abide by the procedures for record access in the Privacy Act Notice and the NDIS responsibilities and procedures manuals. The Privacy Act authorizes the same disclosures as the DNA Act (discussed earlier), including release to researchers and by court order.[69] The purpose of such a search is to present a defense and obtain exculpatory information from the government.

Scope of Requests and Eliminating Duplicates

Prosecution and crime labs have also argued that the scope of the request — to find matches of nine or more loci — is excessive. However, even nine loci can generate miniscule RMPs. The defense can also give the court the option of limiting the search to matches of 10 or more loci, if necessary.

The prosecution has previously argued that it needs significant time to sort out duplicates and relatives among the matching profiles. The prosecution has expressed concern that the results of the match will be misleading because the databanks contain some duplicates and do not consist of random samples.[70] However, duplicates are easily identified. First, they would definitely match at all 13 loci. The next step would be to pull the case files to determine the source(s) and whether they are, in fact, duplicates or a true coincidental match. As for the identification of relatives in the databank, it is speculative to say how long this would take. If the government produces evidence showing a high number of matches, then the government can, if it wants, seek to determine if any of the matches are duplicates or relatives.

Finally, the prosecution might argue that such a search is irrelevant. However, a cold hit DNA case hinges on the significance of the DNA match. The match data provide useful fodder for cross-examining the government's witnesses and could create reasonable doubt about the government's statistics. The Arizona, Maryland and Illinois matches are *prima facie* evidence to believe that there are coincidental matches in the state's databank. Performing such a search can be critical to representing the defense's client effectively.

Section 7. Identifying the Theory of Defense: Defenses Specifically Based on a Cold Hit

In addition to normal defenses, counsel should consider the following three defense theories with an eye to the differences in a cold hit case:

- The RMP statistic does not address the correct question because the defendant was identified through a cold hit.
- The cold hit was the product of contamination or innocent presence.
- The government did not do enough testing and databank searching to prove that the defendant was the perpetrator.

Section 8. Statistics

Defense counsel should consider making a motion to prevent the prosecution from introducing the RMP as evidence in a cold hit case. Depending on the test used in the jurisdiction, counsel can argue that the prosecution's RMP statistic is not generally accepted or not reliable in a cold hit case.

Appellate courts in both *Frye* jurisdictions that have considered this argument have rejected it and upheld admission of the RMP in cold hit cases.[71] There have been no *Daubert*-based decisions.

If the RMP is admitted, the defense can seek to discredit the statistic through the methods available in all DNA cases: The number is based on datasets of 150–200 people.

The independence of genetic locations is an assumption. The databases relied upon are not truly in Hardy-Weinberg equilibrium, and the RMP does not account for relatives. RMP's applicability in a cold hit case can also be challenged. Use the information (discussed earlier) on how RMP answers a different question than DMP.

Also, the defense can introduce its own statistics. Apply the NRC I (confirmatory loci) or NRC II (DMP/Np) methodologies to the client's databank match and argue for the admission of one or both of these statistics. The NRC II method is admitted as evidence more easily, as more scientists are willing to vouch for its reliability, but NRC I may produce a more exculpatory number.

Furthermore, defense counsel can cross-examine the state's expert by citing scientific journal articles and NRC publications to demonstrate the relevance — or lack thereof — of the RMP in a cold hit case.

When cross-examining witnesses who report the random match probability, consider challenging their qualifications — they are often forensic scientists and not population geneticists or statisticians. Scientists will agree there are alternative methods for calculating the relevant statistic. For example, the DAB, an FBI-convened group, approved use of the NRC II method. Ask the witness about ascertainment bias and use articles, treatises, and the NRC I and II reports. Point out that the NRC II report states that the NRC I confirmatory loci approach is acceptable. Establish that the lab could have used the NRC I approach but chose not to (it could have run a databank search, one location at a time, until there was only one match in the databank). Use the Arizona match results to show that, despite a small RMP, matches at many loci are expected. If defense counsel is able to get Arizona-style databank search match information from the government for your state, use it. If not, consider establishing the defense's request, in writing, and the government's refusal to provide the data. Without such a search, the government cannot say with certainty that there are no coincidental 10-, 11-, 12- or even 13-loci matches in the databank.

Section 9. Contamination

There have been several cold hit cases worldwide involving contamination. Use these examples to see if there is a plausible contamination theory in the defense's case.

In 1998, an Australian toddler was found dead after having been preserved in icy water for months.[72] In 2003, there was a DNA databank hit based on DNA from the victim's bib and pants. The "match" was a young woman from the opposite coast with no apparent connection to the victim. It turns out that the matching "perpetrator" was a rape victim, and the same lab that handled the DNA from the murder victim's clothing had processed her DNA.

The samples from the rape victim were tested in the lab during February 2–5, 1998. The bib and pants were examined on February 2 or 3; the DNA was extracted on February 4, 1998. The rape victim was not charged in the toddler's death.[73]

- There have also been false cold hits due to laboratory contamination in the United States. A Washington state police lab contaminated samples in a rape case. The error was caught because the suspect had been a young child when the rape occurred. The lab admitted that "the felon's sample was being used as a training sample by another analyst" when the rape case was being analyzed.[74] In a separate case, a California law enforcement crime lab accidentally cross-contaminated samples from two different rape cases being processed at the same time.[75]

- Questions remain about the 1969 murder of a Michigan woman. In December 2003, police received a DNA match based on a cold hit of an evidence sample, but the matched person was only 4 years old at the time of the woman's death. A sample from another convicted offender was tested at the same lab and matched another item of evidence in the case. Police have failed to come up with an explanation for the first match; there was no obvious evidence of laboratory contamination.[76] The second match was eventually tried and convicted on the basis of his DNA match. The case is currently under appeal.
- In an Australian rape case, after numerous inquiries by defense counsel, the crime laboratory withdrew an alleged cold hit match and admitted "there may have been a contamination event or a laboratory error during the DNA extraction process."[77]

When faced with a cold hit case, investigate how the defendant's profile originally got into the databank.

As mentioned previously, some local and state databanks upload suspect profiles that the FBI does not allow in NDIS. For example, the crime lab in Erie County, N.Y., collected and entered DNA profiles from crime victims into its local databank.[78] If the defendant's profile that resulted in the hit was uploaded into a local database but would not have met the FBI's standards for uploading into NDIS, the defense should pursue this line of questioning during cross-examination.

Counsel also must check whether the same facility processed both the evidence sample and the defendant's offender DNA sample. If so, verify the facilities layout, staffing procedures, workflow and timing to see if there could have been contamination.

If the lab processed the defendant's DNA sample before or at the same time as the evidence sample, or if the two samples could have come into contact with each other (directly or through secondary transfer, such as an analyst who is assigned to both the databank and the casework analysis units), then consider contamination as a possibility.

Section 10. When the Government Cannot Produce Certain Evidence

In a cold hit case, the prosecution will often lack more traditional evidence, such as eyewitness identifications or fingerprints. Along with an attack on the reliability of the proposed cold hit evidence (whether by claiming contamination or a misstatement of the match's statistical significance), defense counsel can highlight how much evidence the prosecution *cannot* produce for the jury.

Counsel should challenge the claim that a cold hit match between the defendant's DNA and an evidence sample means that the defendant committed a crime — particularly when there is no corroborating evidence. In a St. Louis case, prosecutors dismissed two rape-murder cases where the DNA evidence against the suspect showed that the suspect had sex with the women, both of whom were murdered. After a jailhouse informant died, no other evidence linked the suspect to the crime. In fact, one of the victim's sisters was adamant that the suspect was not the man she saw choking her sister the night before she was found dead.[79]

Section 11. Cases in Which No DNA Evidence Was Tested

When Can the Defense Request Initial DNA Testing?

Rules differ from state to state about a defendant's right to test evidence. Defense counsel's first step should be to determine the jurisdiction's rules and limitations governing defense testing.

There may be a case in which DNA testing was not done and the defense thinks it should be. An example of this might be when a victim claims she had never met the defendant before the alleged rape, but the defendant asserts that not only did he know her but also they also had drinks and cigarettes in his apartment the night before. Testing the cigarettes for DNA profiles might be the prudent thing to do.

The lawyer should never be the one to collect evidence; either an investigator knowledgeable in evidence collection or a law enforcement officer should collect the evidence. The decision of whether to involve the police will hinge on several factors: First, is the evidence in a place where an investigator can access it? If not, the choice is easy. However, if the police and an investigator have equal access, careful consideration must be given to who is going to collect the evidence. Counsel cannot be certain the client is telling the truth; to be found to be lying about such a thing could be devastating to the defense. As such, it may be best to have an investigator collect the evidence and send it to a private lab for testing.

Alternatively, the evidence the defense wishes to test may be in state custody. In this case, the first question is, "Should the defense ask the jurisdictional lab to perform testing or send the item to a private lab?" Counsel may wish to contact the lab or the prosecution and ask about the lab's policy — does the lab do testing at the request of the defense?

If items are to be sent out for testing, the defense should consider what items it will request from the lab or investigating agency. Asking for a single item of evidence will, of course, let the investigator know what the defense is testing. It may be wise to ask for a number of items — or all of the items — so that the testing of the item of interest is done without highlighting the goal. Additionally, the defense should consider to whom it requests the items be released. If the agency that has custody of the evidence sends the items directly to the private lab, that agency naturally knows who is doing the testing. In some jurisdictions, this could result in the prosecution calling the lab scientist or investigator to testify on behalf of the state. Counsel should consider having their investigator pick up the items from the custodial agency and then sending them out for testing. It is a good idea to speak with the prosecutor about chain-of-custody issues before the items leave the agency that has custody of the evidence.

It is important to be aware that a private lab (a non-CODIS lab) will not have the ability to upload and/or search any profiles that are generated using CODIS if this is of interest once the results of testing are reviewed.

When and What Evidence Can Be Retested?

There may be cases in which DNA testing was already performed by the state but the defense chooses to have items retested. Counsel should know whether the jurisdiction allows confidential retesting. If the retest shows the same result as the state lab, the defense may be bolstering the state's case. Will the jury be made aware that retesting was a possibility? In many cases, in order to render effective assistance of counsel, an expert must be consulted.

Alternatively, the defense may want to have additional testing performed on items of evidence. There may be cases where it would be helpful to attempt to identify other contributors. Consider a situation in which the state performs STR testing on items of evidence, obtains partial profile information, and does not exclude the defendant — but it also observes or reports additional DNA types that are perhaps below the lab's sensitivity threshold. This information may not be in the analytical report, but it should be found as raw data in the case file, which can be obtained through a discovery request. Testing the evidence with different DNA markers (such as Y-STRs, mtDNA or miniSTRs) may help identify the additional contributors or show that the defendant is actually excluded. A qualified expert should help counsel decide whether to have additional testing conducted and new results interpreted. Be aware that a private lab (a non-CODIS lab) will not have the ability to upload or search any profiles that are generated using CODIS, should a foreign or nonattributed DNA profile be generated. See Chapter 5, Section 14, for a discussion regarding this issue.

What if There is no DNA Evidence?

The lack of DNA analysis may be important to the defense's case. Counsel should determine whether evidence not collected or tested could have provided probative information, such as a sexual assault case in which the police did not collect, or collected but did not test, the victim's clothing. Jurors have come to expect DNA evidence to be presented in certain types of cases.

It may also be possible that DNA was unavailable because the crime was not reported immediately. One argument that could be made is that the victim knew about DNA testing and delayed reporting because DNA evidence would not support his or her claim. In a case where police officers fail to collect evidence, it is important to establish that the particular officer who failed to collect the items was knowledgeable in evidence collection procedures. For example, a used glass at the crime scene could have identified someone who handled or drank from the glass. Or, if a cigarette butt at the crime scene was not collected, does the defendant smoke cigarettes?

End Notes

[1] United States v. Marion, 404 U.S. 307, 322 (U.S. 1971).

[2] Stogner v. California, 123 S. Ct. 2446, 2452 (U.S. 2003).

[3] Stogner v. California, 123 S. Ct. 2446, 2461 (U.S. 2003) ("a law enacted after expiration of a previously applicable limitations period violates the Ex Post Facto Clause when it is applied to revive a previously time-barred prosecution").

[4] State v. Garcia, 169 P.3d 1069 (Kan. 2007).

[5] Stogner v. California, 123 S. Ct. 2446, 2454 (U.S. 2003).

[6] See National Conference of State Legislatures, Statutes of Limitations for Sexual Assaults (April 2007), www.ncsl.org/programs/cj/IncludingDNA. htm.

[7] See, e.g., Commonwealth v. Laventure, 894 A.2d 109, 118 (Pa. 2006) (finding inadequate a warrant describing the perpetrator as "John Doe Steve, having an unknown address, and who was a white male, in his thirties" but citing to cases where a partial description coupled with a location were deemed sufficient). Professor Wayne R. LaFave has taken the position that a John Doe arrest warrant satisfies the particularity requirement if it

describes the person's "occupation, his personal appearance, peculiarities, place or residence or other means of identification," LaFave, X., 3 W AYNE R. Search and Seizure §§ 5.1(g) (3d ed. 1996 & Supp. 2003) (footnote citations omitted).

[8] People v. Robinson, 156 Cal. App. 4th 508, 520 (Cal. Ct. App. 2007); People v. Robinson, 156 Cal. App. 4th 508: Subsequent appellate history contains negative analysis. State v. Danley, 853 N.E.2d 1224, 1227 (Ohio 2006); State v. Dabney, 663 N.W.2d 365 (Wis. App. 2003).

[9] State v. Dabney, 663 N.W.2d 366, 374 (Wis. Ct. App. 2003) ("a defendant is not entitled to specific notice that the state is issuing a complaint and seeking an arrest warrant").

[10] United States v. Gouveia, 104 S. Ct 2292, 2300 (1984).

[11] See, e.g., Commonwealth v. Scher, 803 A.2d 1204, 1221-1222 (Pa. 2002) (allowing the claim where "the evidence shows that the delay was the product of intentional, bad faith, or reckless conduct by the prosecution").

[12] Goldfarb, "When Judges Abandon Analogy: The Problem of Delay in Commencing Criminal Prosecutions," 31 WM. & M ARY L. R EV. 607 (Spring 1990).

[13] See www.fbi.gov/hq/lab/pdf/codisbrochure.pdf.

[14] FBI statistics on NDIS are available at http:// www.fbi.gov/about-us/lab/codis/ndis-statistics.

[15] See, e.g., 42 U.S.C. § 14132.

[16] See 42 U.S.C. § 14135(b), National DNA Index System DNA Acceptance Standards.

[17] Updated information on the coverage of each state's DNA statute can be found at www.dnaresource.com.

[18] See generally, Whiteley v. Warden, 401 U.S. 560 (1971) (judicial officer issuing warrant must be "supplied with sufficient information to support an independent judgment that probable cause exists"). Whiteley v. Warden, 401 U.S. 560: Overruled in part as stated in: Thompson v. Wagner, 2008 U.S. Dist. LEXIS 75066 (W.D. Pa. Sept. 29, 2008).

[19] Sweeney, A., and F. Main, "Botched DNA Report Falsely Implicates Woman: Case Compels State to Change How It Reports Lab Findings," CHICAGO S UN-TIMES (Nov. 8, 2004), at 18.

[20] "'It could call into question the integrity of DNA evidence,' one source said. 'Do we have probable cause to take someone into custody for questioning based on an 'investigative lead' from a partial DNA match? It raises serious legal issues.'" (Id.)

[21] See P.J. Neufeld, Co-Director of the Innocence Project, Member of N.Y. State's Forensic Science Review Board, Testimony at the Subcommittee on Crime, Terrorism, and Homeland Security (July 17, 2003); see also EPIC 5th Circuit Amicus Brief on DNA Dragnets.

[22] Courts agree that the taking and analysis of a person's DNA is subject to Fourth Amendment protections. Nicholas v. Goord, 430 F.3d 652, 658 (2d Cir. 2005); Nicholas v. Goord, 430 F.3d 652: United States v. Sczubelek, 402 F.3d 175, 182 (3rd Cir. 2005); Jones v. Murray, 962 F.2d 302, 306 (4th Cir. 1992); Groceman v. United States Dep't of Justice, 354 F.3d 411, 413 (5th Cir. 2004) (per curiam); Green v. Berge, 354 F.3d 675, 676-77 (7th Cir. 2004); United States v. Kincade, 379 F.3d 813, 821 (9th Cir. 2004); United States v. Kimler, 335 F.3d 1132, 1146 (10th Cir. 2003); Padgett v. Donald, 401 F.3d 1273, 1277 (11th Cir. 2005); Johnson v. Quander, 440 F.3d 489, 493 (D.C. Cir. 2006).

[23] Remember to challenge the client's inclusion in the databank on the basis of the state constitution in addition to the federal Constitution.

[24] The search of a DNA profile and its continued retention in a databank fall under the Fourth Amendment, just like the initial extraction of a person's DNA, even if there was nothing wrongful in the initial extraction of a person's DNA. See United States v. Stewart, 468 F. Supp. 2d 261, 264-65 (D. Mass. 2007) ("the first expectation of privacy concerns the physical penetration of the person to extract the blood. The second expectation is implicated when the blood is tested and the information contained in DNA is revealed") (citing Skinner v. Railway Labor Executives' Ass'n, 489 U.S. 602, 616 (though urine samples do not require surgical intrusion of the body, "chemical analysis of urine, like that of blood, can reveal a host of private medical facts about an employee, including whether he or she is epileptic, pregnant, or diabetic")); United States v. Stewart, 468 F. Supp. 2d 261. See also, Ferguson v. City of Charleston, 532 U.S. 67, 76 (2001) (test of a urine sample implicates the Fourth Amendment).

[25] See, e.g., United States v. Amerson, 483 F.3d 73, 79 (2d Cir. 2007); United States v. Sczubelek, 402 F.3d 175, 184 (3d Cir. 2005); Jones v. Murray, 962 F.2d 302 (4th Cir. 1992); Groceman v. U.S. Department of Justice, 354 F.3d 411 (5th Cir. 2004); Green v. Berge, 354 F.3d 675 (7th Cir. 2004); United States v. Kraklio, 451 F.3d 922, 924-25 (8th Cir. 2006); United States v. Kriesel, 508 F.3d 941 (9th Cir. 2007); Padgett v. Donald, 401 F.3d 1273, 1280 (11th Cir. 2005); Johnson v. Quander, 440 F.3d 489, 496 (D.C. Cir. 2006).

[26] See Stewart, 468 F. Supp. 2d at 261; In Re: C.T.L., 722 N.W.2d 484 (Minn. Ct. App. 2006); Stewart, 468 F. Supp. 2d at 261: Reversed by: United States v. Stewart, 532 F.3d 32, 2008 U.S. App. LEXIS 14469 (1st Cir. Mass. 2008). Vermont v. Watkins (Vt. Dist. Ct. App. 24. 2006) (No. 6805-2-04).

[27] See generally, www.dnaresource.com/documents/statequalifying offenses 2011.pdf (as of the end of 2007, 34 states included some misdemeanors as qualifying offenses, and 45 states include all convicted felons).

[28] Id.

[29] Id.

[30] Unlike the totality of the circumstances test, if the special needs test is applied, the government must prove that the search is justified by a special need beyond the ordinary need for law enforcement. Griffin v. Wisconsin, 483 U.S. 868, 873 (1987). Thus, it may be beneficial to argue that the special needs test applies in the case. See United States v. Amerson, 483 F.3d 73, 79 n. 6 (2d Cir. 2007) (special needs test, as applied in the 2nd Circuit, is "more stringent" than the general balancing test because it requires a threshold step before engaging in a balancing test, unlike the "totality of the circumstances" test) (citing Nicholas v. Goord, 430 F.3d 652, 664 n. 22 (2d Cir. 2005)); Nicholas v. Goord, 430 F.3d 652. See also, United States v. Kincaide, 379 F.3d 813 (9th Cir. 2004) (in 6-5 decision, five of the judges in the majority applied "totality of the circumstances" and one applied "special needs," whereas the five dissenting judges applied "special needs"). However, see United States v. Sczubelek, 402 F.3d 175 (3rd Cir. 2005) (stating, without explanation, that the "totality of the circumstances" test is "more rigorous" than the "special needs test").

[31] 547 U.S. 843 (2006). Although some federal courts (2nd and 7th Circuits) still use a special needs test for some categories, the U.S. Supreme Court's decision in Samson, plus the weight of the majority of courts addressing the issue, means that counsel probably will be working with the totality of the circumstances balancing test when dealing with persons with reduced expectations of privacy. However, the Second Circuit argues that "nothing in Samson suggests that a general balancing test should replace special needs as the primary mode of analysis of suspicionless searches outside the context of the highly diminished expectation of privacy presented in Samson." United States v. Amerson, 483 F.3d 73, 79 (2d Cir. 2007) (discussing that probationers (at issue in Amerson) have a greater expectation of privacy than parolees (at issue in Samson)).

[32] Id. at 2197-2200 & n. 2.

[33] Hudson v. Palmer, 468 U.S. 517, 530 (1984).

[34] United States v. Kriesel, 508 F.3d 941, 949 (9th Cir. 2007) (quoting United States v. Kincade, 379 F.3d 813, 841 (Gould, J., concurring)); see also, Green v. Berge, 354 F.3d 675, 679-81 (7th Cir. 2004) (Easterbrook, J., concurring) (noting that "[f]elons whose terms have expired" form a different category of individuals than supervised releasees for the purposes of a Fourth Amendment inquiry).

[35] See United States v. Stewart, 468 F. Supp. 2d 261 (D. Mass. 2007) (finding unconstitutional the DNA Analysis Backlog Elimination Act of 2000 as applied to a probationer convicted of Social Security fraud); United States v. Stewart, 468 F. Supp. 2d 261: Vermont v. Watkins, (Vt. Dist. Ct. App. 24. 2006) (No. 6805-2-04) (invalidating, on state constitutional grounds, the "suspicionless collection and banking" of DNA samples from all convicted nonviolent felons); however, see, e.g., Green v. Berge, 354 F.3d 675 (7th Cir. 2004) (upholding Wisc. Stat. Ann. § 165.77 (West 1999)); Jones v. Murray, 962 F.2d 302 (4th Cir. 1992) (upholding Va. Code Ann. § 19.2-310.2 (1990)); Va. Code Ann. § 19.2-310.2. Doles v. State, 994 P.2d 315 (Wyo. 1999) (upholding Wyo. Stat. Ann. § 7-19-403 (1997)).

[36] In re Calvin S., 150 Cal. App. 4th 443 (Cal. App. 2007) (recognizing juvenile's stronger privacy rights but still finding balance of interests to favor requiring entry of juvenile DNA into databank).

[37] In re C.T.L., 722 N.W.2d 484 (Minn. Ct. App. 2006) ("Because Minn. Stat. § 299C.105, subd. 1(a)(1) and (3) (Supp. 2005), direct law-enforcement personnel to conduct searches without first obtaining a search warrant based on a neutral and detached magistrate's determination that there is a fair probability that the search will produce contraband or evidence of a crime, and because the privacy interest of a person who has been charged with a criminal offense, but who has not been convicted, is not outweighed by the state's interest in taking a biological specimen from the person for the purpose of DNA analysis, the portions of Minn. Stat. § 299C.105, subd. 1(a) (1) and (3), that direct law-enforcement personnel to take a biological specimen from a person who has been charged but not convicted violate the Fourth Amendment to the United States Constitution and Article I, Section 10, of the Minnesota Constitution."), Minn. Stat. § 299C.105.

[38] See 42 U.S.C. § 14135a(a)(1)(A) (2006).

[39] See Walker, S., "Police DNA 'Sweeps' Extremely Unproductive: A National Survey of Police DNA 'Sweeps,'" POLICE P ROFESSIONALISM INITIATIVE, Department of Criminal Justice, University of Nebraska (2004).

[40] See, e.g., State v. Kinder, 942 S.W.2d 313 (Mo. 1996); United States v. Porter, 618 A.2d 629 (D.C. 1992); People v. Axell, 235 Cal. App. 3d 836 (Cal. App. 1991).

[41] See, e.g., Budowle. B., et al., Clarification of Statistical Issues Related to the Operation of CODIS, Pub. No. 07-01, Laboratory Division, Federal Bureau of Investigation (2006) (unpublished manuscript) [hereinafter Budowle et al. (2006)], http://www.promega. com/~/media/files/resources/ conference%20proceedings/ ishi%2017/oral%20 presentations/ budowle.pdf?la=en.

[42] See, e.g., National Research Council. The Evaluation of Forensic DNA Evidence, at 134 (1996) (NRC II) ("There is an important difference between [a standard DNA case] and one in which the suspect is initially identified by searching a databank to find a DNA profile matching that left at a crime scene. In the latter case, the calculation of a match probability or LR [likelihood ratio] should take into account the search process."); DNA Advisory Board, Statistical and Population Genetics Issues Affecting the Evaluation of the Frequency of Occurrence of DNA Profiles Calculated From Pertinent Population Databanks (Feb. 23, 2000) (clarifying that random match probability and the statistical significance of databank matches are distinct concepts requiring different calculations); Stockmarr, A., "Likelihood Ratios for Evaluating DNA Evidence When the Suspect Is Found Through a Database Search," 55 B iOMETRiCS 671, 671 (1999) (observing "distinction" between differing statistical considerations necessary to derive statistical significance in two types of cases); Devlin, B., "The Evidentiary Value of a DNA Database Search," 56 B iOMETRiCS 1276 (2000) (echoing Stockmarr's position).

[43] NRC I was composed of a professor of medical genetics; a professor of medicine, biochemistry and cell biology; two directors of forensic science laboratories; a chemical engineer; a professor of epidemiology and genetics; a professor of biology; a professor of law and sociology; a molecular geneticist; a J.D.-M.D. ethicist; a professor of forensic sciences and biomedical sciences; and a U.S. district court judge. See National Research Council, The Evaluation of Forensic DNA Evidence (1992) at 173-76 (NRC I).

[44] See www.cstl.nist.gov/biotech/strbase/ multiplx.htm; Krenke, B.E., et al., "Validation of a 16-Locus Fluorescent Multiplex System," 47(4) J. F ORENSiC S Ci. 773 (2002); Collins, P.J., et al., "Developmental Validation of a Single-Tube Amplification of the 13 CODIS STR Loci, D2S1338, D19S433, and Amelogenin: The Amp-FlSTR Identifiler PCR Amplification Kit," 49(6) J. F ORENSiC S Ci. 1265 (2004).

[45] Butler, J.M., et al., New Autosomal and Y-Chromosome STR Loci: Characterization and Potential Issues, presented at PROM EGA S YMPOSiUM ON HUMAN iDENTiFiCATiON (October 2007).

[46] National Institute of Standards and Technology, New STR Loci Under Development by the NIST Forensics/Human Identity Project Team, available at www.cstl.nist.gov/biotech /strbase/ newSTRs.htm (last visited Jan. 17, 2008).

[47] NRC II (1996), supra note 43, at 134.

[48] Id. at 32.

[49] Id.

[50] See DNA Advisory Board (2000), supra note 43.

[51] See Balding, D.J., and P. Donnelly, "Evaluating DNA Profile Evidence When the Suspect Is Identified Through a Databank Search," 41(4) J. FORENSiC S Ci. 603 (1996).

[52] Id. at 605.

[53] See Budowle et al. (2006), supra note 42.

[54] See, e.g., United States v. Jenkins 887 A.2d 1013 (D.C. 2005); People v. Nelson, 48 CAL. R PTR. 3d 399 (Cal. App. 2006), review granted, 147 P.3d 1011; People v. Nelson, 48 CAL. R PTR. 3d 399: Subsequent appellate history contains negative analysis. Review granted, Depublished. See also, People v. Johnson, 139 Cal. App. 4th 1135 (2006).

[55] See Devlin, B., "The Evidentiary Value of a DNA Database Search," 56 B iOMETRiCS 1276 (2000); Amicus Brief in People v. Nelson; Letter Brief of California Attorneys for Criminal Justice in People v. Johnson, Exhibit B, Letter to the Honorable Frederick K. Ohlrich, signed by 25 scientists. (It is a scientific manner to answer "the central and oft-presented question in a DNA database match case, 'What is the likelihood that a match found between a known DNA profile and a profile found by a search of a DNA database is merely coincidental.'" Though not always agreeing on the direction of the effect, the scientists agreed that "[t]he fact that a suspect is first identified by searching a database unquestionably changes the likelihood of the match being coincidental.") See also, Stockmarr, A., "Likelihood Ratios for Evaluating DNA Evidence When the Suspect is Found Through a Database Search," 55 Biometrics 671 (1999).

[56] See, e.g., Transcript, People v. Robinson, J.N.6 p. 1581; J.N.3, p. 634 (testimony of prosecution experts); Letter Brief of California Attorneys for Criminal Justice in People v. Johnson, Exhibit B, Letter to the Honorable Frederick K. Ohlrich, signed by 25 scientists.

[57] See generally, NRC II at 122 (1996) (setting forth Recommendation 4.1 for calculation of profile frequency).

[58] Kathryn Troyer, K., et al., A Nine STR Locus Match Between Two Apparently Unrelated Individuals Using AmpflSTR Profile Plus and COfiler, PROMEGA 12T H iNTERNATiONAL S YMPOSiuM (2001), available

at http://www.promega.com/~/media/files/resources/conference%20proceedings/ishi%2017/oral% 20 presentations/budowle. pdf?la=en.

[59] See National Institute of Justice, U.S. Department of Justice, The Future of Forensic DNA Testing, at n. 13 (Nov. 2000) (reporting 10 six-locus DNA profile matches in the New Zealand databank of 10,907 records, in which eight matches were brothers, and two matches were unrelated persons); Willing, R., "Mismatch Calls DNA Tests Into Question," uSA TODAY (Feb. 8, 2000), at 3A (a coincidental six-locus match in a United Kingdom databank).

[60] See Ungvarsky, E., "What Does One in a Trillion Mean?" 20(1) GENEW ATCH (Jan.-Feb. 2007).

[61] Arizona Department of Public Safety Crime Laboratory, 9+ Locus Match Summary Report (Oct. 2005).

[62] For the original nine-loci match, RMPs were 1 in 754,100,000 in Caucasians and 1 in 561,500,000 in African-Americans.

[63] See Budowle et al (2006), supra note 42, at 7.

[64] See Paoletti, D.R., et al., "Assessing the Implications for Close Relatives in the Event of Similar but NonMatching DNA Profiles," 46 J uRiMETRiCS J. 161 (2006); Weir, B., "Matching and Partially-Matching DNA Profiles," 49 J. F ORENSiC S Ci. 1009 (2004).

[65] See Weir, B., "The Rarity of DNA Profiles," 1(2) ANN. A PPL. S TATS. 358-370 (2007), available at http://projecteuclid.org/euclid.aoas/ 119643 8022; see also Buckleton, J.S., C.M. Triggs, and S.J. Walsh, eds., Forensic DNA Evidence Interpretation: Biology, Technology, and Genetics of STR Markers, at 460-462 (CRC Press, 2005).

[66] See Ungvarsky, E., "What Does One in a Trillion Mean?" 20(1) GENEW ATCH 12 (Jan.-Feb. 2007).

[67] See, e.g., Order of the Honorable Vincent Gaughan, Circuit Court of Cook County, Illinois (July 11, 2006); Order of the Honorable Steven I. Platt, Circuit Court for Prince George's County, Maryland (Aug. 4, 2006).

[68] Dolan, M., and J. Felch, "The Verdict Is Out on DNA Profiles," L.A. T iMES (July 20, 2008).

[69] For a discussion, see Murphy, E., "The New Forensics: Criminal Justice, False Certainty, and the Second Generation of Scientific Evidence," 95 CAL. L. R EV. 721, 783 (2007).

[70] See, e.g., Budowle et al. (2006), supra note 42.

[71] See People v. Nelson, 185 P.3d 49 (Cal. 2008); United States v. Jenkins, 887 A.2d 1013 (D.C. 2005).

[72] www.dbc.uci.edu/~mueller/pdf/leskie_ decision.pdf.

[73] See Thompson, W., "Tarnish on the 'Gold Standard': Recent Problems in Forensic DNA Testing," CHAM PiON M AGAZiNE (Jan./Feb. 2006), available at www.nacdl.org/ public.nsf/0/6285f6867724e1e68525712 4006f9177; see also, www.bioforensics.com/ articles/Kranereport.pdf.

[74] Id.

[75] Id.

[76] Id.; Murphy, E., "The New Forensics: Criminal Justice, False Certainty, and the Second Generation of Scientific Evidence," 95 CAL. L. R EV. 721, 755 n. 151 (2007).

[77] See Banks, A., "DNA Lab Admits Rape Case Bungle," THE AUSTRALIAN (March 16, 2006), available at www.dbc.uci.edu/ ~mueller/pdf/ australia_pathwest_false_match2.pdf

[78] See Precious, T., "Crime Lab Lambasted Over DNA Database," THE B UFFALO NEWS (April 25, 2006).

[79] See Bryan, B., and R. Patrick, "Snitch's Death Frees Murder Suspect," S T. L OUIS POS T-DIS PATCH (Jan. 11, 2006).

10. Proactive Uses of DNA

Section 1. Using DNA to Establish Third-Party Guilt

Recent advances in DNA technology — particularly the standardized and widespread use of STR typing and Y-STR DNA testing — have made it possible to acquire and test DNA from incredibly minute samples of biological material, including transferred skin cells, traces of saliva and cells contained in sweat. Practically any item handled or used by the perpetrator of a crime can be subjected to DNA analysis in an effort to obtain his or her DNA profile. These items include weapons, hats, bandanas, masks, eyeglasses, facial tissues, toothpicks, cigarettes, tape, ligatures, bottles, cans, glasses, swabs of bite marks, fingernail clippings or scrapings, and even half-eaten food.[1] A DNA profile can be generated from testing seemingly invisible sweat and skin cells from the inside of a baseball cap worn by an assailant or on a knife an assailant used to inflict fatal stab wounds. In cases where a perpetrator forcefully removed a victim's underwear or pants, DNA testing can be performed on the waistband or cuffs that he or she grabbed when pulling off the clothing.

As a result, DNA testing is now performed on a wide range of evidence and in cases that go beyond the framework of a rape case, where semen and hair are collected from the victim.[2] Just as DNA is a powerful tool for the prosecutor, it can be a powerful tool for the innocent defendant. DNA test results showing the presence of DNA from someone other than the accused on probative evidence can provide powerful scientific support that someone else may have committed the crime.

Exclusion from a Probative Item

Depending on the facts of a case, a single *exclusionary result* may be sufficient to establish that the defendant did not commit the crime — for example, testing a cigarette butt the police collected from the crime scene in a case where the assailant was seen smoking, or testing saliva from a bite mark in a rape case where the perpetrator bit the victim.

When testing a single piece of evidence, it may be helpful to test several key areas of the item to help place the results in context and thereby show that the DNA belongs to someone other than the defendant. If, for example, a shirt left at the crime scene is attributed to the assailant, DNA testing can be performed on areas that would contain the sweat and skin cells of someone who wore the shirt, including the inside surfaces of the neck, cuffs and underarms. Test results that exclude the defendant and establish that the same unknown person contributed the DNA profiles from each area of the shirt would form powerful evidence that the person whose DNA profile was found is the person who wore the shirt — not someone who came into casual contact with it. In a similar fashion, testing a ski mask worn by an assailant and discarded at the crime scene can generate a DNA profile from sweat, skin cells and dandruff on the inside surfaces of the mask's head area and from saliva around the mouth.

Redundancy

In some cases, testing only a single piece of evidence will be insufficient to argue third-party guilt because the evidentiary significance of the evidence — that is, whether the DNA on the item came from the perpetrator or some other source — is unclear. Take, for example, a case in which the victim was bludgeoned to death with a hammer. DNA testing can be performed on skin cells and sweat from the hammer's handle. However, because there is no way to know for certain whether the biological material on the hammer comes from the assailant or someone who previously used the hammer, the prosecution may easily explain away the finding of DNA from someone other than the defendant.

In such a case, redundant results — results showing that the same DNA profile is on a number of items of crime scene evidence — may be critical to the argument that the DNA found on the various items, which does not belong to the defendant, comes from the person who committed the crime. In the previous example, the probative value of the DNA from the hammer handle could be transformed by results that show this profile is consistent with DNA found under the victim's fingernails. The more redundant the profiles are from the evidence, the stronger the argument will be that the DNA came from the perpetrator and not someone unrelated to the crime.

Case example: Larry Peterson. In 1987, a victim's partially clad body was found in a southern New Jersey soybean field. She had been manually strangled, and her legs were spread apart with her jeans and underwear pulled down toward her ankle on one leg. Police learned that, in the hours before her murder, the victim had consensual sex with two men.

A local man, Larry Peterson, was arrested within weeks of the crime and later convicted of felony murder based on:

- Testimony from four individuals who claimed that Peterson confessed (the alleged confessions contained nonpublic details of the crime).
- Testimony from a state forensic expert that seven hairs found on the victim's body and a stick that was used as a weapon were microscopic matches to Peterson.
- Testimony from three witnesses who claimed that, in the days after the murder, they saw fresh fingernail scratches on Peterson's arm.
- Peterson allegedly threatened a witness and tried to borrow money to leave town.

During the original investigation of the crime, semen and sperm were found on the victim's pants, but none was detected on any of the victim's body orifice swabs. DNA testing was not performed before the trial, as forensic DNA testing was in its infancy at the time. The state opposed postconviction DNA testing, arguing that DNA would not shed light on the perpetrator's identity because the victim had consensual sex with at least two people immediately before her rape-murder. The Appellate Court of New Jersey subsequently ordered DNA testing under the state's DNA testing statute (N.J.S.A. 2A: 84-32a).

The DNA results showed that the hairs that had been microscopically matched to Peterson actually belonged to the victim. DNA from someone other than Peterson — an "unknown male" — was found under the victim's fingernails. STR testing of the victim's oral and vaginal swabs showed sperm from two males, which had been overlooked during the original examination. The majority of the sperm from the victim's vaginal swab, as well as the sperm from inside the victim's mouth, matched the profile of the unknown male whose DNA was found under the victim's fingernails. *Reference samples* from the victim's two prior consensual sex partners were tested. One partner was identified as the minor source of sperm from the victim's vaginal swab, and he was the only source of the sperm on the victim's underwear and pants. The "unknown male's" sperm — although in the victim's vaginal swab sample — was not on her underwear or pants.

Because semen drains from the vaginal cavity after intercourse, the fact that there was sperm from the unknown male in the victim's vagina, but not on her underwear or pants, provided powerful proof that (1) he was the last person to have vaginal intercourse with the victim before she was killed, and (2) the victim did not put her clothing back on after intercourse with the unknown male. As the unknown male's sperm was also found in the victim's mouth and his DNA was found under her fingernails, the test results provided powerful scientific evidence that the unknown male — not Peterson — vaginally and orally raped the victim and strangled her. On the basis of these redundant results, Peterson's conviction was vacated and the charges against him dismissed.[3]

Note: When testing for redundancy, it is critical to take into account that different DNA tests generally use different genetic markers. STR and *Mini-STR* results share significant genetic markers and can be compared with one another. On the other hand, Y-STR test results can be compared only with Y-STR test results, and *mtDNA* test results can be compared only with mtDNA test results. Competing goals are sometimes encountered when testing DNA — for example, the desire to obtain an STR profile that can be searched in *CODIS,* the goal of getting redundant test results, and the need to use Y-STR testing on samples with a high female-to-male DNA ratio. This may require testing all of the evidence with the type of test needed for one item or subjecting a certain item to more than one type of test.

Take, for example, a case in which a female victim is murdered by ligature strangulation. A DNA profile will be obtained from fingernail scrapings and the ligature collected at the scene. If the laboratory's quantitation shows a low level of male DNA in the fingernail samples, Y-STR testing may be required to yield a profile. Because one goal of testing is to obtain a profile searchable in CODIS, and Y-STR results cannot be searched in CODIS, the ligature could be subjected to STR testing. However, to support the theory that the person whose DNA was found under the victim's fingernails is the person who strangled her with the ligature, the male DNA found underneath the fingernails will need to be compared to the male DNA on the ligature. One solution is to perform both types of testing on the ligature: STR testing to get a CODIS-searchable profile and Y-STR testing to obtain results that can be compared with the Y-STR profile from the victim's fingernails.

Matching Crime Scene Evidence to a Specified Third Party

It is well-established that a defendant is entitled to introduce evidence that another person committed the crime with which the defendant is charged. Both before trial and after conviction, DNA test results can link crime scene DNA to a third party by comparing the DNA profile of an alternate suspect with crime scene evidence or through a CODIS hit that matches crime scene evidence to another unsolved crime or to an individual whose profile is in the databank.

Case example: Clarence Elkins. Clarence Elkins was convicted in 1999 for the rape and murder of his mother-in-law and the rape of his 6-yearold niece. The young victim reported that the attacker looked like her Uncle Clarence. Results from pretrial mtDNA testing of pubic hairs found on the victims' bodies excluded Elkins as the contributor. Nevertheless, Elkins was convicted and sentenced to life in prison on the basis of the young victim's identification.

In 2004, postconviction Y-STR DNA testing was performed. The results showed a male DNA profile on the victim's vaginal swab that matched DNA on portions of the girl's underwear, which the assailant had touched when pulling it off. Elkins was excluded, but his motion for a new trial based on the DNA results was denied.

Elkins' wife, working with a private investigator, learned that a convicted rapist named Earl Mann had been living near the victim's house at the time of the crime. Mann was coincidentally transferred to Elkins' cell block in 2005. Elkins picked up a cigarette butt that Mann dropped in the yard and mailed it to his wife. When the cigarette butt was tested, it matched the crime scene DNA. Elkins was exonerated in 2005 after serving six and a half years in prison. Mann pled guilty to charges related to the case.[4]

The following examples show how DNA can provide essential corroboration of a third-party confession previously deemed unreliable:

- Ryan Matthews was convicted and sent to death row in Louisiana for the 1997 shooting of a store owner by a masked assailant. Another inmate, incarcerated for a different murder, bragged about committing the murder for which Matthews was falsely convicted. DNA testing of the mask matched the profile of the other inmate, and Matthews was exonerated.[5]
- Marvin Anderson spent 11 years in a Virginia prison for a rape he did not commit. Years before the postconviction DNA testing took place, Otis Lincoln had confessed at Anderson's habeas corpus proceeding that he — not Anderson — had committed the crime, but the reviewing court found Lincoln's confession unreliable. It was only after postconviction DNA testing excluded Anderson and matched Lincoln in the DNA databank that Anderson was exonerated and Lincoln was tried and convicted for the rape.[6]
- Christopher Ochoa falsely confessed and pled guilty to the 1988 rape-murder of a woman in Texas. He also falsely implicated his friend Richard Danziger. Ochoa and Danziger were exonerated more than a decade after their convictions when Achim Marino wrote to then-Governor George W. Bush, confessing that he alone committed the murder. Marino's confession was corroborated by DNA testing that linked him to semen found in the victim's body.[7]
- Five youth in the infamous "Central Park Jogger" case falsely confessed to beating and raping a woman in New York's Central Park in 1989. Thirteen years after the crime, Matias Reyes admitted that he alone had committed the terrible crime, and DNA testing showed that the semen recovered from the crime scene did, in fact, belong to Reyes. The convicted youth were then exonerated.[8]

Laboratory policies differ as to whether the lab will perform a CODIS search at the request of the defense. If a lab will not run a search at the defense's request — or the lab policy, state laws, or CODIS user regulations prevent such a search — counsel should seek prosecutorial cooperation or a court order.

At least three states — Georgia,[9] Illinois[10] and North Carolina[11] — have statutory provisions that explicitly allow a defendant to obtain a court order to search an unknown DNA profile in the DNA databank upon showing that access to the databank is material or relevant to the defense.

Although a defendant's right to demonstrate third-party guilt through a CODIS search is a novel issue, at least one appellate court has addressed it in the context of an application for DNA testing under New Jersey's DNA access statute (N.J.S.A. 2A:84A-32a). In *State* v. *DeMarco,*[12] the Appellate Division of New Jersey ruled that a convicted defendant was entitled to retest semen evidence — from which he was excluded before trial — with STR DNA technology for the purpose of conducting a CODIS search to prove third-party guilt.

When obtaining a court order for a CODIS search, it is important to determine the scope of the search. Will the crime scene evidence be permanently uploaded into CODIS, or will a keyboard search be performed? A keyboard search is a one-time search that compares the crime scene DNA profile with all profiles on file at that particular time in the databank. Every day, new samples from convicted offenders and unsolved crimes are added to the databank, increasing the possibility of getting a hit. A sample must be permanently uploaded for it to be compared with new samples as they are added to the system.

Full profiles, partial profiles (deemed allowable) and data from mixed samples (deemed allowable) can be searched in CODIS (see Chapter 7, Section 12). The FBI governs searches of the national databank, whereas state law governs statewide databanks — the requirements may not be the same.

Even if a profile is not eligible to be permanently uploaded (in either the national or a state databank), it may still be possible to perform a one-time keyboard search depending on the suitability of the sample for searching. Similarly, if a profile does not qualify for uploading to the national or state databanks, it may still be possible to search it against a local databank.

Section 2. When to Seek Postconviction DNA Testing

Postconviction DNA testing can yield results that scientifically establish a defendant's innocence, entitle him or her to a new trial, or mitigate the sentence. Postconviction testing should be considered in cases where:

- A defendant was convicted prior to the routine use of DNA testing.
- The conviction occurred after the advent of DNA generally, but new DNA technology may enable more meaningful results or testing of previously unsuitable evidence (such as an assailant's sweatshirt).
- Testing might link DNA from the crime scene to an alternate suspect.

In deciding which items to test, it is important to consider all of the items that the perpetrator would have used or touched.

Case example: Stephan Cowans. In 1997, a Boston police officer pursuing a man acting suspiciously was shot after the two scuffled and the officer lost his gun. The assailant forced his way into a nearby house, where he asked the occupant for a glass of water and removed his sweatshirt before fleeing. Stephan Cowans was convicted of the shooting on the basis of eyewitness identification by the surviving victim/officer and testimony by two fingerprint analysts, who stated that a fingerprint taken from the mug from which the assailant drank matched Cowans' fingerprint.

In 2004, STR DNA testing was performed on saliva from the mug and on sweat and skin cells from both the sweatshirt and the brim of a baseball cap that fell off the assailant's head and was found in a nearby yard. The testing yielded the same STR DNA profile on all three items. The DNA results conclusively excluded Cowans as the source and led officials to review and reject the earlier, erroneous fingerprint analyses. In 2004, Cowans was released from prison upon a joint motion by the prosecution and defense.[13]

In a postconviction case in which the defendant maintains he or she is not the person who committed the crime, it will generally be insufficient to simply show that the defendant's DNA was not at the scene. The defense will be looking for testing to yield a DNA profile from the crime scene evidence that is attributable to the perpetrator and does not match the defendant's profile.

Take, for example, a rape case in which the defendant maintains he was misidentified and the victim's rape kit was not examined before the trial. DNA test results from the rape kit that yield only the female victim's DNA profile would be of little probative value. After a conviction, the absence of the client's DNA alone will be insufficient. (Note that, before the trial, such results may have been helpful to the defense theory.) To help the defendant, testing will need to show a male DNA profile in the rape kit that does not match the defendant or any other previously identified consensual partners. For example, in *In re Pers. Restraint of Bradford,*[14] the defendant's conviction for a 1996 attack was vacated in 2005 after DNA testing performed on a piece of tape — used by the perpetrator to adhere a mask covering the victim's face — generated a male DNA profile that excluded the defendant.

However, in cases where the state used biological or physical evidence at trial against the defendant, DNA test results that show the defendant's DNA is not on the item attributed to the assailant may warrant a new trial on the basis of newly discovered evidence. For example, consider a rape case in which the state's evidence consisted primarily of the victim's identification of the suspect. The state also used microscopic hair comparison analysis that linked the defendant to a hair fragment from the victim's rape kit. Postconviction mtDNA analysis excludes the defendant from the hair fragment and shows instead that the hair's mtDNA profile is consistent with the victim. Although the results do not establish innocence — in the sense that they do not exclude the defendant from evidence belonging to the assailant — they do show that the defendant was convicted on the basis of false evidence and may be entitled to a new trial.

In some cases, exclusionary test results can provide key support for defense theories — other than factual innocence — that could minimize the defendant's culpability or mitigate the sentence. For example, the defendant is charged with murder for allegedly entering the victim's home with an accomplice and shooting the victim. The defendant maintains that he only gave his friend a ride to the victim's house, but never entered the house and was not the shooter. Pretrial testing of the defendant's shirt was presumptive for blood, and the blood was attributed to the victim and used to establish the defendant as the shooter. DNA results that exclude the victim as the source of the DNA on the defendant's clothing would provide support for the defense theory that he did not enter the house and was not the shooter.

Pretrial Examination, Testing and DNA Analysis

It is extremely important to keep in mind how items may have been handled if they were used as exhibits in court. For example, if seeking to demonstrate that the client's DNA is not on items of evidence from the victim, but the court record indicates that the defendant was handed each item to review before it was published to the jury, pursuit of the DNA testing may have a detrimental effect on the postconviction process. Using this example, finding trace amounts of the defendant's DNA during postconviction testing on these items wouldn't be particularly surprising, but would be extremely hard to overcome once detected.

The handling of evidence after crime lab processing without taking care to avoid transfer of material between suspect, crime scene and victim-associated evidence is common — particularly during pretrial preparation by both sides and in the courtroom.

Individuals handling evidence have not been required to wear a particle mask to avoid inadvertently depositing their DNA on an item, nor have individuals been required to change gloves prior to handling each new item of evidence. Most scientists will not handle evidence in court without wearing gloves.

When considering what items to submit for testing, it is important to understand the scope of the original examination of the evidence, if any, and the implications of any negative findings.

For example, in the pre-STR DNA era, testing was not typically done for sweat or skin cells on clothing or weapons. (Clothing may have been deemed significant and tested if it contained blood.) There is no way to know whether clothing or weapons contain skin cells or sweat for DNA testing until they are subjected to DNA analysis.

Also, before the advent of DNA testing, there would have been no way to tell if material under the victim's fingernails contained foreign DNA unless there was ample blood or tissue, which could be tested to determine blood or enzyme types or obtain an *RFLP* DNA, or other DNA profile using a now-obsolete DNA testing technique.

It is not uncommon for re-examination and testing of evidence to detect semen, sperm or other important biological evidence that was undetectable or simply overlooked during the original examination of the evidence. The National Institute of Justice (NIJ) has endorsed the re-evaluation of evidence, noting:

> It may be important to re-evaluate/analyze previously collected evidence samples to determine if there are: (1) other relevant evidence samples that could be tested... ; (2) samples containing stains or other biological samples that had not been detected previously; or (3) samples that were unsuitable for testing with previous techniques but may give conclusive results with currently available DNA tests (e.g., very small blood or semen stains, hair shafts).[15]

The No-Semen Fallacy

Due to the traumatic nature of sexual assault, a rape victim's perception and recollection of events may be impaired. Many times, rape victims believe that the perpetrator did not ejaculate when, in fact, ejaculation did occur.

It is important to consider the crime scene evidence anew and not rely on a previous negative finding that indicates there is no biological evidence to test. Both the scientific literature and individual DNA exoneration cases have confirmed the importance of re-examining evidence in sexual assault cases. Early technology used to detect the presence of semen frequently produced false negatives; newer tests are far more discerning.[16] For example, in a rape case in which no semen or sperm were detected during the original examination of the rape kit specimens, it is possible that the semen or sperm were simply overlooked. Despite an original negative finding, male DNA could still be present in skin cells or from pre-ejaculate left behind during sexual contact, which may contain low levels of semen or sperm cells. Conventional serologic tests, such as the *p30* test, have a sensitivity threshold that may not be low enough to detect semen at low levels. Additional tests for the presence of male DNA are now available, such as using the quantitation kits that target human male DNA during the quantification step of DNA typing, that provide a more definitive answer as to the presence or absence of male DNA on swabs and other samples contained within a sexual assault kit.

Case example: Michael Mercer. Because testimony at Michael Mercer's 1992 rape trial stated that vaginal swabs taken from the victim had tested negative for sperm, Mercer's motion for postconviction DNA testing was denied on the grounds that additional testing would be pointless, as there was no semen from the perpetrator to test. However, in early 2003, unknown to Mercer, the rape kit swabs from Mercer's case were sent to a private

DNA laboratory with whom New York City had contracted to test all rape kits in its possession as part of a "backlog project" to solve open cases through the DNA databank. Despite the police chemist's "negative" test for sperm in 1992, the DNA lab was able to obtain a full STR DNA profile from *spermatozoa* detected on the vaginal swabs collected from the victim in Mercer's case. When the profile was entered into the databank, it produced a hit to a convicted serial rapist. Mercer was exonerated and released from prison in May 2003.[17]

Case example: Ronnie Taylor. Ronnie Taylor was convicted of raping a Texas woman in her home in 1993. Police responded to the scene less than an hour after the crime and identified a "wet spot" on the bed sheet where the rape occurred. A serologist testified for the state at Taylor's trial that testing of the bed sheet and other evidence yielded "negative" results for the presence of semen. In 2007, evidence from Taylor's case was submitted for postconviction DNA testing, and semen was found on the bed sheet. Testing of the sperm yielded a profile that not only excluded Taylor but also matched a man, Chili Charlie, who was already in prison for other sex crimes.[18]

Even in sexual assault cases where the assailant did not ejaculate, traditional *autosomal STR DNA testing* coupled with Y-STR testing may still be able to generate the perpetrator's DNA profile, even if it is only a Y-STR *haplotype,* from skin cells or pre-ejaculate left behind on clothing during intercourse or in a rape kit.

Case example: Dean Cage. In 1994, a 15-yearold girl was raped on her way to school in Chicago. She identified Dean Cage as the perpetrator when police took her to a meat market and asked her to identify the attacker. She also identified Cage in a line-up by the sound of his voice but was not entirely sure he was the man who had raped her. The rape kit and other bodily swabs as well as her clothing were tested for DNA, but no sperm cells were revealed in the initial examination. Postconviction Y-STR testing revealed that the same male DNA profile was on both the clothing and rape kit swab and that this profile did not match Cage. He was exonerated after serving 12 years of a 40-year sentence.[19]

Pretrial Exclusion

Generally, postconviction DNA testing has little use in cases where an older or less informative forensic science methodology excluded the defendant as the source of the biological material before the trial. For example, if the defendant was excluded through microscopy as the source of a crime scene hair, it would be of little use to perform mtDNA testing to exclude the defendant again (albeit with more informative and, in some instances, reliable technology). However, there are important circumstances in which testing evidence from which the defendant was excluded can be the key to exoneration.

In some cases, law enforcement discount the significance of crime scene evidence after the prime suspect is excluded as the source of that biological material. This commonly occurs in murder cases in which investigators initially believe that the victim was also sexually assaulted but, after excluding the accused as the source of semen evidence, change their theory of the crime, attribute the semen to prior consensual sex, drop charges related to the sexual assault, and instead focus solely on a prosecution for homicide. In such cases, it can be critical to link the evidence deemed insignificant by the state to an alternative suspect. When prior *serology* (ABO and enzyme typing) or older forms of DNA testing (for example, *RFLP* and *PM/DQ_)* excluded the defendant before trial, counsel should consider retesting the evidence with autosomal STR technology to generate a profile that could be searched in

CODIS. Retesting could yield a hit to a person convicted of a similar crime or to an unsolved crime that was committed at a time when the defendant could not have been involved.

Case example: Jeffrey Deskovic. In 1989, the body of a missing high school student was found in the woods in Westchester, N.Y. The victim had been strangled and beaten and was partially nude. During the investigation, sperm was recovered from the victim's body and submitted for DQ_ testing. When the lead suspect, Jeffrey Deskovic — a classmate of the victim who had (falsely, but convincingly) confessed to the crime — was excluded as the source of the sperm, investigators deemed the biological material unrelated to the murder, speculating that it originated from a prior act of consensual sex. Despite the pretrial exclusion, Deskovic was convicted. In 2006, the sperm evidence was retested with STR technology, and the profile matched a man in prison for a similar murder. On the basis of these results, the Westchester County District Attorney's Office moved to vacate Deskovic's conviction.[20]

Case example: Douglas Warney. In 1996, a Rochester, N.Y., man was found stabbed to death in his home. There were drops of blood leading from the room where the victim's body was found to a nearby bathroom. Inside was a bloody towel as well as blood in the sink. Douglas Warney was convicted of the crime almost exclusively on the basis of a confession he gave to police shortly after the crime (which contained details known only to the killer and police). Serologic testing performed before the trial excluded Warney as the source of nonvictim blood at the scene. Nevertheless, prosecutors successfully argued at trial that the blood could have come from an accomplice, and Warney was convicted. In 2006, the crime scene blood was retested with STR technology, and the DNA profile was entered into CODIS. There was a hit to a man named Eldred Johnson, who was incarcerated for various similar stabbing offenses. Johnson confessed to committing the crime alone and has since pled guilty.[21]

Aside from matching evidence to a specific third party, defense counsel may want to test to establish redundancy. For instance, a defendant may have been convicted despite *exclusion* from biological evidence before the trial because there were alternate explanations for the significance of the biological testing results. Current DNA technology may not only confirm the exclusion of the defendant but may also show that the same DNA profile is on a separate piece of evidence.

Inconclusive or No Results

Retesting with newer technologies should be considered in cases where DNA testing was inconclusive or failed to yield a result. For example, consider a case in which previous DNA testing — even autosomal STR testing — of a female murder victim's fingernails showed only the victim's profile. Retesting with Y-STR technology could yield a haplotype profile of a foreign male DNA source.

Limitations of non-DNA Forensic Testing

It is important to understand the limitations of prior testing. For example, ABO blood typing could include an individual only in a relatively large segment of the population that could be the source of the biological material. In general, 40 percent of the Caucasian population have type A blood, 45 percent have type O blood, 11 percent have type B blood, and 4 percent have type AB blood. Although the approximate percentages of each blood type vary among population groups, an overwhelmingly large percentage of each population group has type A or type O blood. However, even if someone has type AB blood, statistically speaking that is roughly 1 in every 25 people having that "rare" blood type. In comparison to autosomal DNA typing results, ABO testing, while useful in its day, provided a limited ability to discriminate among people in the population.

A trace evidence examination, such as microscopic hair comparison, is subjective and based on class characteristics. Although still used to screen hairs for DNA testing, microscopic hair comparison analysis has never been considered a conclusive form of identification. The appropriate interpretation of hairs that exhibit the same microscopic characteristics is only association.[22] A recent study of microscopic hair comparisons found that, out of 80 "microscopic associations" made independently by two top FBI examiners, nine were demonstrated to actually be exclusions when later subjected to DNA testing (approximately 11 percent of the cases).[23] Given the availability of DNA testing, the current approach should be to support any associative microscopic hair comparison results with nuclear autosomal STR DNA typing when there is cellular material on the root of the hair, or via mtDNA testing when only a hair shaft exists or when autosomal STR DNA testing fails to yield results.

Case example: Ron Williamson and Dennis Fritz. In the cases of Ron Williamson and Dennis Fritz, mtDNA testing contradicted the microscopic hair comparison analysis performed before trial. Williamson and Fritz were convicted of the 1982 rape and murder of a woman in her Oklahoma home. Williamson received a death sentence; Fritz received life in prison. A key element of the state's case was forensic expert testimony that 17 hairs from the crime scene microscopically "matched" either Fritz or Williamson. The state's case appeared to be so compelling that Williamson's appeals were quickly exhausted. He came within five days of execution. However, in 1999, Williamson received a last-minute stay of execution and obtained DNA testing, which proved that none of the 17 hairs that were deemed matches at the time of trial actually belonged to Fritz or Williamson. These results — combined with results excluding the men as the source of semen collected from the victim — exonerated both men and led to their release from prison. DNA testing also led to the identification of the true assailant: Glen Gore, a key state witness who falsely incriminated Williamson. Gore was convicted of the woman's murder in 2003.[24]

Prior DNA Testing That Was "Inculpatory"

Even if DNA testing was performed that did not exclude the defendant, additional tests may be warranted after conviction. NIJ guidelines for handling postconviction DNA testing requests point out that:

> It is important to understand what the previous test results really mean and whether those results could have been obtained if another individual other than the alleged donor was the source of the sample. For instance, ABO blood testing and/or DQa PCR test results alone are not sufficiently discriminating such that a falsely accused individual would necessarily be excluded with these tests.[25]

For example, one earlier form of PCR-based DNA testing — the DQa test — is akin to basic blood typing in that:

> As with serological tests, an exclusion with [the DQa] test eliminates an individual as the source of the sample; however, an inclusion with this test simply includes an

individual within a set of a large number of individuals that also have the same DNA types. A falsely accused individual may be included as a possible donor of a DNA sample with this test system.[26]

In such cases, requesting testing with more discriminating technology is appropriate. The cases of Josiah Sutton from Texas and Jerry Watkins from Indiana illustrate how indiscriminating DQa testing is and how innocent people can be included through DQa testing even though the biological material did not come from them. Both Sutton and Watkins were included as the source of biological crime scene evidence through DQa testing and convicted of sexual assaults. They were later excluded through the more informative autosomal STR DNA testing and exonerated. See the case profiles of Sutton and Watkins at www.innocenceproject.org.

Cases Involving Multiple Perpetrators

Postconviction DNA testing can yield material exculpatory evidence and even prove innocence in cases with multiple assailants. For example, in a rape perpetrated by two assailants, a range of possible test results could establish a defendant's innocence. DNA testing of semen from the rape kit may reveal two male DNA profiles (belonging to each of the assailants). If two profiles are obtained and the defendant is excluded, such test results would establish innocence.[27]

This same logic holds true for cases where there are more than two assailants. In fact, many of the postconviction DNA exoneration cases have involved crimes committed by multiple perpetrators — as many as five or six assailants. Just because testing may not yield the requisite number of profiles to demonstrate innocence is no reason to forgo testing.

Prior Consensual Sex and the Potential Need for an Elimination Sample

In some cases, after the defendant is excluded during postconviction testing, it will be necessary to obtain an elimination sample to establish that the biological material actually belongs to the assailant. For example, consider a case where the defendant was excluded as the source of sperm found on swabs in which the victim had consensual sex the day of her attack. An elimination sample would be required from the prior sex partner. If the sex partner is eliminated as a source of the sperm, then it can reasonably be inferred that the sperm came from the assailant — therefore, the defendant's exclusion establishes innocence. If the sperm matches the consensual sex partner, then the test results may entitle the defendant to a new trial (depending on how the presence of sperm was used at trial) but would not prove innocence. The potential need for an elimination sample should not be an obstacle to DNA testing.[28] Elimination samples were successfully obtained and used in many of the DNA exoneration cases.

It is important to keep in mind that if the consensual partner is no longer available (perhaps he is deceased), there are a number of possibilities for locating a medical sample or a family member willing to provide a sample for comparison purposes.

Evidence Searches

The most difficult step in the postconviction DNA testing process can be locating the physical evidence from the case to test. Defense counsel may need to track the chain of custody from the time the evidence was collected and check with every agency that possessed it. Evidence can be found decades after a crime in a variety of places, including the hospital, morgue, trial or appellate courts, district or state's attorney's office, police department, sheriff's office, court reporter's office, and the laboratory that did the original or any previous testing in the case. Every jurisdiction has its own procedure for storing evidence.

Finding evidence may depend on learning these policies and procedures as well as where evidence could be if there was a departure from procedure. Counsel will want to learn all of the locations where evidence is stored by agencies that once possessed the evidence.

Ultimately, locating the requested evidence — or determining that it has all been destroyed — will depend on a competent, comprehensive search. The jurisdiction must physically and thoroughly search all facilities and locations that can reasonably be expected to store the evidence, and it must produce *chain-of-custody* documents and other information detailing its efforts to the defense. At a minimum, the state should be made to disclose:

- All agencies that might possess evidence from the case.
- The evidence storage locations under each agency's control.
- Whether physical searches of those locations were conducted or the agency relied only on paper records to determine that evidence no longer exists.
- If physical searches were conducted, who performed the search.
- When this person (or people) performed the search.
- What facilities, locations and areas were searched within each agency.
- An inventory of what evidence items were recovered.
- Contemporaneous business records documenting the destruction of any evidence that the state maintains no longer exists.
- Copies of all other chain-of-custody documents, including evidence logs that reflect the chain of custody from the time the evidence was collected.
- Each agency's procedures and policies for storing evidence from the time of the crime to the present.

If the state is unable to locate the physical evidence, the defense should seek a hearing to determine the adequacy of the state's search and the availability of evidence.

Counsel can start the evidence search by calling or writing each agency to request voluntary cooperation. The custodians of evidence or records should be willing to look for evidence upon request and share any documents pertaining to chain of custody. (Counsel can also make a request for this information under the jurisdiction's public records law.) Each agency has its own protocol for numbering cases; when requesting a search for evidence or confirmation that evidence exists, it is important to include that number. For example, when calling the crime laboratory, provide the laboratory's case number (as opposed to the indictment number, which in most circumstances would be useless to lab personnel looking for the evidence).

The defense must be clear in what it asks the agency to look for and ensure that the agency does not simply rely on paper records but actually physically searches for the requested evidence. For example, it is routine for laboratories to return original evidence, such as a rape kit or the victim's clothing, to the submitting agency when the examination is complete. If counsel contacts a lab and asks for evidence from the case, the lab may say (and provide paperwork indicating) that the evidence was returned to the submitting agency many years ago. However, labs frequently retain, or separate out, cuttings, slides, swabs, extracts and other materials during the testing. Therefore, be clear in what you ask for — it is always better to be overly inclusive because you do not know what evidence might be available.

A flowchart that shows each step in evidence processing may also be of value. You would need to create it showing the type of evidence category and the practices used when the case was originally processed.

See below for some tips for various types of agencies that may be of value when requesting a search for evidence.

Hospital and Medical Examiners' or Coroners' Offices

After a rape, the victim is generally taken to a hospital, where she is treated and a kit is used to collect potential evidence. In addition to collecting samples for the kit, hospitals collect samples for their own testing, such as for sexually transmitted diseases. Hospitals also have been found to retain slides from their exam of the victim (their patient) — in places such as the hospital's laboratory and pathology departments — even decades after the original examination. In a similar manner, medical examiners' and coroners' offices may have also retained biological materials that are potentially suitable for analysis.

Case example: Bernard Webster. Bernard Webster was convicted of a 1982 Maryland rape on the basis of three eyewitness identifications. In 2002, Webster was exonerated by postconviction DNA testing. Although the Baltimore County Police Department had destroyed the case's biological evidence by the time Webster sought postconviction testing, three slides from the victim's rape examination were found at the pathology department of the Greater Baltimore Medical Center, where the victim had been treated immediately following her attack. DNA testing of the 20-year-old hospital slides yielded an autosomal STR DNA profile that not only exonerated Webster but also identified, through a search of the DNA databank, the true perpetrator of the crime.[29]

When asked whether they keep rape kits in general or in a specific case, hospitals will almost always say no and explain that the evidence was released to the police. The defense must make clear that it is not asking for the rape kit but rather duplicate slides or specimens made during the exam for the hospital's own use (as opposed to law enforcement's use). Due to privacy laws, counsel will likely need a court order or subpoena to require a hospital to search for any evidence from the case.

In murder cases, an autopsy would have been performed, and evidence from the victim's body and clothing would have been collected and, typically, later released to law enforcement. As with hospitals, the medical examiner's or coroner's office may have retained slides or other specimens.

Investigating Law Enforcement Agencies

After the rape examination, the investigating agency (police or sheriff's office) usually retrieves the evidence from the hospital. In addition to the rape kit, police frequently collect the clothing the victim wore to the hospital. Police also collect physical or biological evidence from the crime scene. Typically, the law enforcement agency submits some or all of the evidence to the crime lab; evidence not submitted stays in the agency's custody. After the lab completes its examination, the submitting agency often retrieves the evidence for long-term storage in its facility or for transport to court for a legal proceeding. Evidence not used at trial may stay with the investigating agency indefinitely. Alternatively, the law enforcement agency may follow a protocol for destroying stored evidence on a prescribed schedule.

Case example: Alan Newton. On July 6, 2006, Alan Newton of the Bronx, N.Y., was freed after 22 years of wrongful imprisonment for rape. Newton had been seeking DNA testing for 12 years, and both he and the court were told that repeated searches of police evidence facilities — where the rape kit was last held in 1985 — had yielded no evidence. In July 2005, a prosecutor newly assigned to the case asked the commanding officer of the police warehouse to make a renewed search. The prosecutor provided the officer with documentation indicating the specific bin where the evidence had been stored 20 years ago. When a hand search of that bin was performed, the rape kit was found inside it. DNA testing of that evidence thereafter conclusively proved Newton's innocence, and he was exonerated by a joint motion by the defense and the Bronx District Attorney's Office.[30]

Crime Laboratories

Although crime labs typically return the original evidence (such as a rape kit or clothing) to the submitting agency after the examination is complete, they frequently will preserve and store cuttings, slides and swabs of evidence used during testing. Such critical evidence has been found at crime labs decades after the original exam. It is important to have a crime lab check its evidence vaults for retained items as well as all relevant case files for evidence or chain-of-custody information. Crime labs that do not store these biological samples that are suitable for subsequent DNA testing in their facility may have created a packet or container that holds the samples and submitted that to the investigating agency as a newly created "item" of evidence.

Some crime labs will retain portions of the evidence for only a fixed number of years following the original examination. Their policy or storage limitations may have resulted in this evidence being returned to the submitting agency. In recent years, more and more crime labs have been returning swabs, slides and cuttings that had been retained at the lab. These portions of the original items are often returned in bulk, along with similar biological evidence from other cases, to submitting law enforcement agencies. This separate avenue for locating biological crime scene evidence must be explored because the packages containing the biological evidence cuttings, swabs and slides may not have been re-filed with the original case evidence.

Case example: Marvin Anderson. When Marvin Anderson of Virginia sought DNA testing, he was told that the rape kit from his case had been destroyed. A special search in 2001 by the director of the Virginia Division of Forensic Science revealed that the criminalist who had performed the conventional serology tests in Anderson's case in 1982 had broken lab protocol and, instead of returning the slides containing semen samples to the rape kit, had scotch-taped them into the lab notebook. The combination of this breach of procedure and the dedicated search led to the discovery of evidence that, once subjected to DNA testing, exonerated Anderson.

Clerk of the Court/Trial Court Exhibits

Depending on the jurisdiction, physical evidence used at trial may be with the trial court, reviewing courts, prosecutor's office or sometimes in the custody of the court reporter, bailiff or judge.

Prosecutors' Offices

Evidence used in postconviction testing has been found in district attorneys' evidence storage rooms as well as the district attorney's trial file.

Postconviction Evidence Search — Treatment by Courts

As noted above, when an agency asserts that the evidence was destroyed, the agency should provide contemporaneous records documenting the evidence destruction. Even then, a physical search must be conducted to ensure that the order was carried out and all of the evidence was, in fact, destroyed.

Absent conclusive proof that each item of potential biological evidence was destroyed, one or more items capable of resolving the petitioner's guilt or innocence beyond any doubt may still be in existence. In several DNA cases described earlier, evidence that had been reported as lost or destroyed was later discovered intact after a more diligent search. Indeed, when evidence has, in fact, been destroyed, there will usually be specific documentation of the destruction of every item, even if the evidence was destroyed in bulk. If there is not, one simply cannot be confident that the evidence is truly gone.

Evidence has been located in the back of a storage closet, a trial judge's locker, and between a wall and a prosecutor's desk. Sometimes evidence is labeled under the victim's name rather than the defendant's name, or it is simply misfiled among unrelated case evidence boxes. In other cases, it became clear after evidence was located in the original storage location that prior searches were half-hearted or perhaps not performed at all.

Although counsel can begin by contacting facilities in the chain of custody and requesting information, it may be necessary to litigate the evidence search to locate the evidence. Counsel can make a motion under the jurisdiction's access statute. Many state statutes either explicitly provide for evidence searches or require the court to make a factual determination regarding the existence of the evidence. Counsel can file a motion under the statute, requesting an evidence search and testing of any evidence located.

Many state courts have ordered law enforcement agencies to conduct comprehensive evidence searches. These cases place significant weight on NIJ's 1999 report, *DNA Testing: Recommendations for Handling Requests,*[31] which notes that "[f]inding the evidence is the most difficult part of the [DNA testing] process."[32] The report cautions prosecutors against concluding too hastily that evidence no longer exists and advises prosecutors to "use their best efforts to locate the crime scene samples."[33] The report also notes that "[m]any times all parties believe that the evidence has been destroyed, when in fact it has not."[34] The report further states:

> If, from initial contact with the investigating officer or review of case files, it appears that evidence suitable for DNA analysis was never collected, or has since been destroyed, it may prove impossible to continue with the rest of this guideline. ... However, no final decision or notification should be made until it has been carefully verified that evidence did not or does not still exist.[35]

In ordering state agencies to conduct comprehensive evidence searches, courts have rejected both unsworn affidavits that the evidence no longer exists and reliance on paper records without physical searches, even when a comprehensive physical search would require the state to inventory large evidence storage rooms.[36] In *Arey* v. *State,* a Maryland appellate court ruled that the state had the burden of establishing that the evidence no longer existed. The state failed to meet its burden by submitting an unsworn affidavit after it briefly searched only one location for the evidence. The court found that relying on computer records and paper files alone — absent a physical search — was insufficient to conclude that the evidence no longer existed. According to the court, "the [s]tate needs to check any place the evidence could reasonably be found, unless there is a written record that the evidence had been destroyed in accordance with then-existing protocol."[37]

In *Blake* v. *State,* the court held that an "unsworn memorandum, stating that the [s]tate merely requested the police to look in the evidence control unit, is insufficient to establish this critical fact," adding that "[s]imply asking a police officer to check an evidence unit locker is not sufficient. There are many other likely places where the evidence may have been stored."[38] Similarly, in a New York case, it was held that the state is "the gatekeeper of the evidence, who must show what evidence exists and whether the evidence is available for testing," and the "mere assertion that the evidence no longer exists based on a phone call to a police [p]roperty [c]lerk's office is insufficient as a matter of law" to summarily dismiss a DNA testing motion.[39]

Moreover, the court in *Arey* described what a search should entail. Simply put, "a court should not conclude that evidence no longer exists until the [s]tate performs a reasonable search for the requested evidence."[40] The court explained:

> At a minimum, a reasonable search ... would have required the [s]tate to look in the crime lab referred to in Detective Russell's testimony, if the lab is still in existence, for any slides used to test the blood evidence used against appellant or for pieces of the clothing he requested; the property room, if it was different from the ECU; and because the testimony at trial was that the evidence had been stored in the Judge's chambers, as unlikely as it is that it would be there after all these years, an inquiry as to that location.[41]

Additionally, the appellate court held that the state had a duty to identify whether the various state agencies "had protocols in place for the destruction of evidence" and, if so, whether these agencies followed their respective protocols:[42]

> [I]t is reasonable to assume that police departments, sheriff departments, clerk offices of the court, and like departments had protocols in place for the destruction of evidence, even before the enactment of [the DNA testing statute]. The [s]tate should identify the protocol that was in place from the time of the trial to the time of the request for testing, if possible, and see if that protocol was followed.[43]

The appellate court explained that the inquiry into the protocols might lead to other locations to search.[44]

Section 3. When Are You Entitled to Postconviction DNA Testing?

Seeking State Consent

In many cases, the state will consent to a defendant's request for postconviction DNA testing, obviating the need for unnecessary litigation. In fact, the NIJ guidelines recommend that prosecutors avoid unnecessary litigation and consent to testing in cases where an exclusionary result would prove innocence.[45] Even when a prosecutor agrees, it is important to proceed with a consent or stipulated order entered by the court that:

- Provides for chain of custody of the evidence.
- Details the procedures to be used to collect a new reference sample from the defendant and disseminate the test results.
- Addresses the circumstances under which the laboratory can use up the evidence during testing.

State Statutes on Postconviction DNA Access

Postconviction testing can be obtained through an application under the state's postconviction DNA-testing access statute. Currently, more than 40 states and the federal government have statutes that entitle defendants to access to postconviction DNA testing.[46] As many of these statutes were enacted after 2000, prosecutors frequently consent either to testing or to a motion under the statute, and courts routinely order testing on opposed motions under state statutes. However, there is not an abundance of case law interpreting these laws. In this sense, the DNA exoneration cases can provide powerful support for a motion for DNA testing by illustrating how DNA can provide evidence of innocence in a given case.

The requirements of state postconviction DNA statutes vary. They use differing burdens of proof requiring a defendant to show that favorable test results would most likely change the verdict and establish innocence.

Identity-At-Issue Requirement

Many DNA access statutes contain a requirement that the perpetrator's identity was or should have been an issue at trial. Some prosecutors and lower courts have misconstrued this requirement to mean that the applicant must show that the state's identification evidence was weak or the conviction was based mainly on eyewitness identification testimony. However, the identity-at-issue requirement has the same meaning in the DNA context as it does elsewhere in criminal law.

The identity issue is present in every criminal case in the sense that, to warrant conviction, the evidence must establish beyond a reasonable doubt the identity of the defendant as the person who committed the crime. When asserting a defense such as consent, self-defense, necessity, or insanity, the defendant admits that he or she participated in the charged acts. In such cases, the state still must prove the defendant's identity, but because the defendant admits participation, identity is not a genuinely contested issue. On the other hand, when the defense is misidentification, alibi or any other defense in which the defendant disputes that he or she was the perpetrator, identity is a significant issue.[47]

Whether the identity-at-issue requirement is met in a particular case depends entirely on whether the applicant for postconviction DNA testing put forth a defense at trial that he or she was not the person who committed the crime. The identityat-issue requirement is a common-sense limitation on postconviction DNA testing requests.[48] The very purpose of the postconviction DNA statute is to use more informative scientific technology to test the state's identification proof — proof a jury has already determined to be beyond a reasonable doubt — to determine if a wrongful conviction occurred. Generally, there would be little point to test in cases where the defendant admits that he was the perpetrator but maintains nevertheless that he is legally innocent. For example, in a rape case where the defense is consent, DNA testing would not resolve the primary question: whether the victim consented. Rather, DNA testing would be expected simply to confirm a fact not in dispute: that the defendant had sexual contact with the victim.

In *State* v. *Peterson,* Peterson was convicted of felony murder based on "strong evidence of defendant's guilt," including:

- Testimony by three people that Peterson had described the crime to them "in lurid detail" only a few hours after the crime, before police had released any detailed information to the public.
- Testimony of an inmate in the jail where Peterson was incarcerated before trial. Peterson allegedly admitted to the inmate that he had committed the crime.
- In the days after the crime, Peterson had fresh scratch marks that looked like fingernail marks on his arms.
- After the crime, Peterson had asked several people for money so he could travel to Germany, and he threatened several potential witnesses.[49]

The trial court initially found that the identityat-issue requirement was not met because the case did not involve a stranger's eyewitness identification. The Appellate Division found that the trial court erred in concluding that Peterson had failed to show that identity was a significant issue at trial, noting:

Eyewitness identification is simply one method of proving a perpetrator's identity. The identification of the defendant also may be established by various other forms of evidence including, as in this case, the defendant's inculpatory statements and efforts to avoid apprehension and physical evidence found at the crime scene.[50]

The Appellate Division further stated:

> [D]espite the strong evidence of his guilt, defendant's identity as the perpetrator was the only issue at trial. Defendant took the stand and denied that he was the one who raped and murdered the victim. Moreover, defendant presented his girlfriend's testimony that he was with her in a motel room at the time of the crime. Although this alibi was discredited, defendant's only defense was that he was not the perpetrator of this horrific crime.[51]

Reasonable Likelihood of A Different Outcome

All statutes require an applicant to make some sort of showing regarding the probative value of favorable DNA test results in the case. New York's statute (C.P.L. 440.30(1-a)) requires the applicant to show that if a DNA test had been conducted on the requested evidence and the results had been admitted in the trial resulting in the judgment, there exists a reasonable probability that the verdict would have been more favorable to the defendant. Louisiana's DNA statute (La. C. Cr. P., article 926.1) requires an applicant to show there is "a reasonable likelihood that the requested DNA testing will ... establish the innocence of the petitioner." Some statutes employ a stricter standard. For example, Ohio's DNA statute (R.C. § 2953.74(B) & (C)) requires that the DNA results be "outcome determinative with respect to the question of [applicant's] guilt."[52]

The "reasonable likelihood" prong ensures that the evidence to be tested has probative value. Because it is impossible to know the outcome of DNA testing in advance of actual testing, this inquiry requires the court to presume favorable test results and determine the significance of those favorable test results — that is, whether it is "reasonably likely" that favorable test results would be probative enough to establish the applicant's innocence.[53]

The "reasonable probability" prong requires the court to consider the probative value of favorable test results on the case — not whether it thinks it is likely, as a matter of fact, that the applicant is actually innocent.

Overwhelming Evidence of Guilt

As of July 22, 2010, there have been 255 DNA exonerations in the United States. In many of these cases, the trial evidence against the defendant could be characterized as overwhelming — including cases where the defendant gave a detailed confession or pled guilty; the state's evidence consisted of "certain" or multiple eyewitnesses; or other forensic evidence, even prior DNA testing, that linked the defendant to the crime.[54] It is important to understand that properly conducted DNA testing is more accurate and reliable than practically all other types of identification proof.[55] The existence of overwhelming evidence of guilt in a particular case should not stand in the way of DNA testing.

Eyewitness Identification

Eyewitness misidentification played a role in more than 75 percent of convictions overturned through DNA testing.

Case example: Michael Williams. Michael Williams from Chatham, La., was convicted of aggravated rape in 1981 on the basis of what seemed to be an unassailable identification: the eyewitness testimony of the victim, who had known Williams since he was a little boy and tutored him months before she was attacked. The 22-year-old victim stopped tutoring 16-yearold Williams after he became infatuated with her. Williams harassed her, broke a window at her home, and was arrested as a consequence. Several months later, when a man entered her home in the middle of the night and raped her, the victim immediately reported that she was able to see the assailant and identified Williams. However, in 2005, testing of semen from the clothing the victim wore during the attack excluded Williams, and he was exonerated after 24 years of wrongful imprisonment.[56]

Case example: Kirk Bloodsworth. Kirk Bloodsworth was convicted in 1985 in Maryland for the sexual assault and brutal killing of a 9-year-old girl. The victim had been strangled, raped and beaten with a rock. Bloodsworth's conviction rested on five eyewitnesses, who testified at trial that they had seen him with the victim before she was murdered. Additional evidence against Bloodsworth included:

- Testimony that he had said he did something terrible that day that would affect his relationship with his wife.
- Bloodsworth's mention of a bloody rock to police before that information had been made public.
- A shoe impression that matched his size, found near the victim.

In 1992, the prosecution agreed to DNA testing of the victim's shorts and underwear, a stick found at the scene, and an autopsy slide. Testing of sperm on the victim's panties excluded Bloods-worth. After spending more than eight years in prison — two of those years facing possible execution — Bloodsworth was released from prison and pardoned. However, even though prosecutors deemed Bloodsworth's exclusion from the semen evidence sufficient to overturn his conviction, it was not until 2003, when the evidence was retested with autosomal STR technology and matched through CODIS to a convicted sex offender named Kimberly Shay Ruffner, that the state finally acknowledged Bloodsworth's innocence. At the time of the DNA hit, Ruffner was in prison for an attempted rape and murder he committed just three weeks after the 9-year-old girl's murder.[57]

Case example: Kevin Green. In 1980 in California, Kevin Green was convicted of second-degree murder for the death of an unborn fetus and of the attempted murder and assault of his pregnant wife. The case against Green rested entirely on the testimony of his wife, who had been attacked, raped and beaten into unconsciousness while alone in their apartment. Green maintained that he had left their home to get fast food and that she was attacked in the brief time he was gone. His wife maintained that he was her attacker and claimed that he beat her after she refused to have sex with him. After the creation of CODIS, the California Department of Justice laboratory found that the DNA profile from sperm collected from the wife's rape kit matched Gerald Parker, a serial killer known as the "Bed-room Basher" for breaking into women's bedrooms to rape and kill them. Parker confessed to the attack on Green's wife and was connected to five other rapes and murders in the area.[58]

Confessions and Guilty Pleas

In approximately 25 percent of DNA exoneration cases, innocent defendants pled guilty or made incriminating statements or confessions.

Case example: Chris Ochoa. In 1988, Chris Ochoa was an employee of the Pizza Hut restaurant chain in Austin, Texas. After a young woman was found raped and murdered in another Pizza Hut restaurant, Ochoa was brought in for questioning under the theory that a master key had been used to gain access to the premises. After several hours of interrogation, Ochoa gave a detailed confession that contained important details of the crime not available to the public. He described in graphic detail how he and a friend and fellow employee, Richard Danziger, raped the victim before Ochoa shot her in the head. Unlike many defendants who confess to crimes while in police custody, Ochoa did not recant his statements before the trial. Instead, he pled guilty to the crime and went on to testify in detail about the events of that night at Danziger's trial.

Danziger was convicted on the basis of that testimony, in addition to expert testimony that a pubic hair found near the victim's body was microscopically similar to Danziger's hair. However, in 1998, a man named Achim Marino wrote to then-Governor George W. Bush, confessing to the murder and stating that he could no longer bear responsibility for the fact that two innocent men were in prison for his crimes. Postconviction DNA testing subsequently showed that the single male DNA profile obtained was a perfect match to Marino, confirming Marino's claim and exonerating both Ochoa and Danziger by excluding them as the source of semen found in the victim's body.[59]

Possession of Crime Scene Evidence

In several cases where DNA testing has brought to light a wrongful conviction, the defendant's guilt seemed solidified by the fact that he possessed an item taken from the victim during the crime.

Case example: Gene Bibbins. In 1986, police in Louisiana arrested Gene Bibbins when they found him with the radio that a man had stolen from a teenage girl after raping her less than an hour earlier. His clothing also resembled the assailant's outfit. Bibbins, who lived in the same apartment complex as the victim, claimed he found the radio between buildings. The radio, along with the victim's immediate identification of Bibbins, led to his conviction.

In addition, conventional serology testing could not exclude him as a source of the semen from the victim's rape kit. However, in 2002, Bibbins was exonerated after postconviction DNA testing excluded him from the rape kit evidence.[60]

Case example: Robert Clark. In 2005, Robert Clark was exonerated nearly 24 years after his conviction for the rape of a woman in her car in Georgia. Days after the crime, the victim saw a man driving her car, which had been stolen during the attack. This man was Clark, who maintained that he received the car from a friend. Nonetheless, the victim identified Clark as her attacker, testifying that there was no doubt in her mind. More than two decades later, postconviction DNA testing on a vaginal slide excluded Clark. A CODIS hit matched the man whom Clark had named as the source of the car, who by then was in prison for subsequent crimes (but was on the verge of release).[61]

Constitutional Right to DNA Testing

In states that do not have specific postconviction DNA testing statutes, access to DNA evidence may be obtained through the general postconviction statute dealing with newly discovered evidence. However, many of the traditional post-conviction statutes have time limits or other bars to applications for DNA testing.

The issue of whether the U.S. Constitution provides a sentenced prisoner a right to DNA testing independent of state statute was largely but not entirely resolved in *DA's Office for the Third Judicial Dist.* v. *Osborne.*[62] Although federal law and the laws of 46 states provide for at least some access to DNA testing, Alaska did not. In *Osborne,* the U.S. Supreme Court declined to hold that an Alaskan inmate had a constitutional right to testing enforceable under civil rights laws as a due process claim. The Supreme Court majority concluded that Alaska law provided sufficient procedures for presenting "new" evidence to satisfy due process concerns — procedures the inmate had not invoked, the majority contended. *Osborne* is a case of limited applicability. The Supreme Court majority's conclusion that Alaska state law provided an avenue for the inmate to seek testing did not answer the question of when a state statute might be unreasonably narrow or strict in its conditions, so that the denial of access to evidence for DNA testing might violate federal constitutional rights. What is clear from *Osborne* is that any incarcerated person who wants testing must first attempt to use state law procedures before seeking relief in federal court under a claim of a denial of due process rights.

Useful Resources

More than 40 projects and organizations in the United States that work on postconviction cases are affiliated with the Innocence Network. Consider contacting one of them for assistance, referrals, or resources at www.innocencenetwork.org.

End Notes

[1] National Institute of Justice, U.S. Department of Justice, *Using DNA to Solve Cold Cases: Special Report* (July 2002), NCJ 194197.

[2] President's DNA Initiative, *Advancing Justice Through DNA Technology,* www.dna.gov/uses/ solving-crimes/property_crimes/; Zedlewski, E., and M.B. Murphy, "DNA Analysis for 'Minor' Crimes: A Major Benefit for Law Enforcement," 253 NIJ J OURNAL (January 2006), available at www.ojp.usdoj.gov/nij/journals/253/dna_analysis. html.

[3] See *State* v. *Peterson,* 364 N.J. Super. 387, 397 (App. Div. 2003); *New Jersey Death Penalty Study Commission Report,* Jan. 2007, at 52; Possley, M., "Convict Seeks New Trial on Basis of Flawed Hair Analysis," CHICAGO TRIBUNE (July 29, 2005); Mansnerus, L., "Case Dropped Against New Jersey Man After 18 Years," N.Y. T IMES (May 27, 2006).

[4] See "Man Tells of Ordeal of 7 Years in Prison," DAYTON NEWS DAILY (Feb. 15, 2006); *The Innocence Project: Know the Cases — Clarence Elkins,* www.innocenceproject. org/Content/92. php.

[5] Bragg, R., "DNA Clears Louisiana Man on Death Row, Lawyer Says," N.Y. T IMES (April 22, 2003).

[6] Glod, M., "Cleared Va. Man to Be Pardoned," THE W ASHINGTON P OST (Aug. 21, 2002).

[7] Donald, M., "Lethal Rejection," DALLAS OBSERVER (Dec. 12, 2002).

[8] *People* v. *Wise,* 752 N.Y.S.2d 837 (N.Y. Sup. Ct. 2002).

[9] Ga. Code Ann., § 24-4-63(b)(3) states, in pertinent part:

Upon a showing by the defendant in a criminal case that access to the DNA data bank is material to the investigation, preparation, or presentation of a defense at trial or in a motion for a new trial, a superior court having proper jurisdiction over such criminal case shall direct the bureau to compare a DNA profile which has been generated by the defendant through an independent test against the data bank, provided that such DNA profile has been generated in accordance with standards for forensic DNA analysis adopted pursuant to 42 U.S.C. Section 14131, as amended.

[10] 725 ILCS 5/116-5 (2003) states, in pertinent part:

(a) Upon motion by a defendant charged with any offense where DNA evidence may be material to the defense investigation or relevant at trial, a court may order a DNA database search by the Department of State Police. Such analysis may include comparing: (1) the genetic profile from forensic evidence that was secured in relation to the trial against the genetic profile of the defendant, (2) the genetic profile of items of forensic evidence secured in relation to trial to the genetic profile of other forensic evidence secured in relation to trial, or ... cured in relation to trial, or (3) the genetic profiles referred to in subdivisions (1) and (2) against: ... (ii) genetic profiles, including but not limited to, profiles from unsolved crimes maintained in state or local DNA databases by law enforcement agencies.

[11] N.C.G.S.A. § 15A-267 (2001) states, in pertinent part:

(c) Upon a defendant's motion made before trial in accordance with G.S. 15A- 952, the court may order the SBI to perform DNA testing and DNA Database comparisons of any biological material collected but not DNA tested in connection with the case in which the defendant is charged upon a showing of all of the following: (1) That the biological material is relevant to the investigation. (2) That the biological material was not previously DNA tested. (3) That the testing is material to the defendant's defense.

[12] 387 N.J. Super. 506, 521-22 (N.J. App. Div. 2006).

[13] Saltzman, J., and M. Daniel, "Man Freed in 1997 Shooting of Officer," BoSToN G LoBE (Jan. 24, 2004).

[14] 165 P.3d 31 (Wash. Ct. App. 2007).

[15] National Institute of Justice, U.S. Department of Justice, Postconviction DNA Testing: Recommendations for Handling Requests, Report From the National Commission on the Future of DNA Evidence, NCJ 177626 (September 1999), at 23.

[16] For example, one test claims to be at least 10 times more sensitive than the presumptive acid phosphatase test frequently used in laboratories. See Denison, S.J., et al., "Positive Prostate-Specific Antigen (PSA) Results in Semen-Free Samples," 37 CAN. S oC. F oRENSiC S Ci. J. 197, 200 (2004). In 16.8 percent (17 cases) of the forensic casework samples tested in one recent study, positive PSA levels were found in samples that had previously tested negative for PSA and where there were no visible spermatozoa (Id.).

[17] See McFadden, R.D., "DNA Clears Rape Convict After 12 Years," N.Y. T iMES (May 20, 2003); "Wrong Man is Set Free by DNA," N.Y. PoST (May 20, 2003). Similarly, a New Jersey court vacated the 1989 rape-murder conviction of Larry Peterson after postconviction DNA testing showed that sperm on both vaginal and oral swabs from the murder victim did not come from Peterson but rather belonged to another man (and also matched the DNA profile in the victim's fingernail scrapings). See Mansnerus,
L.,"Citing DNA, Court Annuls Murder Conviction from 1989," N.Y. T iMES (July 30, 2005). These same vaginal/oral swabs had been examined and tested by the state's forensic expert in 1989, yet the state lab had failed to detect semen or spermatozoa on any of these items. See Possley,
M.,"Convict Seeks New Trial on Basis of Flawed Hair Analysis," Chicago Tribune (July 29, 2005).

[18] See "Finally Free, Ronald Taylor Has No Grudges," HoUSToN C HRoNiCLE (Oct. 10, 2007), www.chron.com/disp/story.mpl/ metropolitan/falkenberg/5204596.html; The Innocence Project: Know the Cases — Ronald Taylor, www. innocenceproject.org /Content/1124.php.

[19] See "Man Cleared in 1994 Rape Says Faith, Perseverance Got Him Through Prison, Now Free Thanks to DNA Evidence, Dean Cage Looks to Start His Life Over," CHiCAGo TRiBUNE (May 29, 2008); The Innocence Project: Know the Cases — Dean Cage, www.innocence project.org/ Content/1376.php.

[20] Santos, F., "DNA Testing Frees Man Imprisoned for Half His Life," N.Y. T iMES (Sept. 21, 2006).

[21] Zeigler, M., and G. Craig, "Final Vindication," DEMoCRAT AND C HRoNiCLE (March 7, 2007); Dobbin, B., "DNA Tests Free Man Held 10 Years," THE BUFFALo NEWS (May 17, 2006).

[22] Houck and Budowle, "Correlation of Microscopic and Mitochondrial DNA Hair Comparisons," 47 J. F oRENSiC S Ci. 964, 966 (2002).

[23] Id.

[24] Jones, C.T., "DNA Tests Clear Two Men in Prison," DAiLY o KLAHoMAN (April 16, 1999); Yardley, J., "The Innocent Man," THE W ASHiNGToN PoST (Oct. 8, 2006).

[25] National Institute of Justice, U.S. Department of Justice, Postconviction DNA Testing: Recommendations for Handling Requests, at 14 (1999), NCJ 177626, http://www.dna. gov/postconviction.

[26] Id. at 27.

[27] The NIJ guidelines provide the following examples of Category 1 cases (that is, cases in which exclusionary results would be dispositive of a petitioner's innocence *and* "the prosecution should be willing to stipulate to test testing"):
Example 2: Petitioner was convicted of the rape of a woman who reported that she was sexually attacked by two men. Vaginal swabs were taken and preserved. Exoneration of the defendant may depend on whether the DNA test of sperm on the vaginal swabs shows two male DNA profiles, both of which exclude petitioner. (*Id.* at 4, 35)

[28] *Id.* at 14.

[29] "Wrongfully Held 20 Years, Md. Man to be Freed Today, Imprisoned for Rape DNA Shows He Did Not Commit," THE B ALTIMORE S UN (Nov. 7, 2002); "Md. DNA Evidence Yields New Suspect in '82 Rape; Earlier Test Freed Man After 20 Years," THE W ASHINGTON P OST (Nov. 19, 2002).

[30] Williams, T., "Freed by DNA, and Expressing Compassion for Rape Victim," N.Y. T IMES (July 7, 2006); Dwyer, J., "New York Fails at Finding Evidence to Help the Wrongfully Convicted," N.Y. TIM ES (July 6, 2006).

[31] See, e.g., *Arey* v. *State,* 929 A.2d 501, 507 (Md. 2007) ("[W]e carefully considered a cogent report published by the National Commission of the Future of DNA Evidence"); *Blake* v. *State,* 909 A.2d 1020, 1024-25 (Md. Ct. App. 2006).

[32] National Institute of Justice, U.S. Department of Justice, *Postconviction DNA Testing: Recommendations for Handling Requests,* at 45.

[33] *Id.* at 7.

[34] *Id.* (emphasis added).

[35] *Id.* at 36 (emphasis added).

[36] Harris, M., "DNA Search Delayed," B ALTIM ORE S UN (April 18, 2008).

[37] Arey v. State, 929 A.2d 501, 508 (Md. 2007).

[38] Blake v. State, 909 A.2d 1020, 1028, 1031 (Md. Ct. App. 2006).

[39] People v. Pitts, 4 N.Y.3d 303, 311-312 (2005).

[40] Id. at 508.

[41] Id. at 508-09.

[42] Id. at 508.

[43] Id. at 508.

[44] Id. at 508 (emphasis added).

[45] Postconviction DNA Testing: Recommendations for Handling Requests (1999) states:

> A prosecutor should normally agree to testing without opposition in category 1 cases. For example, when a rape case turned solely, or in large part, on eyewitness testimony, where serology at the time was inconclusive or not highly discriminating, and newer, more discriminating tests are now available, the prosecutor should order DNA testing. (*Id.* at 40)

[46] See 18 U.S.C. 3600; Ariz. Rev. Stat. § 13-4240; Ark. Code Ann. § 16-112-201; Cal. Penal Code § 1405; Colo. Rev. Stat § 18-1-411); Conn. Gen. Stat. Ann. § 52-582 (2003); Del. Code. tit. 11, § 4504; D.C. Code Ann. § 22- 4133; Fla. Stat. Ann. § 925.11; Ga. Code Ann. § 5-5- 41(c); Haw. Rev. Stat. §§ 844D 121-133; Idaho Code § 19-4902; 725 Ill. Comp. Stat. Ann. 5/116-3; Ind. Code Ann. § 35-38-7; Iowa Code § 81.10; Kan. Stat. Ann. § 21-2512; Kan. Stat. Ann. § 21-2512; Ky. Rev. Stat. § 422.285; Ky. Rev. Stat. § 422.285; La. Code Crim. Proc. Ann. art. 926.1; Me. Rev. Stat. Ann. tit. 15, § 2137; Md. Code Ann., Crim. Proc. § 8-201; Mich. Comp. Laws § 770.16; Minn. Stat. § 590.01; Mo. Rev. Stat. § 547.035; Mont. Code Ann. §§ 46-21-110, 53-1-214; Neb. Rev. Stat. § 29-4120; A.B. 16, 2003 Leg., 72nd Reg. Sess.; Nev. Rev. Stat. § 176.0918; N.H. Rev. Stat. Ann. § 651-D:1 - D:4; N.J. Stat. Ann. § 2A:84A-32a; N.M. Stat. Ann. § 31-1a-2; N.Y. Crim. Proc. Law § 440.30(1-a); N.C. Gen. Stat. § 15A-269; N.D. Cent. Code Ann. § 29-32.1-15; Ohio Rev. Code Ann. § 2953.71; Or. Rev. Stat. § 138.510 *et seq.;* Pa. Stat. Ann. 42 § 9541 *et seq.;* S.C. Code Ann. §§ 17-28-10; R.I. Gen. Laws § 10-9.1-11; Tenn. Code Ann. § 40-30-403; Tex. Code Crim. Proc. Ann. art. 64.01 *et seq.;* Utah Code Ann. § 78-35a-301; Vt. Stat. Ann. tit. 13, § 5561 *et seq.;* Va. Code Ann. § 19.2-327.1; Wash. Rev. Code § 10.73.170; W. Va. Code Ann. § 15 2B 14; Wi. Stat. Ann. §974.07; Wyo. Stat. Ann. 7-12-302-315; Pa. Stat. Ann. 42 § 9541 *et seq.;* S.C. Code Ann. §§ 17-28- 10; Vt. Stat. Ann. tit. 13, § 5561 *et seq.;* W. Va. Code Ann. § 15 2B 14; Wyo. Stat. Ann. 7-12-302- 315- A.B. 16, 2003 Leg, 72nd Reg. Sess.

[47] State v. Peterson, 364 N.J. Super. 387, 395- 96 (App. Div. 2003) (concluding "the strength of the evidence against a defendant is not a relevant factor in determining whether his identity as the perpetrator was a significant issue" and finding identity was a significant issue under N.J.S.A. 2A:84-32a(d)(3), where defendant's only defense was that he was not the perpetrator of the crime).

[48] People v. Urioste, 736 N.E.2d 706, 711-12 (Ill. App. Ct. 2000).

[49] State v. Peterson, 364 N.J. Super. at 392.

[50] Id. at 395.

[51] Id. at 395-96.

[52] See State v. Emerick, 868 N.E.2d 742, 746 (Ohio Ct. App. 2007) (ordering postconviction DNA testing of evidence, including a hammer and screwdriver used to murder the victims and crime scene blood).

[53] See State v. Peterson, 836 A.2d 821, 827 (N.J. 2003); State v. DeMarco, 387 N.J. Super. 506, 517 (2006) ("Even if a trial court concludes, in light of the overwhelming evidence of a defendant's guilt presented at trial, that it is unlikely DNA testing will produce favorable results, the court may not deny a motion for DNA testing on that basis. Because it is difficult to anticipate what results DNA testing may produce in advance of actual testing, the trial court should postulate whatever realistically possible test results would be most favorable to defendant in determining whether he has established" the statutory requirements for testing.), www. innocenceproject.org.

[54] Aside from the DNA exoneration cases, DNA is yielding scores of pretrial exclusions. To illustrate, the Georgia Bureau of Investigation (GBI) routinely performs DNA testing in homicide, rape and other violent crime cases. Of the more than 700 cases in which GBI has conducted DNA testing, 59 percent resulted in the inclusion of a suspect and 25 percent excluded the suspect. See www.state.ga.us/gbi/fsdna.html. DNA testing performed at the FBI laboratory, similarly, has excluded 20 percent of the primary suspects and resulted in a match with the primary suspect in only about 60 percent of the cases. National Institute of Justice, U.S. Department of Justice, Convicted by Juries, Exonerated by Science: Case Studies in the Use of DNA Evidence to Establish Innocence After Trial, NCJ 161258 (June 1996), at xxviii.

[55] O'Brien, K., "From Jail to Joy," THE TIM ES -PICAYUNE (March 12, 2005).

[56] National Institute of Justice, U.S. Department of Justice, Convicted by Juries, Exonerated by Science: Case Studies in the Use of DNA Evidence to Establish Innocence After Trial, NCJ 161258 (June 1996), at 35-37; Valentine, P.W., "Jailed for Murder, Freed by DNA: Md. Waterman, Twice Convicted in Child's Death, Is Released," THE W ASHINGTON P OST (June 29, 1993); Hanes, S., "'84 Investigation Quick to Overlook the Culprit," BALTIMORE S UN (May 22, 2004); Seigel, A.F., "Taking Felons' DNA in Dispute," BALTIMORE S UN (June 7, 2004).

[57] Associated Press, "Wrongly Convicted Man Finally Sees Justice," THE VIRGINIA PILOT (Oct. 7, 1998); Goodyear, C., and E. Hallissy, "The Other Side of DNA Evidence: An Innocent Man is Freed," THE S AN FRANCISCO CHRONICLE (Oct. 19, 1999).

[58] See Donald, M., "Lethal Rejection," DALLAS OBSERVER (Dec. 12, 2002); Wrolstad, M., "Hair-Matching Flawed as a Forensic Science; DNA Testing Reveals Dozens of Wrongful Verdicts Nationwide," THE DALLAS M ORNING NEWS (March 31, 2002).

[59] See AP, "DNA Tests Free Convicted Rapist," (Dec. 6, 2002), available at www.cbsnews.com/ stories/2002/12/06/national/main532165.shtml. See also Angelette, A., "Judge Officially Overturns Bibbins' Rape Conviction," THE A DVOCATE (March 12, 2003); Barrouquere, B., "Number of Wrongful Convictions in LA Immense," THE A DVOCATE (Nov. 23, 2003).

[60] See AP, "DNA Tests Clear Georgia Inmate of Rape Charges," (Dec. 8, 2005). See also Innocence Project, browse profiles under "Robert Clark," at http://www.cbsnews.com/ stories/2002/12/06/national/ main532165. html (last visited July 20, 2007); see Innocence Project, browse profiles under "Donte Booker," at http://www.innocenceproject. org/Content/ 55.php (last visited July 20, 2007) (Booker was convicted of rape in Ohio on the basis of the victim's identification, the fact he was found in possession of an item that had been taken from the victim's car during the attack, and microscopy testimony. He was exonerated through postconviction DNA testing that excluded him from the rape kit evidence.)

[61] 129 S. Ct. 2308 (U.S. 2009).

GLOSSARY

13 core CODIS loci: The 13 core CODIS autosomal loci are CSF1PO, D3S1358, D5S818, D7S820, D8S1179, D13S317, D16S539, D18S51, D21S11, FGA, THO1, TPOX and VWA.

Accreditation: Procedure used by an authoritative body that gives formal recognition to laboratories that have demonstrated (via production of objective evidence) their competence to conduct specified tasks.

Accredited DNA laboratory: A forensic DNA laboratory that has received formal recognition by an accrediting body that it meets or exceeds a list of standards, including The Quality Assurance Standards for Forensic DNA Testing Laboratories, to perform certain tests.

Accrediting body: An organization that defines the elements needed to demonstrate competence, administers its accreditation program, and grants accreditation.

Acid phosphatase: A chemical substance found in high quantities in semen/seminal fluid. The AP test is a presumptive color test that is used to screen for the presence of semen/seminal fluid by detecting acid phosphatase content. Also referred to as the *seminal acid phosphatase (SAP) test.*

Acrosomal cap: A cap-like structure on the tip (anterior end) of a sperm cell that contains enzymes that aid in egg penetration during fertilization.

Administrative review: An evaluation of a report and the supporting documentation for consistency with laboratory policies and editorial correctness.

Allele calls: Allele calls for STRs are the designated numbers given to each allele detected for a genetic marker. For amelogenin, allele calls correspond to the designated letter(s) — X and Y — that denote the detected DNA fragment or fragments. Allele calls can be generated manually or via a software program.

Allele frequency: The proportion of a particular allele in a population or population category (e.g., Caucasian, African-American, Hispanic or Asian). Allele frequencies are calculated using the number of times an allele is observed in a sampling of persons within a population. Allele frequencies are then used to determine the probability that a particular DNA profile might occur randomly in the larger population from which the sampling was obtained.

Alleles: Different forms of a gene or genetic marker at a particular locus. An allele is described as the characteristics of a single copy of a specific gene or of a single copy of a specific location on a chromosome. For example, one copy of a specific short tandem repeat region might have 10 repeats, while the other copy might have 11 repeats. These would represent two alleles of that short tandem repeat region, designated as alleles 10 and 11, respectively.

Allelic drop-in: An allele not originating from the sample but appearing on the electropherogram, often caused by low-level contamination or use of robotic systems for analysis.

Allelic drop-out: Failure to detect an allele within a sample, or failure to amplify an allele during a polymerase chain reaction (PCR).

Amelogenin: Referred to as the gender differentiation locus, and colloquially as the sex determination locus. It is a gene present on the X and Y sex chromosomes that is used in DNA identification testing to determine the gender of the DNA donor in a biological sample.

Amplicon: A DNA sample that has undergone the polymerase chain reaction (PCR) process. Also referred to as *amplified DNA*.

Amplified DNA: A DNA sample that has undergone the polymerase chain reaction process (PCR). Also referred to as an *amplicon.*

Analytical documentation: The documentation of procedures, standards, controls and instruments used; observations made; results of tests performed; and charts, graphs, photographs and other documentation generated that are used to support the analyst's conclusions for a case.

Analytical procedure: A defined progression of steps designed to ensure uniformity of a testing process by all analysts within a laboratory on a day-to-day basis and over time.

Analytical threshold: The minimum height requirement at and above which detected peaks can be reliably distinguished from background noise on an electropherogram; peaks above this threshold are generally not considered noise and are either true alleles or artifacts.

Artifacts: Any non-allelic products of the amplification process (e.g., stutter or minus A/non-templated nucleotide addition), anomalies of the detection process (e.g., pull-up or spike), or byproducts of primer synthesis (e.g., a dye blob).

Aspermic male: A male who is unable to produce sperm. Colloquially, the terms aspermic and azoospermic are often used interchangeably; however, technically, azoospermia refers to a male who does not have sperm cells in his ejaculate. The cause of aspermia may be sterility, vasectomy, venereal disease or injury.

Assessment: An inspection used to evaluate, confirm or verify activity related to quality. Also called an *audit.*

Audit: An inspection used to evaluate, confirm or verify activity related to quality. Also called an *assessment.*

Autoradiographic film: An X-ray film image showing the position of radioactive substances. Sometimes called an *autorad.*

Autosomal chromosomes: Chromosomes that are not sex chromosomes.

Autosomal STR analysis/locations: DNA analysis that targets autosomal chromosomes for short tandem repeat typing. Autosomal pertains to chromosomes that are not sex chromosomes. Individuals normally have 22 pair of autosomes and one pair of sex chromosomes (X,X in females or X,Y in males).

Azoospermic male: A male who does not have sperm cells in his ejaculate. The cause may be sterility (in this instance, the inability to produce spermatozoa), vasectomy, disease or injury.

Bases/base pairs: Adenine (A), thymine (T), cytosine (C) and guanine (G) are molecular building blocks of DNA, called bases. Each base will only combine with its specific, corresponding base to form base pairs when DNA is in its typical double-stranded form. This predictable pattern of base pairing — specifically, that A pairs only with T (and vice versa) and C pairs only with G (and vice versa) — is exploited during the DNA typing process to make copies of very specific areas on the DNA molecule (using a polymerase chain reaction, or PCR) where differences between people in the population are known to exist.

Bayesian approach: System of probability based on beliefs in which the measure of probability is continuously revised as available information changes.

Buccal sample/swab: A sample obtained from the interior cheek area of the mouth.

Calculated match: A statistical calculation performed on a match.

Calibration: A set of operations that establish, under specified conditions, the relationship between values indicated by a measuring instrument or measuring system, or values represented by a material, and the corresponding known values of a measurement. See *Equipment calibration.*

Capillary: A narrow silica tube containing a polymer solution used to separate out components of a mixture based on their size and/or chemical composition. See *Capillary electrophoresis.*

Capillary electrophoresis (CE): An instrument used to separate fragments of DNA based on size. The platform for CE uses narrow silica capillaries (or tubes) containing a polymer solution through which the negatively charged DNA molecules migrate under the influence of a high-voltage electric field. An important advantage of the multi-channel CE instruments, compared with slab gel electrophoresis, includes easier automation of analyses.

Casework CODIS administrator: An employee at a laboratory performing DNA analysis on forensic and casework reference samples who is responsible for administration and security of the laboratory's Combined DNA Index System (CODIS).

Chain of custody: A continuous log documenting the location of sample(s) of physical evidence collected from a crime scene at every step of the process — from the crime scene to the laboratory to the analyst's workstation, and any other movement or handling in between (including any time that evidence was removed from storage or returned).

Christmas tree stain: A staining method used to improve visualization of cells. In forensics, this type of staining is used in microscopic examination for the presence of sperm cells and/or epithelial cells. Two solutions are used in the process: Kernechtrot solution (also called *nuclear fast red*), a red-staining solution; and Picro Indigo-Carmine solution (also called *picro-indigo-carmine*), a green-staining solution.

Chromosomes: The biological structure by which hereditary information is physically transmitted from one generation to the next. Located in the cell nucleus, chromosomes consist of a tightly coiled thread of DNA with associated proteins and RNA. The genes are arranged in linear order along the DNA.

Coincidental match: A match that occurs by chance.

"Cold hit" DNA match: When the Combined DNA Index System (CODIS) recognizes a match between an offender's DNA profile and a forensic DNA profile, it is referred to as a "cold hit."

Combined DNA Index System (CODIS): CODIS is a collection of databases of DNA profiles administered by the FBI. CODIS contains DNA profiles obtained from evidence samples from solved and unsolved crimes, known individuals convicted of particular crimes, missing persons and relatives of missing persons, unidentified human remains, and in some jurisdictions, individuals arrested for particular crimes. The three levels of CODIS: the Local DNA Index System (LDIS), used by individual laboratories; the State DNA Index System (SDIS), the state's DNA database containing profiles from the LDIS laboratories; and the National DNA Index System (NDIS), managed at the national level by the FBI and containing all of the DNA profiles that have been uploaded from participating states.

Combined probability of exclusion (CPE): A statistic produced by multiplying the probabilities of inclusion from each location (locus) and subtracting the product from 1 (i.e., 1 – CPI). CPE is also defined as the percentage of the population that can be excluded from a mixed DNA profile.

Combined probability of inclusion (CPI): A statistic produced by multiplying the probabilities of inclusion from each location (locus). Also defined as the percentage of the population that can be included in a mixed DNA profile.

Composite profile: A DNA profile generated by combining DNA typing results from different locations (loci) obtained from multiple injections of the same amplified DNA sample and/or multiple amplifications of the same DNA extract. When separate extracts from different locations on a given evidentiary item are combined before amplification, the resultant DNA profile is not considered a composite profile.

Confirmatory test: Testing used to confirm the presence of a body fluid, such as blood or semen.

Contamination: The unintentional introduction of exogenous DNA into a DNA sample or into a polymerase chain reaction (PCR).

Counting method: A statistical approach for estimating the genotype frequency, and thus the potential probative value, of a haplotype DNA testing result generated as a result of mitochondrial DNA (mtDNA) testing or Y-STR DNA testing. This approach involves actually counting the number of times the observed haplotype profile has been observed in the population database(s) being searched.

Critical equipment or instruments: Instruments or equipment requiring calibration or a performance check before use and periodically thereafter.

Critical reagents: Substances used for testing or chemical reactions that have been determined, by empirical studies or routine practice, to require testing on established samples before use on evidentiary or casework reference samples.

Cytoplasm: The viscid, semi-fluid matter contained within the plasma membrane of a cell, excluding the nucleus.

Deconvolute/deconvolution: Separation of the contributors to a mixed DNA profile into major and/or minor contributor profiles. Deconvolution is typically based on quantitative peak height information and may depend on underlying assumptions (e.g., whether the sample has been deemed an intimate sample).

Degradation profile: A DNA typing profile in which higher allele heights may be observed in shorter fragments of DNA. A classic degradation pattern can be said to mimic a ski slope, whereby the peak height diminishes as the DNA fragments increase in length, from left to right on the electropherogram.

Degraded DNA samples: Samples of DNA that have been fragmented, or broken down, by chemical or physical means.

Deoxyribonucleic acid (DNA): Often referred to as the "blueprint of life," DNA is the genetic material present in the nucleus of cells that is inherited, half of which originates from each biological parent. DNA is a chemical substance, contained in cells, that determines a person's individual characteristics. An individual's STR DNA profile is unique, except in cases of identical twins.

Developmental validation: The acquisition of test data, and the determination of conditions and limitations of a new or novel DNA methodology, for use on forensic and/or casework reference samples.

Differential DNA amplification: The selection of one target region or locus over another during the polymerase chain reaction (PCR). Differential amplification can also arise between two alleles within a single locus if one of the alleles has a mutation within a PCR primer binding site, causing the allele with the mutation to be copied less efficiently because of the primer template mismatch.

Differential DNA degradation: A DNA typing result in which contributors to a DNA mixture are subject to different levels of degradation (e.g., due to time of deposition), thereby impacting the mixture ratios across the entire profile.

Differential DNA extraction: A procedure in which sperm cells are separated, or extracted, from all other cells in a sample.

Diploid: A cell (or organism) containing two complete sets of chromosomes. The pair of chromosomes is homologous.

Distinguishable DNA mixture: A DNA mixture in which relative peak height ratios allow deconvolution of, or separation into, the profiles of major and minor contributors.

DNA dragnet: Process of obtaining DNA samples from multiple members of a specific geographical area for comparison with evidence samples to attempt to determine the identity of the perpetrator of a crime. DNA dragnets are often conducted without any specific suspicion of a particular individual's guilt.

DNA extraction: Process where DNA is removed, or isolated, from other present cellular material in an evidence sample in order to conduct DNA type testing.

DNA marker: Refers to a specific chromosomal location that is analyzed in a forensic DNA laboratory. The term *DNA marker* — rather than "gene" — is typically used in forensics because the areas of DNA (loci) being tested, with the exception of amelogenin, do not code for a specific protein.

DNA match: The generation of the same alleles at a locus or loci in an evidence sample and a reference sample. A DNA match will typically refer to both the evidence/crime scene DNA profile and the sample from a known individual having the same DNA typing results at all loci for which results were obtained.

DNA profile: The genetic makeup of an individual at defined locations (loci) in the DNA. A nuclear DNA (nDNA) profile typically consists of one or two alleles at a minimum of 13 STR loci plus the amelogenin locus. A mitochondrial DNA (mtDNA) profile is described in relation to the revised Cambridge Reference Sequence.

DNA sample: An evidentiary sample or a sample from a known source/reference submitted to a laboratory for DNA testing.

DNA sequences: Specific combinations of four bases of the DNA molecule: adenine (A), cytosine (C), guanine (G) and thymine (T).

Electropherogram: The graphic representation of the separation of molecules by electrophoresis or other means of separation.

Electrophoresis: A method of separating large molecules (such as DNA fragments) from a mixture of similar molecules. An electric current is passed through a medium at a different rate, depending on its electrical charge and size. Separation of DNA markers is based on these differences.

Enzyme-linked immunosorbant assay (ELISA): Common serological test for the presence of a particular body fluid using corresponding antigens or antibodies. ELISA tests are rapid immunochemical analyses that involve the use of antigens or antibodies and an enzyme (a protein that catalyzes a biochemical reaction). ELISA tests are typically highly sensitive and specific.

Epithelial cells: Cells from the outer surface of the skin (epithelium) or a body cavity.

Equipment calibration: A test performed on equipment or instruments that perform a particular operation or measurement to ensure accuracy of results.

Evidentiary sample: For the purposes of DNA testing, a biological sample recovered from a crime scene or collected from persons or objects associated with a crime.

Exclusion/exclusionary result: A conclusion that eliminates an individual as a potential contributor of DNA obtained from an evidentiary item, based on the comparison of known and questioned DNA profiles (or multiple questioned DNA profiles with each other). An exclusionary DNA test result indicates that an individual is excluded as the source of the DNA evidence. In a criminal case, however, "exclusion" does not necessarily equate to "innocence." An exclusion results when one or more loci from the DNA profile of a known individual are not present in the questioned DNA profile generated from an evidence sample.

Exculpatory peaks/exculpatory values: A conclusion that excludes a suspect on the basis of a DNA typing analysis, depending on the specific threshold set by each laboratory; even a one-allele difference in a full, single-source DNA profile can exclude a suspect as a possible perpetrator.

Expectation bias: Having a strong belief or mindset toward a particular outcome.

Forensic unknown: A DNA profile, obtained from a crime scene evidence sample, that does not match the DNA profile of a known individual.

Four bases of a DNA molecule: Adenine (A), cytosine (C), guanine (G) and thymine (T).

FST (Fst, F*st*) value: A statistical value that measures the amount of variance in allele frequency in a sampled population relative to the maximum possible amount of variance in the population as a whole. The FST value may also be considered to be the proportion of the diversity in a sampled population that is due to allele frequency differences among populations. More simply, the FST is used to determine whether the variances between two populations are significantly different. Typically, the lower the FST value, the better, and the more populations included (i.e., the more global coverage), the more the FST value stabilizes. However, there will always be population groups that are outliers.

Gamete: In humans, gametes are sperm cells and egg cells. Gamete cells are haploid and combine during fertilization to form a new, diploid organism.

Gene pool: A population of interbreeding individuals who share a common set of genes and genetic markers; the system of Mendelian genetics is used for classification.

Genotype: The genetic constitution of an organism, as distinguished from its physical appearance (its phenotype). The designation of two alleles at a particular locus is a genotype.

Haplogroup(s): A group of similar haplotypes that share a common ancestor with a single nucleotide polymorphism mutation.

Haplotype(s): The term for denoting the collective genotype of a number of closely linked loci on a chromosome that are inherited together or the sequence of the control regions of mitochondrial DNA that pass from a mother to her offspring unchanged.

Haploid: A cell (or organism) containing only one complete set of chromosomes. In humans, sperm cells and egg cells are haploid.

Hardy-Weinberg equilibrium: In a large random intrabreeding population not subjected to excessive selection or mutation, the gene and genotype frequencies will remain constant over time. The sum of $p^2 + 2pq + q^2$ applies at equilibrium for a single allele pair, where p is the frequency of the allele A, q is the frequency of a, p^2 is the frequency of genotype AA, q^2 is the frequency of aa, and $2pq$ is the frequency of Aa.

Heteroplasmic/heteroplasmy: The presence of more than one mitochondrial DNA type within a single individual.

Heterozygous/heterozygote: An individual having different alleles at a particular locus; usually manifested as two distinct peaks for a locus in an electropherogram. If two alleles are different at one locus, the person is heterozygous at that genetic location.

Homologous: Having similar characteristics and structure. Diploid cells contain a set of homologous chromosomes.

Homozygous/homozygote: A homozygote is an individual having the same (or indistinguishable) alleles at a particular locus, manifested as a single peak for a locus in an electropherogram. If two alleles at a locus are indistinguishable, the person is homozygous at that genetic location.

Hypervariable: An area on the DNA that can have many different alleles in differing sequences.

Inclusion: A conclusion for which an individual cannot be excluded as a potential contributor of DNA obtained from an evidentiary item, based on the comparison of known and questioned DNA profiles (or multiple questioned DNA profiles with each other). The inability to exclude an individual as a possible source of a biological sample occurs when all the DNA typing at a specific location in the DNA profile of a known individual is the same typing as in the DNA profile from an evidence sample.

Inconclusive or uninterpretable results: Interpretations or conclusions for which the DNA typing results are insufficient, as defined by the laboratory, for comparison purposes. A situation in which no conclusion can be reached regarding testing performed can be due to a number of situations, including no results obtained, uninterpretable results obtained, or no exemplar or standard being available for testing.

Indistinguishable DNA mixture: A DNA mixture in which relative peak height ratios are insufficient to attribute alleles to individual contributor(s).

Internal validation: Evaluation of the methods of DNA analysis used in a specific laboratory to ensure accurate measurements, equipment calibration, and adherence to standard protocols.

Intimate sample: Definitions of an intimate sample can vary, but the term most commonly refers to a biological sample or swab recovered directly from the interior or exterior of the body of an individual, for example, from the vaginal, perianal or buttocks area, or breast. An intimate sample is generally expected to contain DNA from the person from whom the sample was collected.

Known sample: Biological material, for which the donor's identity is established, that is used for comparison purposes.

Likelihood ratio (LR): The ratio of two probabilities of the same event under different hypotheses. In DNA testing, typically the numerator contains the prosecutor's hypothesis and the denominator the defense's hypothesis. This is often expressed as the ratio between the likelihood that a given profile came from a particular individual and the likelihood that it came from a random unrelated person.

Linkage equilibrium: When two or more genetic loci appear to segregate randomly in a given population. The genotypes for each locus appear randomly with respect to each other.

Local DNA Index System (LDIS): The local DNA index system of the Combined DNA Index System (CODIS) is the entry point for casework profiles being uploaded into the DNA databank. Profiles that meet the criteria for entry into LDIS can then be submitted to the State CODIS Administrator for consideration.

Locus (pl. loci): The specific physical location(s) of gene(s) on a chromosome.

Low copy number (LCN) DNA testing: Typically refers to either the examination of less than 100 picograms (0.1 nanogram) of input/template/ sample DNA, or the analysis of any results below the stochastic threshold values used for normal interpretation. Also called *low-level* or *low-quality template DNA.*

Major contributor(s): The individual(s) who account for the major portion of DNA in a mixed biological sample.

Masked allele: An allele of a minor contributor that may not be readily distinguishable from the alleles of the major contributor or an artifact.

Match: Genetic DNA profiles are said to "match" when they have the same allele designations at every location on their corresponding chromosomes.

Mini-STRs: Reduced-size amplicons for short tandem repeat (STR) typing. They are created by designing polymerase chain reaction (PCR) primers that bind closer to the targeted repeat unit. This type of DNA typing is used when typing degraded DNA samples.

Minor contributor(s): The individual(s) who account for the lesser portion of DNA in a mixed biological sample.

Mitochondria: Structures found outside the nucleus of a cell whose function is to produce energy. The DNA in mitochondria is genetically distinct from the DNA in the nucleus of a cell. Mitochondrion is the singular of mitochondria.

Mitochondrial DNA (mtDNA): The DNA found in the many mitochondria in each cell of a human body, except for red blood cells. The sequencing of mtDNA can link individuals descended from a common female ancestor.

Mixture: A DNA typing result originating from two or more individuals.

Mixture ratio: The relative ratio of the DNA contributions of multiple individuals to a mixed DNA typing result, using quantitative peak height information. A mixture ratio may also be expressed as a percentage.

Modified random match probability (MRMP) statistic: Not typically used for mixed DNA samples, even when major contributor(s) can be isolated and tested separately; other contributing DNA profiles in the sample may not be obtainable using this method.

Molecular Xeroxing: Refers to polymerase chain reaction, an enzymatic process in which specific regions of the DNA strand are replicated over and over again.

National DNA Index System (NDIS): Commonly referred to as the national DNA databank. Authorized by the DNA Identification Act of 1994, NDIS is administered by the FBI. NDIS compares evidence DNA profiles with DNA profiles collected from known convicted offenders and, in some states, with arrestees as well as with other evidence profiles that have been deemed acceptable for upload to NDIS. When DNA profiles are uploaded to NDIS, they are searched against all casework, convicted offender and arrestee sample profiles submitted by all participating states.

No results obtained: No allelic peaks are detected above the analytical threshold values previously established by the testing laboratory.

Noise: Background signal detected by a capillary electrophoresis or other data collection instrument.

Nuclear DNA (nDNA): The DNA found in the nucleus of a cell. Nuclear DNA testing includes both autosomal STR DNA typing and Y-STR DNA typing.

Nucleated cells: Cells having a nucleus.

Nucleus (pl. nuclei): The structure in a cell that contains most of the DNA.

Oligozoospermic male/oligospermic male: A male who produces less than 20 million spermatozoa per milliliter of ejaculate. Oligospermia has many possible causes, and the effects of these causes may be temporary or permanent.

p30: A protein, also called *prostate-specific antigen (PSA)*, found in high quantities in semen/seminal fluid that is male specific. In forensics, detection of p30/PSA can be used to confirm the presence of semen/seminal fluid. Produced by cells of the prostate gland, p30, when at elevated levels, can indicate potential prostate cancer.

Partial (incomplete) DNA profile: A DNA profile for which typing results are not obtained at all tested loci, typically due to DNA degradation, inhibition of amplification and/or low-quantity template DNA.

Peak height/peak height ratio (PHR): The relative ratio of two alleles at a given locus, as determined by dividing the peak height of an allele with a lower relative fluorescent unit (RFU) value by the peak height of an allele with a higher RFU value, and then multiplying the value by 100 to express the PHR as a percentage; used to indicate which alleles may be heterozygous pairs and also in mixture deconvolution.

Phylogeographic/phylogeography: The study of historical processes that are believed to be responsible for the contemporary geographic distributions of individuals.

Polymarker/DQ alpha (PM/DQ_): Early polymerase chain reaction (PCR) testing procedure, often referred to as dot blot testing. PM/DQ_ DNA markers are used to discriminate between individuals, but they are less discriminating than autosomal STR typing.

Polymerase chain reaction (PCR): A process used in DNA testing in which one or more specific small regions of the DNA are copied, using a DNA polymerase enzyme so that a sufficient amount of DNA is generated for analysis.

Polymorphic/polymorphism: Having multiple alleles that match at a specific gene within a population.

Population genetics: The study of the distribution of genes in populations and how the frequencies of genes and genotypes are maintained or changed.

Presumptive tests: A screening test used to indicate the possible presence of the named body fluid.

Primer binding site: Site at which the primer binds to the DNA strand.

Primers: Short DNA sequences which precede the region to be copied that are added to the polymerase chain reaction.

Proficiency: The demonstration of technical skills and knowledge necessary to perform forensic DNA analysis successfully.

Proficiency test(s): Written, oral and/or practical test, or series of tests, designed to establish that an individual has demonstrated achievement of technical skills and met minimum standards of knowledge necessary to perform forensic DNA analysis.

Prostate-specific antigen (PSA): A protein, also called p30, found in high quantities in semen/ seminal fluid that is male specific. In forensics, detection of PSA/p30 can be used to confirm the presence of semen/seminal fluid. PSA/p30 is produced by cells of the prostate gland; elevated levels in the blood indicate potential prostate cancer.

qPCR data: Quantitative data of a polymerase chain reaction (PCR), also called real-time PCR.

Quality assurance: The overall program used to ensure the accuracy and reliability of the testing performed and the results reported by a laboratory.

Quality assurance review (QAR): A program conducted by a laboratory to ensure accuracy and reliability of the tests performed.

Quality control: Each process or step used by the laboratory to ensure the accuracy and reliability of the testing performed and the results reported by a laboratory. Collectively, the quality control steps, or measures, comprise the quality assurance program

Quantitation slot blots: Method used to determine the quantity of "x" in a given sample. In this context, it refers to the quantity of DNA in a sample and is usually reported as nanograms per microliter (ng/μl).

Questioned sample: A biological sample recovered from a crime scene or collected from persons or objects associated with a crime.

Random match/man probability (RMP): The probability of randomly selecting an unrelated individual from the population whose DNA is a potential contributor to an evidentiary DNA profile.

Rapid Stain Identification of Human Blood (RSI-Blood): Test used to detect the presence of human blood.

Raw data: Data generated from DNA testing before analysis and interpretation.

Reagents: Chemicals and test substances that are added to a system to bring about a reaction or to see whether a reaction occurs.

Real allele peaks versus stutter: Peak on an electropherogram that is representative of a true DNA allele versus a peak or spike that is an artifact or anomaly that, although it may appear as a true peak, it does not, in fact, represent actual DNA.

Reference sample: Biological material for which the identity of the donor is established and used for comparison purposes.

Relative fluorescence unit (RFU): A unit of measurement used in electrophoresis methods involving fluorescence detection. Fluorescence is detected on the charge coupled device (CCD) array as the labeled fragments, separated in the capillary by electrophoresis and excited by the laser, pass the detection window. The software interprets the results, calculating the size and relative quantity of the fragments from the fluorescence intensity at each data point.

Resolvable DNA mixture: Mixture of two or more individuals' DNA detected from an item of evidence in which the ratios, and therefore potentially the alleles, of major and minor contributors can be deduced due to the proportion of one versus the other.

Restriction fragment length polymorphism (RFLP): Variation in the length of a stretch of DNA.

Revised Cambridge reference sequence (rCRS): The rCRS is a modified version of the original Cambridge Reference Sequence (a "master template" of the HVR-1 region of mitochondrial DNA); (GenBank #J01415.o gi:337188), from Anderson, S., A.T. Bankier, B.G. Barrell, et al., "Sequence and Organization of the Human Mitochondrial Genome," 290 Nature 457–465 (1981).

RNA: The abbreviation for ribonucleic acid, a nucleic acid molecule similar to DNA. RNA contains ribose sugar within the structure, rather than the deoxyribose in DNA.

Scientific Working Group on DNA Analysis Methods (SWGDAM): A group of approximately 50 scientists representing federal, state and local forensic DNA laboratories in the United States and Canada. SWGDAM generates and promulgates interpretation guidelines for use by forensic DNA testing laboratories.

Semenogelin: A protein found in high quantities in human semen, produced by seminal vesicles.

Seminal acid phosphatase (SAP): A chemical substance found in high quantities in semen/seminal fluid. The SAP test is a presumptive color test that is used to screen for the presence of semen/seminal fluid by detecting acid phosphatase content. Also referred to as the acid phosphatase (AP) test.

Serology: In forensics, serology typically refers to the initial examination of items of evidence for the presence of blood, semen and/or other biological materials and/or the recovery of portions of samples for DNA testing. In the general scientific community, serology refers to the study of serums, particularly the properties and immunological (antigen-antibody) reactions of blood serum.

Short tandem repeat (STR) DNA analysis/ typing: A method of DNA analysis that targets regions on the chromosome that contain multiple copies of a short DNA sequence in succession.

Signal-to-noise ratio: An assessment used to establish an analytical threshold to distinguish allelic peaks (signals) from background/instrument noise.

Single nucleotide polymorphisms (SNPs): DNA sequence variations that occur at a single nucleotide (A, T, C or G).

Single-source DNA sample/profile: A DNA profile in which only one individual has contributed biologic material.

Source attribution: A declaration that identifies an individual as the source of an evidentiary profile to a reasonable degree of scientific certainty, based on a single-source or major contributor DNA profile.

Spermatozoa: Sperm cells; a male reproductive cell.

Spermatozoon: A single sperm cell.

State DNA Index System (SDIS): Contains qualifying casework and suspect DNA profiles/records uploaded from local laboratory sites within the state as well as the convicted offender and arrestee samples for the state. SDIS is the state's repository of DNA identification records, under the control of state authorities. Convicted offender and arrestee profiles are entered into CODIS at the SDIS level. The SDIS laboratory serves as the central point of contact for access to the National DNA Index System. The DNA Analysis Unit serves as the SDIS laboratory for the FBI.

Stochastic effects: The observation of peak imbalance and/or allelic drop-out within a given locus resulting from random, disproportionate amplification of alleles in low-quantity template, also called low level DNA, samples.

Stochastic threshold: The peak height value above which it is reasonable to assume that, at a given locus, allelic drop-out of a sister allele has not occurred.

STR DNA analysis: See Short tandem repeat (STR) analysis.

Stutter: A minor peak typically observed one repeat unit smaller than a primary STR allele resulting from strand slippage during amplification.

SWGDAM: See Scientific Working Group on DNA Analysis Methods.

The product rule: The product rule calculates the expected chance of finding a given short tandem repeat (STR) profile within a population by multiplying the frequency of occurrence of the combination of alleles (genotype) found at a single locus, by the frequency of occurrence of the genotype found at the second locus, and by the frequency of occurrence, in turn, of each of the other genotypes at the remaining STR loci.

Theta (Θ) correction: A theta adjustment is a mathematical correction applied to a frequency calculation when both alleles at a locus are the same (known as a homozygous state). It is not applied when alleles are different at a locus (known as a heterozygous state). This correction adjusts the frequency slightly upward to account for the presence of subpopulations in a general population database that might otherwise cause the genotype frequency to be underestimated at that locus.

Touch DNA: DNA that is left behind, typically from skin (epithelial) cells, when a person touches or otherwise comes into contact with an item or person.

True match probability: A formula for determining the uniqueness of a DNA profile in a large population, assuming that it is more or less 10-fold, based on guidelines established by the National Research Council (NRC II, 1996) in The Evaluation of Forensic DNA Evidence (National Academies of Science). A confidence interval of 95-99 percent means that an individual is expected to have a unique DNA profile in a population of 300 million, i.e., a true match probability of 30 billion to 1. This probability is then compared with the random match probability (RMP) of the evidence sample matching the reference sample.

Validation/validation studies of DNA analyses: The process of extensive and rigorous evaluation of DNA methods before acceptance for routine use.

Y-STR DNA analysis/profile/typing: DNA typing in which short tandem repeats (STR) are analyzed on the Y, or male, chromosome. This is one variant on the pair of chromosomes (also called sex chromosomes) in a DNA sequence that define the sex of an individual.

Chapter 2

DNA TESTING IN CRIMINAL JUSTICE: BACKGROUND, CURRENT LAW, GRANTS, AND ISSUES*

Nathan James

SUMMARY

Deoxyribonucleic acid, or DNA, is the fundamental building block for an individual's entire genetic makeup. DNA is a powerful tool for law enforcement investigations because each person's DNA is different from that of every other individual (except for identical twins). DNA can be extracted from a number of sources, such as hair, bone, teeth, saliva, and blood. As early as the 1980s, states began enacting laws that required collecting DNA samples from offenders convicted of certain sexual and other violent crimes. The samples were then analyzed and their profiles entered into state databases. Meanwhile, the Federal Bureau of Investigation (FBI) Laboratory convened a working group of federal, state, and local forensic scientists to establish guidelines for the use of forensic DNA analysis in laboratories. The group proposed guidelines that are the basis of current national quality assurance standards, and it urged the creation of a national DNA database. The criminal justice community began to utilize DNA analyses more often in criminal investigations and trials, and in 1994 Congress enacted legislation to authorize the creation of a national DNA database.

Federal law (42 U.S.C §14132(a)) authorizes the FBI to operate and maintain a national DNA database where DNA profiles generated from samples collected from people under applicable legal authority and samples collected at crime scenes can be compared to generate leads in criminal investigations. Statutory provisions also authorize the collection of DNA samples from federal offenders and arrestees, District of Columbia offenders, and military offenders. State laws dictate which convicted offenders, and sometimes people arrested for crimes, will have profiles entered into state DNA databases, while federal law dictates the scope of the national database. Increasing awareness of the power of DNA to solve crimes has resulted in increased demand for

* This is an edited, reformatted and augmented version of a Congressional Research Service publication, CRS Report for Congress R41800, prepared for Members and Committees of Congress, from www.crs.gov, dated December 7, 2011.

DNA analysis, which has resulted in a backlog of casework. Some jurisdictions have started to use their DNA databases for familial searching, which involves using offender profiles to identify relatives who might be perpetrators of crimes. In addition to solving crimes, DNA analysis can help exonerate people incarcerated for crimes they did not commit.

Congress has authorized several grant programs to provide assistance to state and local governments for forensic sciences. Many of the programs focus on providing state and local governments with funding to reduce the backlog of forensic and convicted offender DNA samples waiting to be processed and entered into the national database. Since FY2006, Congress has appropriated approximately $785 million for backlog reduction and laboratory capacity enhancement programs. However, other grant programs provide funding for related purposes, such as offsetting the cost of providing post-conviction DNA testing.

In the 1990s and the early part of the last decade, most of the debate in Congress focused on the scope of DNA databases, reducing the backlog of DNA casework, and providing access to post-conviction DNA testing. Most of the debate about the scope of DNA databases faded away with the enactment of the Violence Against Women and Department of Justice Reauthorization Act of 2005 (P.L. 109-162), which expanded federal collection statutes to include anyone arrested or detained under the authority of the United States. The act also expanded the scope of the national database to include DNA profiles of individuals arrested for state crimes. However, concerns about the backlog of DNA casework and access to post-conviction testing have persisted. In addition, new issues related to the use of DNA in criminal justice have emerged, including whether (1) DNA databases should be used to conduct familial searches, (2) sexual assault evidence collection kits (i.e., "rape kits") should be standardized, and (3) there should be national accreditation standards for forensic laboratories.

INTRODUCTION

Deoxyribonucleic acid, or DNA, is the fundamental building block for an individual's entire genetic makeup. DNA is a powerful tool for law enforcement investigations because each person's DNA is different from that of every other individual (except for identical twins). By analyzing selected DNA sequences (called loci), a crime laboratory can develop a profile to be used in identifying a suspect.[1]

DNA can be extracted from a number of sources, such as hair, bone, teeth, saliva, and blood. Because the human body contains so many copies of DNA, even a minuscule amount of bodily fluid or tissue can yield useful information. Obtaining a DNA sample is not necessarily invasive; it can be as simple as a swab of the inside of the mouth to obtain saliva.

State and federal DNA databases have proved instrumental in solving crimes, reducing the risk of convicting the wrong person, and establishing the innocence of those wrongly convicted. DNA evidence is used to solve crimes in two ways:

- In cases where a suspect is known, a sample of that person's DNA can be compared to biological evidence found at a crime scene. The results of this comparison may then help establish whether the suspect was at the crime scene or whether he/she committed the crime.[2]
- In cases where a suspect is not known, biological evidence from the crime scene can be analyzed and compared to offender profiles contained in existing DNA databases

to assist in identifying the perpetrator.[3] Through the use of DNA databases, biological evidence found at one crime scene can also be connected to other crime scenes, linking them to the same perpetrator or perpetrators.[4]

Early congressional debate on DNA testing centered on whose profiles should be included in DNA databases. In recent years, concerns have been raised about the backlog of DNA casework and access to post-conviction testing. In the future, policymakers may also focus their attention on other issues related to the use of DNA in criminal justice, including whether (1) DNA databases should be used to conduct familial searches, (2) sexual assault evidence collection kits (i.e., "rape kits") should be standardized, and (3) there should be national accreditation standards for forensic laboratories.

This report provides an overview of how DNA is used to investigate crimes and help protect the innocent.[5] It also reviews current statutory law on collecting DNA samples, sharing DNA profiles generated from those samples, and providing access to post-conviction DNA testing. The report also includes a summary of grant programs authorized by Congress to assist state and local governments with reducing DNA backlogs, provide post-conviction DNA testing, and promote new technology in the field. It also reviews select issues Congress might consider should it legislate or conduct oversight in this area.

BACKGROUND

Federal law authorizes the Federal Bureau of Investigation (FBI) to operate and maintain a national DNA database where DNA profiles generated from samples collected from people under applicable legal authority and samples collected at crime scenes can be compared to generate leads in criminal investigations. Statutory provisions also authorize the collection of DNA samples from federal offenders and arrestees, District of Columbia offenders, and military offenders. State law dictates which convicted offenders and persons arrested for crimes will have profiles entered into state DNA databases, but federal law dictates which profiles entered into state databases can be uploaded into the national DNA database.

Increasing awareness of the power of DNA testing to solve crimes has increased demand for DNA analysis, which has resulted in a backlog of casework. The demonstrated ability of DNA testing to generate leads in criminal investigations has led some jurisdictions to use their DNA databases for familial searching, which involves using offender profiles to identify relatives who might be perpetrators of crimes. In addition to solving crimes, DNA analysis can also help exonerate people incarcerated for crimes they did not commit.

The National DNA Index System (NDIS) and the Combined DNA Index System (CODIS)

As early as the 1980s, states began enacting laws that required DNA samples from those offenders convicted of certain sexual offenses and other violent crimes. The samples were then analyzed and their profiles entered into state databases. Meanwhile, the FBI Laboratory convened a working group of federal, state, and local forensic scientists to establish guidelines for the use of forensic DNA analysis in laboratories. The group proposed guidelines that are the basis of current national quality assurance standards, and it urged the creation of a national DNA database.[6] In 1994, Congress authorized the FBI to establish and oversee the National DNA Index System (NDIS). When the NDIS launched in 1998, only nine states participated.[7] Currently, laboratories in all 50 states, the District of Columbia, the federal government, Puerto Rico, and the U.S. Army Criminal Investigation Laboratory participate in the NDIS.[8] The NDIS contains the DNA profiles provided by federal, state, and participating local crime laboratories.[9] As of January 2011, there are 198 laboratories in the United States participating in the NDIS.[10]

DNA profiles generated by laboratories operated by local law enforcement agencies are stored in Local DNA Index Systems (LDIS). DNA profiles generated by state laboratories, along with authorized profiles stored in participating LDIS, are uploaded into State DNA Index Systems (SDIS). Each state has its own laws specifying which profiles can be included in the SDIS. DNA profiles generated by federal laboratories, along with authorized DNA profiles in participating SDIS, are uploaded into the NDIS.[11] Federal law dictates which DNA profiles can be stored in the NDIS (see below). The NDIS allows participating laboratories to compare DNA on the national level while the SDIS allows each state to compare DNA profiles stored at the state level. Federal, state, and local laboratories upload and compare DNA profiles using the Combined DNA Index System (CODIS) software produced and distributed by the FBI.[12]

CODIS searches three indexes (convicted offenders, arrestee, and forensic) to generate investigative leads. The convicted offender index contains DNA profiles developed from samples collected from convicted offenders; the arrestee index contains DNA profiles developed from samples collected from arrested but *not yet convicted* individuals; and the forensic index contains DNA profiles developed from samples collected at crime scenes. CODIS searches across these indexes to look for potential matches (also referred to as “hits”).[13] Matches can occur between either the convicted offender or arrestee indexes and the forensic index, thereby providing law enforcement with the identity of one or more suspects.[14] Also, matches can occur between DNA profiles in the forensic index, thereby linking crime scenes to each other and identifying serial offenders.[15] Matches between multiple samples in the forensic index can allow law enforcement agencies in different jurisdictions to coordinate their efforts and share leads. No names or other personal identifiers for offender and arrestee DNA profiles are stored in the NDIS, so when a match is made in CODIS, the laboratories that submitted the DNA profiles to the NDIS are notified of the match and they contact each other to verify the match and coordinate their efforts.[16]

DNA Profiles

DNA profiles entered into CODIS are based on 13 core short tandem repeat (STR) loci selected by the FBI.[17] The FBI chose 13 core loci to ensure that all forensic laboratories can establish uniform DNA databases and share forensic information through the NDIS.[18] Currently, the 13 STR loci used by the FBI are non-coding, meaning that they have not been shown to be associated with human attributes such as height, eye or skin color, or susceptibility to a particular disease.[19] Each locus has two alleles, and it is these 13 pairs of alleles that are compared to match samples in the forensic index with profiles in either the offender or arrestee indexes. The 13 core loci chosen by the FBI provide a high level of discriminatory power. The probability that two unrelated individuals would share all 13 pairs of alleles is estimated to be one in several hundred billion.[20] Two random Americans will, on average, share two or three alleles.[21]

It is important to ensure the quality of the DNA profiles entered into the NDIS. If the profiles are not accurate, they are of little use for making matches between forensic and offender or arrestee profiles. The FBI helps ensure the quality of DNA profiles included in the NDIS by signing memorandums of understanding with state laboratories whereby the laboratory agrees to adhere to the FBI's Quality Assurance Standards (QAS, see below).[22] Laboratories submitting DNA profiles to the NDIS must be accredited and audited annually.[23] Annual audits can be conducted by either an internal or external auditor, but laboratories must be audited by an external agency at least once every two years.[24] Laboratories that do not pass the annual audit can be prevented from entering DNA profiles in CODIS.[25] Currently, most labs in the United States are audited by the American Society of Crime Laboratory Directors and its Laboratory Accreditation Board (ASCLD/LAB) and Forensic Quality Services (FQS). In addition, DNA analysts must undergo semiannual proficiency testing.[26] DNA analysts who do not pass their semiannual proficiency tests are not allowed to enter profiles into CODIS.[27] Laboratories are also required to conduct two reviews of all DNA profiles before they are entered into CODIS.[28]

Currently, as prescribed by federal law (see below), only public laboratories that comply with the QAS can submit DNA profiles to the NDIS. However, public laboratories are allowed to outsource casework to private laboratories. Data from the Bureau of Justice Statistics (BJS) of the Department of Justice show that public laboratories are outsourcing more work to private laboratories. BJS reports that 28% of public laboratories reported that they outsourced DNA casework in 2005, up from 19% in 2002.[29] All private laboratories that conduct DNA testing for public laboratories must be accredited, be audited annually, and adhere to the requirements of the QAS.[30] Public laboratories are required to conduct an initial site visit to each private laboratory it contracts with to conduct DNA analyses.[31] If the public laboratory signs a contract with a private laboratory that is longer than one year, the public laboratory must conduct an annual site visit.[32] Public laboratories are also required to review all outsourced DNA profiles generated by private laboratories.[33] The review by the public laboratory is in addition to the two reviews private laboratories are required to conduct per the QAS.

An offender or arrestee profile in a DNA database consists of 26 numbers representing each of the two alleles for the 13 STR loci, an agency identification number, a sample identification number, and an identifier for the analyst that entered the information.[34] However, most jurisdictions retain the DNA sample used to generate the profile placed in CODIS.[35] DNA samples are usually retained for quality assurance purposes, such as confirming a hit made using the NDIS, and it allows jurisdictions to retest the sample if new technology is developed in the future.[36] Privacy advocates are concerned that stored DNA samples include a wealth of genetic information that could be misused.[37] States and the federal government have sought to prevent the unauthorized use of DNA samples. Some states have criminal penalties in place for individuals who misuse DNA samples collected for law enforcement purposes.[38] Under current law, anyone who misuses a DNA sample collected under federal authority is subject to a fine of up to $250,000, or imprisonment for up to one year.[39]

The number of offender profiles included in the NDIS has increased as Congress has allowed states to include DNA profiles from a broader range of convicted offenders and persons arrested for certain crimes to be included in the database. States have also amended their DNA collection laws to reflect this expanded authority. As shown in *Table 1*, over 8 million new convicted offender and arrestee profiles have been added to NDIS over the past decade. This is in part because more forensic profiles have been added to the NDIS as state and local governments have started to work their way through backlogs of forensic casework. As also shown, over 300,000 new forensic profiles have been included in the NDIS since 2000. Data presented show that the expansion of the NDIS has mostly been driven by laboratories processing and entering offender profiles. The additional offender and forensic profiles have increased the number of investigate leads generated by DNA databases. Since 2000, the NDIS has aided in the investigation of nearly 120,000 crimes. Data in Table 1 indicate that most matches occur between forensic and offender profiles stored in SDIS rather than the NDIS.

Table 1. Number of Profiles in the NDIS, Investigations Aided, and Hits Generated by Searches of NDIS

Year	Convicted Offender and Arrestee Profiles	Forensic Profiles	Investigations Aided	Forensic Hits	National Offender Hits (NDIS)	State Offender Hits (SDIS)	Total Offender Hits
2000	460,365	22,484	1,573	507	26	705	731
2001	750,929	27,897	3,635	1,031	167	2,204	2,371
2002	1,247,163	46,177	6,670	1,832	638	4,394	5,032
2003	1,493,536	70,931	11,220	3,004	1,151	7,118	8,269
2004	2,038,514	93,956	20,788	5,147	1,864	11,991	13,855
2005	2,826,505	126,315	30,455	7,071	2,855	18,664	21,519
2006	3,977,433	160,582	43,156	9,529	4,276	28,163	32,439
2007	5,372,773	203,401	62,059	11,750	6,508	43,305	49,813
2008	6,539,919	248,943	80,948	14,122	8,479	58,304	66,783
2009	7,688,286	298,369	101,766	17,636	10,969	75,186	86,155
2010[a]	8,646,417	328,067	119,764	19,940	12,791	89,598	102,389

Source: U.S. Department of Justice, Federal Bureau of Investigation, CODIS Brochure.

Notes: Amounts shown are cumulative starting with 2000. "Forensic hit" refers to cases where a match is made between two or more forensic profiles in the database. "Offender hit" refers to cases where an offender profile is matched to one or more forensic profile in the database.

[a] Through July 2010.

One limitation of the data in *Table 1* is that they do not describe how the investigations were aided, the outcomes of the investigations, or whether any of the hits solved the alleged crimes.[40] Database hits do not always generate a new investigative lead; investigators, if they have already identified a suspect and they know that the suspect's profile is already in the database, may enter a forensic profile into the database and wait for a hit to be returned before investigating further. In addition, not all hits generated by the DNA databases are probative; just because someone's DNA is found at a crime scene does not always mean that the person who left the DNA is the perpetrator. Also, it is possible that one forensic or offender hit might lead to several arrests or aid in multiple investigations. The data published by the FBI provide a measure of the output generated by the NDIS, but the "hits" and "investigations aided" metrics are poor indicators of whether DNA databases aided in resolving criminal investigations.[41]

For example, the data provide no indication of whether the hits generated by the NDIS resulted in a conviction or how many investigations resulted in an arrest.

A study of database hits in San Francisco suggests that there is a need for more expansive data collection in order to properly to evaluate the effectiveness of DNA databases.[42] The study measured the outcomes of 198 DNA database hits in cold cases[43] generated by the San Francisco Police Department Forensic Biology Unit between 2001 and 2006. The researchers report that 90% of the cold hits were probative and provided investigators with substantive leads.[44] Probative hits led to judicial resolution (i.e., conviction, guilty plea, or parole revocation) 40% of the time.[45]

Another 28% of the cases involving probative hits were either awaiting jury trial or the investigation was ongoing at the time the article was written. The researchers note that they found that nearly 70% of the probative hits could result in some form of judicial resolution. There were varying rates of success for database hits for different types of offenses. Nearly 9 in 10 probative hits in homicide and burglary cases either reached judicial resolution or could be resolved.

However, judicial resolution or potential resolution was lower for sex offenses (approximately 1 in 2). In nearly half of the cases where a probative hit was made for a sex offense, either the prosecutor (17%) or the victim (31%) declined to move the case forward.[46]

Collection of DNA Samples from State Offenders

Since the late 1990s, state laws have changed to require a greater number of offenders, and in some cases arrestees, to submit a DNA sample for analysis and inclusion in states' databases. When the NDIS was first started in 1998, most states only collected DNA samples from individuals convicted of violent felonies and felony sexual offenses. Over the years, more states passed laws requiring collection of DNA samples from a broader range of offenders. In 1999, only five states collected DNA samples from all convicted felons.[47] By 2004, the number of states that collected DNA samples from all convicted felons increased to 37, and four years later, 47 states collected DNA samples from all convicted felons.[48]

Currently, all states require all convicted felons to submit a DNA sample and 37 states require individuals convicted of misdemeanor sexual offenses to submit a DNA sample (see Appendix).[49]

Moreover, 34 states collect DNA samples from juveniles adjudicated for certain offenses and all states collect DNA samples from offenders on probation.[50] More recently, some states have enacted laws to require individuals arrested (not just convicted) for certain offenses to provide a DNA sample for analysis and inclusion in the NDIS. Currently, 25 states require individuals arrested for murder and sexual offenses to provide a DNA sample.[51]

Of these states, 22 require individuals arrested for burglary to submit a DNA sample.[52] More than half (13) of the 25 states require individuals arrested for all felony offenses to submit a DNA sample.[53]

DNA Backlog

Delays in processing DNA evidence can result in delays in apprehending or prosecuting violent or serial offenders or it can result in wrongfully convicted individuals serving time in prison for crimes they did not commit. In addition, persistent backlogs can result in crime laboratories prioritizing DNA analysis for violent offenses, such as homicide or sexual assault, over other offenses, such as property crimes, or it can result in law enforcement agencies establishing policies stating that biological evidence is not to be collected for minor offenses.[54] Not analyzing or collecting DNA samples for minor offenses could prevent law enforcement from apprehending offenders before they commit more serious crimes. Data indicate that many violent offenders start off with committing property crimes.[55]

Forensic Casework

In a February 2011 report, the National Institute of Justice (NIJ) published estimates of the forensic casework backlogs in state and local laboratories in 2005, 2007, 2008, and 2009 (presented in *Figure 1*).[56]

Different methodologies used to collect the data and survey response rates differed slightly, but the data show a pattern: the backlog of forensic casework continues to increase as the demand for forensic DNA casework continues to outpace the crime laboratory capacity to conduct such analyses.[57]

It is important to remember that data presented in *Figure 1* present a national picture of the forensic DNA backlog; it is likely that some crime laboratories have little or no backlog, while other laboratories have significant backlogs.[58]

Data also show that the backlog of forensic casework is not the result of a glut of old samples awaiting analysis; rather, the forensic casework backlog is mostly driven by increasing demand for DNA analysis in new cases. NIJ reports that the growing demand for DNA analysis is increasing for many reasons, including

- a growing awareness of the potential of DNA evidence to solve cases;
- more DNA samples are being collected from property crimes;
- advances in DNA technology allow tests to be conducted on smaller samples of DNA;
- more DNA testing in old, unsolved cases where the evidence was collected before DNA testing became widespread; and
- post-conviction DNA testing.[59]

Another contributing factor to the forensic casework backlog is the labor-intensive nature of processing forensic casework. NIJ reports that processing forensic evidence is time-consuming because the evidence must be screened to determine if, and what kind of, biological materials are present before DNA testing can begin.[60]

Convicted Offender and Arrestee Samples

In addition to the backlog of forensic casework, there is a backlog in the processing of samples collected from convicted offenders and arrestees. Backlogs of convicted offender and arrestee samples means that there are fewer DNA profiles in CODIS to match forensic profiles to, and this could result in a delay in indentifying suspects in cases where DNA evidence was collected.

Data collected by NIJ show that submission of new DNA samples from convicted offenders and arrestees increased between 2007 and 2009 (see *Figure 2*).[61]

At the same time, the total number of convicted offender and arrestee samples analyzed decreased between 2007 and 2008, but the number of samples analyzed remained at approximately 1 million in both 2008 and 2009.[62] Therefore, the increased backlog of convicted offender and arrestee samples between 2008 and 2009 was the result of states collecting more offender samples, especially from arrestees, while the ability to analyze those samples remained flat.

Data indicate that there are more convicted offender and arrestee DNA samples for laboratories to process compared to forensic samples; NIJ notes, however, that offender and arrestee samples are easier and faster to analyze because they are collected on a standard, consistent medium.[63] The standardized collection method makes it possible to use automated analysis on robotic platforms that can process approximately 96 samples and controls simultaneously.[64]

This suggests that laboratories might be better able to process the backlog of convicted offender and arrestee samples if they increase their technological capacity.

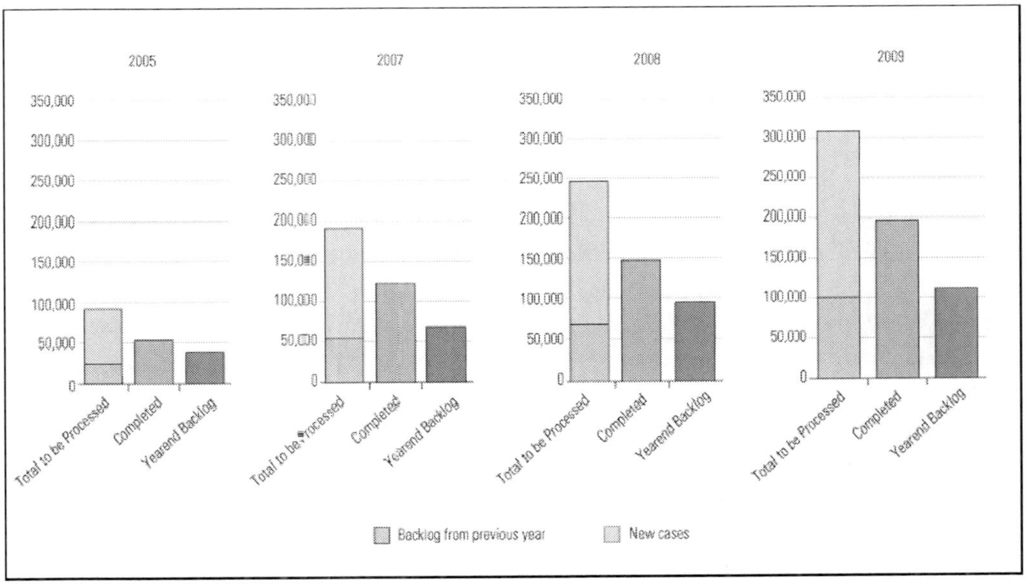

Source: CRS reproduction of a figure from the U.S. Department of Justice, Office of Justice Programs, National Institute of Justice, "Making Sense of DNA Backlogs—Myths vs. Reality," p. 3.

Notes: Backlog data for 2005 were collected as a part of the Bureau of Justice Statistic's 2005 Census of Publicly Funded Forensic Crime Laboratories. Data for 2007 were collected as a part of a NIJ-funded study of DNA backlogs. Data for 2008 and 2009 were reported by applicants for NIJ's FY2009 and FY2010 DNA Backlog Reduction Program grant solicitations. Data are not available for 2006.

Figure 1. DNA Forensic Casework: Supply, Demands, Backlogs.

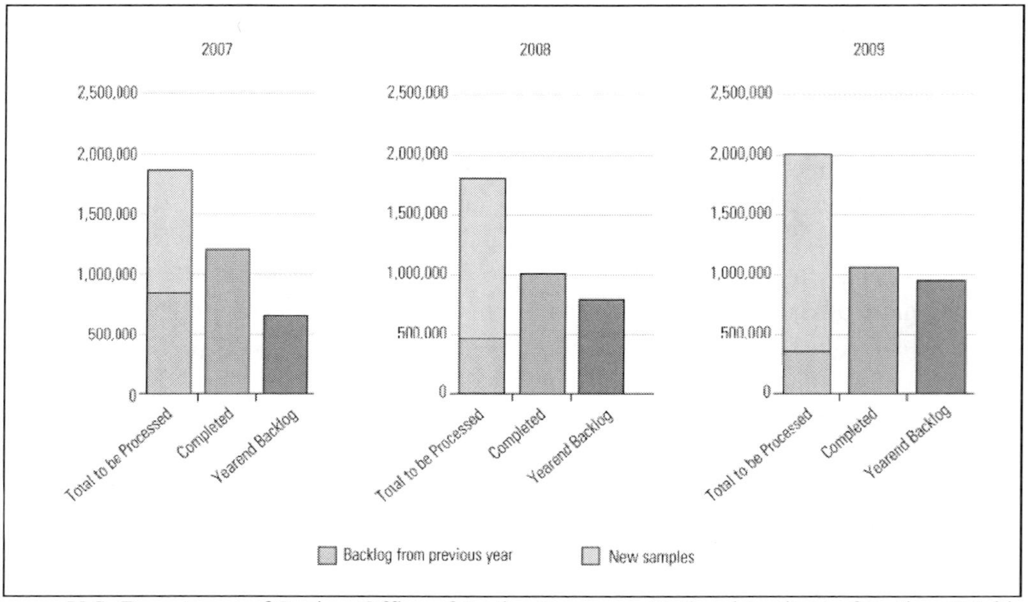

Source: U.S. Department of Justice, Office of Justice Programs, National Institute of Justice, "Making Sense of DNA Backlogs—Myths vs. Reality," p. 7.

Notes: Data for 2007 were collected as a part of a NIJ-funded study of DNA backlogs. Data for 2008 and 2009 were reported by applicants for NIJ's FY2009 and FY2010 DNA Backlog Reduction Program grant solicitations.

Figure 2. Convicted Offender and Arrestee Backlog Trends, 2007-2009.

Evidence in the Possession of Law Enforcement

One limitation to the backlog data discussed above is that they only include samples in the possession of crime laboratories. Samples from evidence still in the possession of law enforcement agencies and not yet transferred to laboratories are not counted as a part of the backlog. While there is no current count of the total amount of unanalyzed evidence in the possession of law enforcement agencies, one group of researchers that surveyed over 2,000 law enforcement agencies in 2007 found that law enforcement agencies had forensic evidence that had not been submitted to a crime laboratory for analysis in 14% of all unsolved homicide cases and 18% of unsolved rape cases.[65] The researchers estimated that nearly 40% of unanalyzed murder and rape cases contained DNA evidence.[66] The results of the survey indicate that there are many reasons why law enforcement agencies chose not to submit evidence for analysis, including that subsequent investigation may have shown that the evidence would not be probative; charges against an alleged perpetrator may have been dropped; or the suspect may have pled guilty.[67] However, data collected by the researchers also suggest that law enforcement agencies may not fully understand the potential value forensic evidence can have in generating leads in cases where they have not identified a suspect. Nearly half of the responding law enforcement agencies reported that they did not submit evidence for analysis because a suspect had not been identified. Also, nearly one in five agencies reported that they did not submit evidence because they felt it would not be useful to the case.68 However, the survey does not reveal how many open cases with unanalyzed evidence would be solved or yield investigative leads if evidence were to be sent to the laboratory.

Processing Time for DNA Analyses

While many policymakers are interested in the size of the backlog of forensic casework and offender and arrestee samples, another important consideration is the time it takes for laboratories to complete new requests for DNA analysis (i.e., turnaround time). Researchers studying the size of the forensic casework backlog in 2007 in publicly funded, accredited DNA laboratories found that 14 of the 145 laboratories (9.7%) responding to their survey reported that they did not have a backlog, meaning that they were able to process all requests for DNA analysis within 30 days of receipt.[69] However, nearly three-quarters (111) of responding laboratories reported completing DNA analysis requests within 119 days (four months) or less. Of the remaining laboratories, 24 (16.6%) reported that turnaround time was more than 180 days (six months) and another 20 (13.8%) laboratories reported turnaround times of 270 days (nine months) or more. The researchers found that the turnaround time for cases of violent crimes was, in general, shorter than the turnaround time for cases of non-violent crimes.[70] The researchers also found that the turnaround time for analysis of offender and arrestee samples tended to be shorter than the turnaround time for forensic casework. Approximately 30% of responding laboratories reported having turnaround times of 30 days or less for offender and arrestee samples.[71] Approximately half of all responding laboratories reported processing offender and arrestee samples within 90 days of receipt. However, one-quarter of laboratories reported turnaround times of more than 270 days (nine months) for these samples.[72]

Sexual Assault Evidence Collection Kits

Recently, the backlog of sexual assault evidence collections kits—also referred to as "rape kits"— has been the focus of a congressional hearing and several pieces of legislation.[73] While there have been several estimates of the backlog in some cities,[74] NIJ reports that currently there is no comprehensive data on the number of unanalyzed sexual assault evidence kits in the United States.[75] NIJ reports that it is currently funding research to better understand why some sexual assault evidence collection kits are not submitted to a crime laboratory for analysis.[76] The backlog of sexual assault evidence collection kits has raised concerns that additional victimizations could have been prevented had the evidence from any given kit been tested and the perpetrator apprehended in a timely manner.[77]

Sexual assault evidence collection kits are collections of tools used by a nurse examiner or another trained professional to collect evidence during a forensic medical exam conducted after someone has reported a sexual assault and consents to the exam.[78] Many jurisdictions have developed their own sexual assault evidence collection kits, or they purchase them from a commercial vendor. As such, the content of a kit can vary from jurisdiction to jurisdiction.[79] In general, sexual assault evidence collection kits include (1) instructions; (2) bags, sheets, and envelopes for evidence collection; (3) swabs for collecting fluids or secretions that could contain the perpetrator's DNA; (4) a comb for collecting hair samples; (5) blood collection devices; and (6) documentation forms.[80] An exam involves collecting a complete medical history from the victim and completing a full-body physical examination.[81] This may include

- collecting blood, urine, hair, and other body secretion samples;
- photo documentation of any injuries sustained during the assault;
- collecting the victim's clothing, especially undergarments; and

- collecting any possible physical evidence that may have transferred onto the victim from the crime scene.[82]

In addition to jurisdictional differences in the content of sexual assault evidence collection kits, procedures for analyzing the evidence collected using the kit can vary from jurisdiction to jurisdiction. In some jurisdictions, all sexual assault evidence collection kits are forwarded to a crime laboratory for analysis.[83] In other jurisdictions, it may be months or even years before the kit is tested, if at all.[84] Some law enforcement agencies might not submit sexual assault evidence collection kits to crime laboratories for various reasons: the identity of the perpetrator was not in question from the beginning of the investigation, detectives identified the suspect through other evidence not included in the kit, or the victim chooses not to proceed with the case.[85] Also, some law enforcement agencies might have a problem working through their backlog of old kits because crime laboratories are operating at full capacity analyzing DNA evidence collected from current cases.[86]

Investigation of Leads Generated from Database Hits

While reducing casework backlogs can help generate new leads in cases without suspects (so-called "cold cases"), law enforcement agencies have to devote time to investigating the leads that result from DNA database matches. Data from a 2009 survey of 235 law enforcement agencies suggest that law enforcement agencies, particularly small agencies, might not have the resources to fully investigate new leads. The survey found that 37% of agencies surveyed had designated "cold case units" (i.e., groups of investigators who are responsible for leads generated from a match between an offender and forensic profile in either the SDIS or the NDIS).[87] In addition, the larger the agency (as measured by the number of sworn officers) the more likely they were to have such a unit. Over two-thirds of law enforcement agencies with 1,000 or more sworn officers reported having a cold case unit.[88] However, less than half of law enforcement agencies with 379- 999 sworn officers reported having this unit, and less than 20% of agencies with 378 or fewer sworn officers reported having such a unit.[89] Even if an agency reported having a cold case unit, the unit was typically small. Three-quarters of law enforcement agencies with cold case units reported that three or fewer staff members were assigned to the unit.[90] Law enforcement agencies that did not have cold case units reported that leads generated from DNA database hits were investigated when resources were available, which usually meant that investigators were paid overtime to follow-up on the new leads.[91] Data suggest that law enforcement agencies would expand cold case units if they had the resources. Surveyed law enforcement agencies were asked to identify, based on their agency's experiences, the resources they needed for DNA-related work. Two-thirds identified cold case unit staffing (both for staffing cold case units or paying overtime if the agency did not have a cold case unit) as a need.[92]

Partial Match Searching

Crime laboratories can use three levels of stringency—high, moderate, and low—when using CODIS to search for matches between an offender or arrestee and forensic profiles.

Searches with high stringency require a match between all 26 alleles,[93] which, as discussed above, indicates that it is highly probable that the identified offender or arrestee was the source of the forensic sample. A moderate stringency search requires all available alleles to match, but the profiles can contain a different number of alleles.[94] Moderate stringency searches can be used to search for matches when the forensic profile contains a mixture of DNA from two or more sources, hence there might be more than two alleles at some loci. Low stringency searches require one allele at each loci to match.[95] Low stringency searches are sometimes required because a degraded sample might not have alleles at all loci.[96]

Crime laboratories can use low stringency searches to make partial matches between and offender or arrestee and forensic profiles. Partial match searching can be used for familial searching, which involves using DNA from known individuals in a database to identify relatives of those individuals as potential suspects in other crimes.[97] There is some debate about whether partial match searching is the same as familial searching. In some states, crime laboratories can release information on partial matches that result from a regular search of the SDIS or NDIS, but they do not consider these partial matches to be familial searches because they were not the result of a *deliberate* search of the database for partial matches between an offender or arrestee and forensic profiles.[98] Others argue that even if the partial match was not the result of a deliberate search of the database, it is still a familial search because it could implicate the relative of someone with a profile in the database.[99] Research indicates that there is a lack of transparency when it comes to policies regarding partial matches. In most cases where a state reports the results of partial matches, it is done without explicit statutory authorization, and in many instances the policy is unwritten or it is not available to the public.[100]

Familial searching is possible because of the way humans inherit genes. Close relatives—especially parents, children, and siblings—who are genetically related are more likely to share alleles used for identification in CODIS than two people who are not closely related.[101] Two unrelated people usually only share a few CODIS alleles, but a parent and his or her child must share no fewer than 13 alleles since children inherit half of their genes from each parent.[102] Parents and children will most likely share between 14 and 16 alleles.[103] It is possible that two siblings will share between 0 and 26 alleles, but on average they will share 16.7 alleles.[104] Familial searching can be conducted by using low stringency searching, but low stringency searches can result in hundreds or even thousands of partial matches, none of which might actually represent a biological relationship. The probability that two unrelated people will share more than 13 alleles with at least one match at each of the 13 loci is about 1 in 2,000.[105] While this probability is low, there are over 8.6 million offender profiles in the NDIS, meaning that a low stringency search for a common genotype could generate thousands of partial matches.

The FBI has been reluctant to allow the NDIS to be used for familial searching without explicit legislative approval,[106] but in July 2006 the FBI issued a policy that permits states, at their discretion, to share identification information with other states in the event that a search of the NDIS turns up a partial match.[107] While the FBI's policy might seem to be at odds with the bureau's reluctance to allow the NDIS to be used for familial searching, a closer review of the FBI's definition of "familial searching" shows how the FBI could allow states to share partial match information without contradicting its stance on familial searching. The FBI defines familial searching "as a 'second deliberate search ... to identify close biological relatives of the perpetrator in the known offender database,' used only after an initial search of the database turns up no candidate matches."[108] The FBI's current policy allows states to share any partial matches; they do not have to be the result of a deliberate search for relatives of individuals with a profile in the NDIS. The FBI's policy means that states have the final say over whether to release identifying information in the case of partial matches.

Post-conviction DNA Testing

In addition to aiding law enforcement in criminal investigations, DNA testing also provides wrongfully convicted prisoners with an avenue to have their convictions overturned. As of April 2011, the Innocence Project reports that 268 prisoners have been exonerated through post-conviction DNA testing.[109] Currently, all states except Massachusetts and Oklahoma have enacted post-conviction DNA testing statutes.[110] However, some states limit who can apply for post-conviction DNA testing. In some states, only individuals sentenced to death or convicted for certain offenses (usually violent offenses) can apply for post-conviction DNA testing. In nearly half the states (24), anyone convicted of a *felony* is eligible to apply for post-conviction DNA testing. However, some states (20) allow *anyone* convicted of a crime to apply for post-conviction DNA testing.

While nearly all states have laws allowing certain individuals access to post-conviction DNA testing, access to testing can be rendered moot if evidence is not preserved or if it is preserved in a manner where there is a chance for it to be contaminated or degraded. Currently, 37 states have statutes that provide for the preservation of evidence after someone is convicted.[111] However, each state has its own standard regarding whether law enforcement agencies are required to retain biological evidence after someone has been convicted of a crime, how long agencies are required to retain evidence, and if there are any requirements for the conditions in which evidence must be stored.

Current Federal Statutory Law

While state law dictates whose profiles will be included in each state's DNA database, federal law provides for the collection of DNA samples from certain federal offenders for analysis and inclusion in the NDIS. Federal law also dictates which profiles included in SDIS can be uploaded into the NDIS. Federal law also states that agencies participating in the NDIS must meet certain specified standards. In addition, federal law provides for post-conviction DNA testing for federal offenders. The following section summarizes current federal law as it pertains to DNA used in a criminal justice capacity.

Quality Assurance and Proficiency Testing Standards

Under current law,[112] the FBI is required to issue (and revise from time to time) Quality Assurance Standards (QAS), including standards for testing the proficiency of forensic laboratories and forensic analysts, in conducting DNA analyses.[113] By law, the QAS must specify the criteria for quality assurance and proficiency tests to be applied to the various types of DNA analyses conducted by forensic laboratories.[114] In addition, the QAS must include a system for grading proficiency testing performance to determine whether a laboratory is performing acceptably.[115] Under current law, FBI personnel who perform DNA analyses must undergo semiannual external proficiency testing by a DNA proficiency testing program that meets the standards set in the QAS.[116]

According to the FBI, the QAS describe the minimum standards for a laboratory's quality assurance program if performing forensic DNA analysis.[117] The minimum standards cover the following areas: organization, personnel, facilities, evidence or sample control, validation, analytical procedures, equipment calibration and maintenance, reports, review, proficiency testing, corrective action, audits, safety, and outsourcing.[118]

Index to Facilitate Law Enforcement Exchange of DNA Identification Information

The Violent Crime Control and Law Enforcement Act of 1994 (P.L. 103-322) authorized the FBI to establish an index of DNA profiles (i.e., NDIS). Under current law,[119] the NDIS can contain the DNA profiles of samples

- taken from individuals convicted of or charged with a crime, or collected under applicable legal authorities (e.g., people arrested for crimes), except for DNA samples that are voluntarily submitted solely for elimination purposes;
- recovered from crime scenes,
- recovered from unidentified human remains; and
- voluntarily contributed from relatives of missing persons.120

The NDIS can only include DNA profiles

- based on analyses performed by or on behalf of a criminal justice agency or the Department of Defense (DOD) in accordance with available standards that satisfy or exceed the FBI's published QAS;
- that are prepared by laboratories that (1) have been accredited by a nonprofit professional organization of persons actively involved in forensic science and nationally recognized within the forensic science community, and (2) undergo external audits, not less than once every other year, that demonstrate compliance with the FBI's QAS;[121] and
- that are maintained by federal, state, and local criminal justice agencies or the DOD pursuant to rules that allow the disclosure of profiles only to other criminal justice agencies for identification purposes, judicial proceedings, criminal defense purposes, and, if personally identifiable information is removed, for research and quality control purposes.[122]

Under current law, the FBI is required to expunge the DNA profile of an individual who had a DNA profile entered into the NDIS on the basis of being convicted for a qualifying federal offense (see below) if the individual provides a certified copy of a final court order showing that the conviction was overturned.[123] Also, the FBI is required to expunge the DNA profile of an individual who had a DNA profile entered into the NDIS on the basis of being arrested under the authority of the United States if the individual provides a certified copy of a final court order that establishes that the charge was dismissed or resulted in an acquittal, or that no charge was filed within the applicable time period.[124] As a condition of having access to the NDIS, states must also have in place a procedure whereby the state will expunge a profile from the state's database based on the same conditions applicable to a profile being expunged from the NDIS.[125] Also, under current law the Department of Defense is required to expunge the DNA profile of an individual who had a DNA profile entered into the NDIS on the basis of being convicted of a qualifying military offense (see below) if the individual provides a certified copy of a final court order showing that the conviction was overturned.[126]

Collection of DNA Samples from Certain Federal, District of Columbia, and Military Offenders

Under current law,[127] the Attorney General is permitted to collect DNA samples from "individuals who are arrested, facing charges, or convicted of a crime or from non-United States citizens who are detained under the authority of the United States."[128] In addition, the Bureau of Prisons (BOP) is required to collect a DNA sample from each federal prisoner who is, or has been, convicted of a felony, a sexual abuse crime under chapter 109A of title 18 of the U.S. Code, a crime of violence,[129] or any attempt or conspiracy to commit any of these crimes.[130]

Federal probation offices responsible for supervising individuals on probation, parole, or supervised release are required to collect DNA samples from individuals who are, or have been, convicted of any of the crimes outlined above.[131] Collected samples are required to be submitted to the FBI for analysis and their resulting DNA profiles are included in the NDIS.[132]

Current law contains similar provisions regarding the collection of DNA samples from District of Columbia offenders. BOP is required to collect a DNA sample from each prisoner who is, or has been, convicted of a qualifying District of Columbia offense.[133] In addition, the Court Services and Offender Supervision Agency for the District of Columbia is required to collect DNA samples from individuals on probation, parole, or supervised release, who are, or have been, convicted of any qualifying District of Columbia offense.[134] The government of the District of Columbia may determine which offenses under the District of Columbia Code are considered qualifying offenses for the purposes of supplying a DNA sample.[135] Collected samples must be submitted to the FBI for analysis and their resulting DNA profiles are included in the NDIS.[136]

Under current law,[137] the DOD is required to collect DNA samples from each member of the Armed Forces who is, or has been, convicted of an offense under the Uniform Code of Military Justice for which a sentence of confinement of more than one year can be imposed, or of any other offense under the Uniform Code of Military Justice that is comparable to the offenses for which a DNA sample can be collected from a federal offender (see above).[138] DOD is required to conduct an analysis of the collected sample and submit the results to the FBI for inclusion in the NDIS.[139]

Post-conviction DNA Testing

The Justice for All Act of 2004 (P.L. 108-405) established procedures for post-conviction DNA testing in federal courts. Under current law,[140] upon a written motion from an individual sentenced for a federal offense (hereafter, "applicant"), the court must order DNA testing of evidence if all of the following apply:

- The applicant asserts, under penalty of perjury, that the applicant is actually innocent of the federal crime for which the applicant was sentenced, or another federal or state offense, if (1) "the evidence was entered during a federal death sentence hearing and exoneration for the offense would entitle the applicant to a reduced sentence or a new sentencing hearing;" or (2) "in the case of a [s]tate offense, the applicant demonstrates that there is no adequate remedy under [s]tate law to permit DNA testing of the ... evidence ... and, to the extent available, the applicant has exhausted all remedies available under [s]tate law for requesting DNA testing of ... evidence."
- The specified evidence to be tested was secured in relation to the investigation or prosecution of the federal or state crime for which the applicant claims to be innocent.
- The evidence to be tested (1) "was not previously subjected to DNA testing, and the applicant did not knowingly and voluntarily waive the right to request DNA testing of the evidence in a court proceeding after the date of enactment of the [Justice for All Act of 2004 (October 30, 2004)] or [did not] knowingly fail to request DNA testing of the evidence in a prior motion for post-conviction DNA testing;" or (2) "was previously subjected to DNA testing and the applicant requests DNA testing using a new method or technology that is substantially more probative that prior testing."
- The evidence to be tested "is in the possession of the [g]overnment and has been subject to a chain of custody and retained under conditions sufficient to ensure that such evidence has not been substituted, contaminated, tampered with, replaced, or altered in any respect" that would affect the DNA testing.
- The proposed DNA testing is "reasonable in scope, uses scientifically sound methods, and is consistent with accepted forensic practices."
- The applicant "identifies a theory of defense that is not inconsistent with an affirmative defense presented at trial and would establish the actual innocence of the applicant."
- If the applicant was "convicted following a trial, the identity of the perpetrator was at issue in the trial."
- The proposed DNA testing may produce new material evidence that would support the affirmative defense theory presented at trial and raise a reasonable probability that the applicant did not commit the crime.
- The applicant certifies that he or she will provide a DNA sample for comparison purposes.
- The motion is made in a timely fashion.[141]

If the court orders DNA testing, the testing is carried out by the FBI.[142] However, the court can order DNA testing to be conducted by another "qualified laboratory if the court makes all necessary orders to ensure the integrity of the ... evidence and the reliability of the testing process and results."[143]

The cost of any DNA testing is borne by the applicant, unless the applicant is indigent; in that case, the cost of the DNA testing is borne by the government.[144]

Test results relating to the DNA sample provided by the applicant are to be included in the NDIS.[145] If the test results ordered by the court are "inconclusive or show that the applicant was the source of the tested evidence, the applicant's DNA profile may be retained in the NDIS."[146] Moreover, if the test results show that the applicant was not the source of the tested evidence, and a comparison of the applicant's DNA profile with other forensic profiles in the NDIS result in a match, DOJ is to contact the appropriate agency and preserve the applicant's DNA sample.[147]

However, if the test results exclude the applicant as the source of the tested evidence, and a comparison between the applicant's DNA profile and forensic profiles in the NDIS does not result in a match, DOJ must destroy the applicant's DNA sample and ensure that the applicant's DNA profile is not stored in the NDIS if there is no other legal authority to retain the profile in the NDIS.[148]

If the results of the DNA test are inconclusive, the court can order further testing, if appropriate, or it can deny the applicant relief.[149] If the results of the DNA test demonstrate that the applicant was the source of the evidence tested, the applicant is denied relief, and on a motion of the government, the court can determine whether the applicant's claim of actual innocence was false. If the court finds the claim was false, it can

- hold the applicant in contempt of court;
- assess against the applicant any cost of DNA testing;
- forward the findings to BOP, who may wholly, or in part, deny the applicant's good conduct time;[150]
- if the applicant is eligible for parole, forward the finding to the U.S. Parole Commission so the commission can deny parole on the basis of the finding; or
- if the test results relate to a state offense, forward the findings to the appropriate state official.[151]

Under current law, if the applicant is convicted for making false assertions relating to post-conviction DNA testing, the applicant is to be sentenced to no less than three years imprisonment, to run consecutively with any other term of imprisonment the applicant is serving.[152]

If the results of the DNA testing demonstrate that the applicant was not the source of the tested evidence presented as a part of the case against the applicant, the applicant can file a motion for a new trial or resentencing, as appropriate, notwithstanding any law that would bar the motion as untimely.[153] Under current law, the applicant would be granted a new trial or resentencing, if the DNA results, when considered with all other evidence in the case (regardless of whether such evidence was introduced at trial), establish by compelling evidence that a new trial would result in an acquittal of the federal offense the applicant is currently sentenced for, or in the case of resentencing, if evidence of a federal or state offense was admitted during a federal death sentencing hearing and exoneration for the offense would entitle the applicant to a reduced sentence or a new sentencing hearing.[154]

Preservation of Biological Evidence

The Justice for All Act of 2004 (P.L. 108-405), among other things, established standards for the preservation of biological evidence by the government. Under current law,[155] the federal government is required to preserve biological evidence[156] that was secured in the investigation or prosecution of a federal offense, if a defendant was imprisoned for the offense, unless[157]

- "the court denied a request or motion for DNA testing [of the evidence] and no appeal is pending";
- the defendant "knowingly and voluntarily waived the right to request DNA testing [of the evidence] in a court proceeding conducted after the date of enactment of the [Justice for All Act of 2004 (October 30, 2004)]";
- "after a conviction becomes final and the defendant has exhausted all opportunities for direct review of the conviction, the defendant is notified that the evidence may be destroyed and the defendant does not file a motion [for post-conviction DNA testing] within 180 days of receipt of notice";
- "the evidence must be returned to its rightful owner, or it is of such size, bulk, or physical character as to render retention impracticable and the [g]overnment takes reasonable measures to remove and preserve portions of the evidence sufficient to permit future DNA testing"; or
- the evidence has been the subject of post-conviction DNA testing (see above) and the results of the testing demonstrate that the defendant was the source of the evidence.

Grants for DNA-Related Programs

Several grant programs provide assistance to state and local governments for forensic sciences. A bulk of the programs focus on providing state and local governments with funding to reduce the backlog of forensic and convicted offender samples waiting to be processed and entered into the NDIS. However, some grant programs provide funding for other purposes, such as offsetting the cost of providing post-conviction DNA testing. This section of the report provides a brief overview of grants for forensic sciences.+

Debbie Smith DNA Backlog Grant Program

The Debbie Smith DNA Backlog Grant Program (hereafter, "Debbie Smith grants") provides grants to state and local governments for five major purposes: (1) conducting analyses of DNA samples collected under applicable legal authority for inclusion in the NDIS, (2) conducting analyses of forensic DNA samples for inclusion in the NDIS, (3) increasing the capacity of state and local laboratories to carry out DNA analyses, (4) collecting DNA samples from people required to submit them and forensic samples from crimes, and (5) ensuring that analyses of forensic DNA samples are carried out in a timely manner. The Attorney General is required to award funds using a formula. The formula distributes funds amongst state and local governments to maximize the effective utilization of DNA technology to solve crimes and protect public safety. The formula must also allocate funding amongst state and local governments to reduce backlogs by considering the number of offender and forensic samples awaiting DNA analysis in the jurisdiction along with the population and number of violent crimes in the jurisdiction. Current law requires DOJ to award not less than 0.5% of the total amount appropriated each fiscal year to each state and the District of Columbia. The territories are to receive 0.125% of the total appropriation. In addition, DOJ is required to award not less than 40% of the total amount appropriated for the program each fiscal year for conducting analyses of forensic casework.

Agencies receiving a grant under the program are required to certify that DNA analyses are conducted in laboratories that satisfy the FBI's QAS and are operated either by a state or local government or by a private laboratory under contract with the state or local government. Grants for conducting analyses of DNA samples collected under applicable legal authority for inclusion in the NDIS, conducting analyses of forensic casework for inclusion in the NDIS, and ensuring that analyses of forensic DNA samples are carried out in a timely manner can be made in the form of a contract or voucher for laboratory services that can be redeemed by nonprofit or for-profit laboratories that satisfy the QAS and have been approved by the Attorney General.

State and local governments receiving funding under the program are required to submit a report to DOJ with a summary of the activities carried out under the grant and an assessment of whether such activities are meeting the needs identified in the grant application, as well as other information the Attorney General may require. DOJ may award not more than 1% of grant funding each fiscal year to states, units of local government, and nonprofit professional organizations of persons actively involved in forensic science and nationally recognized within the forensic science community to help offset the cost of accrediting and auditing laboratories.

Debbie Smith grants were originally authorized under the Justice for All Act of 2004 (P.L. 108- 405). This law amended the DNA Backlog Elimination Act of 2000,[158] authorizing appropriations of $151.0 million for each of FY2004-FY2009.[159] The program was reauthorized under the Debbie Smith Reauthorization Act of 2008 (P.L. 110-360), which includes authorized appropriations of $151.0 million for FY2009-FY2014. A funding history for this program since FY2006 is provided in *Table 2*.

Kirk Bloodsworth Post-Conviction DNA Testing Grant Program

The Kirk Bloodsworth DNA Post-conviction DNA Testing Grant program was authorized by the Justice for All Act of 2004 (P.L. 108-405). The act authorized the Attorney General to make grants to states to help defray the costs of post-conviction DNA testing programs. The act authorized appropriations of $5.0 million for FY2005-FY2009. A funding history for this program since FY2006 is provided in Table 2.

Sexual Assault Forensic Exam Program Grants

The Sexual Assault Forensic Exam Program Grants were authorized under the Justice for All Act of 2004 (P.L. 108-405). The program provides grants for training, technical assistance, education, equipment, and information relating to the identification, collection, preservation, analysis, and use of DNA samples and evidence by medical personnel and those treating victims of sexual assault. Under the program, entities eligible to receive grants include states, units of local government, and sexual assault examination programs. The act authorized appropriations of $30.0 million for each of FY2005-FY2009. P.L. 110-360 extended the same authorized amount through FY2014. A funding history for this program since FY2006 is provided in *Table 2*.

DNA Research and Development Grants

The Justice for All Act of 2004 authorized grants for research and development for improving forensic DNA technology, including increasing the accuracy and efficiency of DNA analysis, decreasing the time and expense of conducting DNA analysis, and increasing its portability. In addition, the law authorized grants for demonstration projects to evaluate the use of DNA technology in conjunction with other forensic analyses. The act authorized funding of $15.0 million for each of FY2005-FY2009. This program has not received any appropriations since FY2006.

DNA Training and Education for Law Enforcement, Correctional Personnel, and Court Officers

Under this program, the Attorney General is required to make grants to provide training, technical assistance, education, and information regarding the identification, collection, preservation, analysis, and use of DNA samples and evidence by law enforcement personnel, court officers, forensic science professionals, and corrections personnel. The program was originally authorized under the Justice for All Act of 2004 (P.L. 108-405), which authorized $12.5 million for each of FY2005-FY2009. P.L. 110-360 extended the same authorized amount through FY2014. This program has not received any appropriations since FY2006.

Appropriations for DNA-Related Grant Programs

Since FY2006, Congress has appropriated over $100 million each fiscal year for DNA analysis and other forensic programs and activities. As shown in *Table 2*, most funding each fiscal year was dedicated to reducing DNA backlogs, enhancing crime laboratory capacity, and other activities related to DNA analysis. In FY2006 and FY2007, Congress gave the Administration discretion in how to award appropriated funding for DNA-related activities. The report to accompany the FY2006 Science, State, Justice, Commerce, and Related Agencies Appropriations Act (P.L. 109-108) stated that the appropriation was for a "capacity enhancement program including eliminating casework backlogs, eliminating offender backlogs, strengthening crime lab capacity, training of the criminal justice community and identifying missing persons."[160] Starting in FY2008, Congress continued to appropriate funding for DNA analysis and capacity enhancement, including Debbie Smith grants.[161] The language in the FY2008-FY2012 appropriations bills did not require DOJ to use all of the funding for DNA analysis and capacity enhancement for Debbie Smith grants, rather, it just had to award a portion of the funding for the programs.

Table 2. Appropriations for Forensic Science Grant Programs (Dollars in millions)

	FY 2006	FY 2007	FY 2008	FY 2009	FY 2010	FY 2011	FY 2012
DNA-Related and Forensic Programs and Activities	107.1	112.1	152.3	156.0	161.0	133.4	125.0
DNA Analysis, Capacity Enhancement, and Debbie Smith Grants	(103.2)	(108.2)	(147.4)	(151.0)	(151.0)	(125.1)	(117.0)
Kirk Bloodsworth Post-conviction DNA Testing	(3.9)	(3.9)	(4.9)	(5.0)	(5.0)	(4.1)	(4.0)
Sexual Assault Forensic Exam		—	—	—	(5.0)	(4.1)	(4.0)

Source: FY2006-enacted appropriations are taken from OJP's FY2008 congressional budget submission; FY2007-enacted appropriations are taken from OJP's FY2009 congressional budget submission; FY2008-enacted appropriations are taken from OJP's FY2010 congressional budget submission; and FY2009- and FY2010-enacted appropriations are taken from OJP's FY2011 congressional budget submission; FY2011-enacted appropriations are based on a CRS analysis of the text of P.L. 112-10; FY2012-enacted appropriations are taken from H.Rept. 112-284.

Notes: Amounts under the DNA-Related and Forensic Programs and Activities might not sum to total due to rounding.

As such, DOJ has awarded funding it received for DNA analysis and capacity enhancement under a variety of programs, including Forensic DNA Backlog Reduction, Convicted Offender DNA Backlog Reduction, Forensic Science Training Development and Delivery, Forensic DNA Unit Efficiency Improvement, Solving Cold Cases with DNA Evidence, DNA to Identify Missing Persons, and DNA Research and Development.[162]

SELECTED LEGISLATIVE ISSUES FOR CONGRESS

In the 1990s and the early part of the last decade, most of the debate about the use of DNA in criminal justice centered on the scope of DNA databases, reducing the backlog of DNA casework, and providing access to post-conviction DNA testing. Most of this debate faded with the enactment of the Violence Against Women and Department of Justice Reauthorization Act of 2005 (P.L. 109-162), which expanded federal collection statutes to include anyone arrested or detained under the authority of the United States. The act also expanded the scope of the NDIS to include DNA profiles of individuals arrested for state crimes. However, concerns about the backlog of DNA casework and access to postconviction testing have persisted. In addition, new issues related to the use of DNA in criminal justice have emerged, including whether (1) DNA databases should be used to conduct familial searches, (2) sexual assault evidence collections kits should be standardized, and (3) there should be national accreditation standard for forensic laboratories. Each of these issues are discussed in more detail below.

The NDIS and Familial Searching

The debate over familial searching is framed by the question of how to balance the desire of law enforcement agencies to use all available tools for solving crimes against the privacy of individuals who might fall under suspicion because they happen to be related to someone convicted, or in some cases arrested, for a crime. The success of familial searching depends on a close relationship between the someone with a profile in the database and the unknown perpetrator.[163] Proponents of familial searching cite research that indicates children of parents who have been convicted of a crime are more likely themselves to be convicted of a crime. Further, siblings who have been convicted are more likely to have other siblings who have been convicted.[164] Proponents also cite data from the Bureau of Justice Statistics (BJS) indicating that 46% of jail inmates reported that they had at least one close relative who had been incarcerated.[165] However, opponents of familial searching challenge its empirical basis. They stress that if offenders are more likely to have *convicted* relatives then their relatives' profiles would already be in the database and there would be no need to expand database searches to include non-convicted relatives.[166] Also, conviction data might not be a perfect indicator of criminal propensity. Biases in the criminal justice system, both racial and geographical, might lead certain groups of people, namely racial and ethnic minorities living in urban centers, to be arrested and convicted more than others with a similar rate of criminality. It could be possible that a law-abiding individual is more likely than a convicted offender to have a convicted relative.[167] On the other hand, many people commit crimes for which they are never arrested or convicted, so it is also possible that convicted offenders have relatives who have committed crimes for which there is no official record.

One of the primary privacy concerns regarding the use of familial searching is that it will put innocent people under "genetic surveillance" because they are related to someone whose profile is in a DNA database.[168] Proponents of familial searching argue that while an offender's family could be questioned by law enforcement as a result of a partial match, they could also come under the scrutiny of law enforcement during the course of an investigation that did not use familial searching.[169] For instance, an eyewitness viewing a lineup may indicate that one face bears a strong resemblance to—but is not actually—the perpetrator; at this point law enforcement might turn attention to a suspect's brother or other relatives. Relatives could also come under suspicion because the investigation turned up evidence of their involvement. Proponents also argue that familial searching might actually *exclude* relatives as the potential perpetrator.[170] Opponents counter that familial searching does not search the entire pool of suspects. Rather, it only subjects relatives of convicted offenders to potential law enforcement scrutiny; relatives of individuals who have not been convicted of, or arrested for, a crime are not at risk of becoming the subject of an investigation through familial searching.[171] Also, some opponents assert, if society has an interest in identifying perpetrators and exonerating the innocent by using a method that would subject non-convicted individuals to law enforcement scrutiny, then the most equitable and optimal mean to achieve this goal is universal DNA collection so that everyone, not just those who are related to someone with a profile in a DNA database, can share in the burdens and benefits of DNA databases.[172]

It has been estimated that using familial searching could increase the cold hit rate (i.e., a match in the database between an offender or arrestee profile and a forensic profile from a case where there is no suspect) in the United States from 10% to 14%.[173] However, opponents argue that familial searching would strain law enforcement's resources because they will have to investigate multiple leads, and in some cases all of the leads might be false-positives. Proponents counter this claim by noting that procedures could be put in place to reduce the number of leads to only the most promising ones. For example, laboratories could use kinship analysis to calculate the probability that a known offender DNA profile in the databases is related to the individual who left a DNA sample at the crime scene.[174] Y-chromosome analysis could, for instance, help determine whether a known offender in the database and the unknown suspect who left a DNA sample at a crime scene are related through male lineage and analysis of mitochondrial DNA (mtDNA) could help determine whether two people are related through maternal lineage.[175] While kinship analysis would make familial searching more efficient, DNA databases are currently not equipped to conduct them.[176] In addition, Y-chromosome and mitochondrial DNA analysis would require additional testing of both the known offender and the forensic samples.

Opponents argue that familial searching would exacerbate existing racial and ethnic disparities in the criminal justice system. African Americans and Hispanics, compared to whites, are disproportionately arrested and convicted, and since the NDIS is comprised of samples collected from individuals arrested and convicted for qualifying offenses, minorities are more likely to have profiles in the NDIS.[177] Disparities in the racial/ethnic compositions of profiles in the NDIS could mean that minority groups would be disproportionately investigated as a result of familial searches.[178] It was estimated that using the NDIS for familial searching could mean that approximately 17% of the African American population in the United States would be "findable" through the database, compared to approximately 4% of the white population (which includes non-African American Hispanics).[179] Disparities in who is investigated as a result of familial searching could also compound existing disparities in the criminal justice system, assuming that more investigations would result in more arrests and convictions.[180] While proponents of familial searching acknowledge that minority groups are more likely to become suspects in investigations resulting from familial searching, they argue that minority groups might benefit from the policy because crime is disproportionately intraracial, meaning that victims of crimes are more likely to be from minority groups.[181] Therefore, minority groups may be more likely to benefit from crimes being solved through familial searching.[182] Proponents have also argued that even if an investigative technique like familial searching does have a disproportionate effect on a specific group, it is not grounds to forgo the technique if it will solve crimes.[183]

Another privacy concern about the use of familial testing is that it could disrupt families. Specifically, opponents of familial searching have voiced concerns that someone might learn that (1) a relative had an unknown criminal history, (2) there is an existing unknown biological relationship between two people, or (3) there is no biological relationship between two people who assume they are related. However, the threat that familial searching poses to families appears to be proportional to how thorough and discreet law enforcement is in investigating leads generated by the search. Proponents argue that family secrets would only be exposed if law enforcement reveals how and why the suspect became a target of the investigation.[184] Since partial matches only suggest that an unknown perpetrator and a known offender might be related, it is unlikely that law enforcement would be able to obtain a warrant to compel the individual to provide a DNA sample.[185] Hence, law enforcement would need to investigate the lead further and develop corroborating evidence that the relative might be the unknown perpetrator. As such, law enforcement would most likely not have to share with the subject of the investigation that the lead was generated through a familial search. In cases where a DNA sample is collected from a suspect, law enforcement should be able to test it against the forensic sample to determine if the suspect committed the crime. They would not have to let the suspect know that they are not related to the person whose profile generated the lead.[186] In addition, law enforcement can legally collect and analyze DNA samples from items the suspect discarded in an area where the suspect had no expectation of privacy (for example, saliva on a soda can the suspect threw away), which would allow law enforcement to surreptitiously exclude a suspect as the possible offender.[187] Proponents have noted that law enforcement could also uncover personal information about a suspect or a suspect's family during the course of an investigation that involved alternative types of surveillance or searches.[188] Yet, despite measures taken by law enforcement to not expose any family secrets, when law enforcement investigates biological relationships, there is the possibility that previously unknown relationships might be revealed.[189]

As outlined above, Congress has delineated the scope of the NDIS in law, but the law is silent as to how the NDIS can be used to make matches.[190] In the conference report for the Commerce, Justice, Science, and Related Agencies Act, 2011 (P.L. 112-55), Congress encouraged the FBI to "undertake activities to facilitate familial DNA searches of the [NDIS] and ... consider the establishment of procedures allowing familial searches only for serious violent and sexual crimes where other investigative leads have been exhausted."[191] Congress also specified that the procedures "should provide appropriate protections for the privacy rights" of individuals with a profile in the NDIS.[192] While Congress has encouraged the FBI to open up the NDIS for familial searching, there are a couple of issues Congress might consider related to this. First, if a potential familial match is made using the NDIS, should law enforcement be required to conduct additional testing before the lead is investigated further (such as Y-chromosome and mtDNA testing of both the forensic and offender DNA samples or searching public records to determine if the offender and the suspect are related)? Second, should there be additional funding for the FBI or the Office of Justice Programs to provide training to law enforcement on how to investigate leads resulting from familial matching so as to decrease any risk of accidently revealing family secrets?

Reducing the Backlog

The reduction of the backlog of DNA samples awaiting analysis has been the subject of concern in this and previous Congresses.[193] If Congress chooses to consider legislation to reduce the size of the DNA backlog, two options it might consider are increasing the capacity of state and local laboratories to conduct DNA analyses and facilitating partnerships between public and private laboratories.

Enhancing the Capacity of State and Local Laboratories

As described above, Debbie Smith grants can be used by state and local laboratories to reduce offender and forensic casework backlogs and enhance their capacity to conduct DNA analyses. This program represents the primary mechanism by which Congress has promoted efforts to reduce the DNA backlog. A study of DNA backlogs in 2007 found that there is a need for Debbie Smith grants. Of the laboratories responding to the survey, 83% replied that they would expect an increase in forensic casework backlogs if they did not receive Debbie Smith grants.[194] Moreover, nearly 90% of laboratories responded that state and local funding would not be sufficient if they did not receive federal support.[195] The study also found that while laboratories are better prepared to process offender and arrestee samples, they still rely on federal funding to conduct these analyses. Nearly 61% of responding laboratories reported that they would expect an increase in offender and arrestee sample backlogs if they no longer received Debbie Smith grants.[196] However, it is possible that respondents to the survey overstated the need for federal support out of fear that federal funding for DNA backlogs could be curtailed or eliminated if there was not a large demand for it. Using data collected as a part of the 2005 census of publicly funded crime laboratories, BJS estimated that based on the average productivity of an analyst in 2005, laboratories performing DNA analyses would have needed 73% more staff to achieve a 30-day turnaround time on all DNA analysis requests they received.[197] While it is possible that crime laboratories have hired additional analysts since 2005 and advances in technology have enabled analysts to process more requests than they once did, the demand for DNA analysis has never been greater and backlogs persist. This means that BJS's findings that crime laboratories have a need for more analysts to complete casework are most likely still valid.

Congress may choose to consider the role the federal government should play in reducing state DNA backlogs. Large backlogs can delay the resolution of criminal investigations, and Congress might have an interest in ensuring that as many criminals as possible are apprehended. As discussed above, it appears that laboratories rely on federal funding to help reduce backlogs, especially forensic casework backlogs. In addition, even though Congress has appropriated approximately $785 million for backlog reduction and laboratory capacity enhancement programs since FY2006 (see *Table 2*), data show that backlogs of offender and forensic casework are persistent. As outlined above, states have expanded the pool of people who are required to submit DNA samples, and law enforcement agencies have started to collect biological evidence in more cases. But the increasing demand for DNA analysis has not been met with increased capacity to conduct the analyses. One policy option policymakers could consider is increasing authorized appropriations for the Debbie Smith grants over the current $151 million per fiscal year.[198] While providing additional funding to states for backlog reduction could allow them to hire more analysts to work on reducing the backlog, it is likely that it would take some time to bring the backlog down from current levels. It can take up to one year to recruit, interview, and hire a new analyst; and it can take another year or two until the analyst is fully trained.[199] In addition, the workloads of more experienced analysts might actually decrease when new analysts are hired because the more experienced analysts will have to train and review the work of the new analysts.[200] Congress might also consider whether to allow Debbie Smith grants to be used for capital investments, such as new construction to expand crime laboratories. Hiring additional staff might ultimately help reduce DNA backlogs, but there could be a limit on the number of staff crime laboratories can hire given space constraints.

However, while Congress might have an interest in helping states resolve their backlogs of DNA samples, there might be some concern that by relying on federal funding to support the operations of state and local laboratories, state and local administrators would be unaware of laboratories' true funding needs, which would prevent them from seeking permanent funding solutions. It could be argued that states have contributed to the backlog problem by expanding DNA collection statutes before increasing their capacity to process the samples they were already collecting. DNA backlogs are not a static phenomenon, they will continue to increase or decrease depending on the demand for DNA analysis and the ability of crime laboratories to conduct the requested analyses. Therefore, the backlog cannot simply be eliminated by providing enough funding to state and local crime laboratories to analyze the number of samples in their backlog for any given point in time; the backlog will only be eliminated when capacity is adequate to meet demand. This means that unless state and local governments make the required investment in crime laboratory capacity, they will likely need continued federal funding to manage backlogs. Congress could consider phasing-out funding for the program to provide states with an incentive to invest in increasing and maintaining their capacity to conduct DNA analyses. Congress could also consider amending the authorizing legislation Debbie Smith grants to make it a matching grant program. Another option may be limiting the number of years states could receive funding under the program so that states would have to consider revenue streams for permanently funding crime laboratories.

Congress might consider whether to modify the Debbie Smith grant program so that a greater proportion of the funding goes to reducing the forensic casework backlog. NIJ reports that demand for grants to assist with reducing the backlog of convicted offender and arrestee samples peaked in FY2007 and has decreased in subsequent fiscal years.[201] As mentioned, currently not less than 40% of the amount appropriated for the program is to be awarded for grants to conduct analyses of forensic DNA samples for inclusion in the NDIS. Given that laboratories appear to be better prepared to process offender and arrestee samples than they are to process forensic casework samples, Congress might consider amending the authorizing statute for Debbie Smith grants so that DOJ is required to award a greater proportion of annual funding for the program to laboratories for reducing the backlog of forensic casework.

Congress could also consider providing additional funding to state and local laboratories for the purchase or development and testing of new technology that will help automate more of the DNA analysis process. Automating some of the more labor-intensive parts of DNA analysis, such as DNA extraction or reviewing the results of the analysis, could help laboratories process more samples in a shorter period of time, thereby reducing the backlog. While automation might help reduce backlogs in the long run, it is unlikely that it would have an immediate effect on reducing backlogs. It can take months to implement and validate new equipment, processes, and procedures, and doing so takes personnel away from working on processing current casework.[202]

Another consideration might be whether to provide funding to help state and local law enforcement to either hire additional investigators or offset the cost of paying overtime to current investigators so they can investigate leads generated by DNA database matches. As discussed, law enforcement agencies have identified funding for either additional staffing for cold case units or paying overtime if the agency does not have a cold case unit as a primary need for DNA-related work. However, there could be a concern that state and local governments will come to rely on federal funding for supporting cold case units rather than identifying permanent revenue streams.

Facilitating Partnerships between Public and Private Laboratories

As discussed above, if a public laboratory outsources some casework to a private laboratory, the public laboratory must review the work of the private laboratory before it is uploaded into the NDIS, even though both public and private laboratories have to be accredited and audited in order to conduct DNA analyses that will generate profiles for inclusion in the NDIS. In many instances, public laboratories do not have additional analysts to review work outsourced to private laboratories. Hence, the requirement that public laboratories check the work of DNA profiles generated by private laboratories might discourage public-private partnerships.[203] Also, in cases where laboratories do not have adequate resources, analysts have to review DNA profiles outsourced to private laboratories in addition to their regular casework, meaning that analysts are paid overtime, which makes the cost of outsourcing the sample more expensive. Further, it takes additional time for the work to be reviewed, resulting in a delayed uploading into the NDIS.[204]

One possible issue before Congress is whether to amend federal policy to facilitate partnerships between public and private laboratories. Congress could consider amending current law so that public laboratories do not have to review the work of private laboratories that are in compliance with the QAS. This might facilitate more public-private partnerships since public laboratories would be able to upload DNA profiles into the NDIS without having to review the private laboratory's work. One potential benefit to outsourcing more samples to private laboratories is that private laboratories might be able to analyze samples at a lower cost than public laboratories. In his testimony before the House Subcommittee on Crime, Terrorism, and Homeland Security, Dr. Jeffery Boschwitz noted that the cost of analyzing DNA samples in private laboratories can be up to 50% less than the cost of comparable analyses conducted by public laboratories.[205] Dr. Boschwitz notes that private laboratories can offer lower per case costs because private laboratories can leverage economies of scale and they have invested in research and development to lower costs in order to remain competitive.[206] However, law enforcement agencies or prosecutors would have to pay private laboratory analysts an expert testimony fee to testify in court, a cost they do not have to bear when an analyst from a public laboratory testifies. This could be a significant cost for local law enforcement agencies or prosecutors in light of the Supreme Court's ruling in *Melendez-Diaz v. Massachusetts* that defendants have a Sixth Amendment right to cross-examine an analyst who conducts an analysis of forensic evidence used in the case against the defendant.[207] However, Dr. Boschwitz testified that, in his experience, most defense attorneys do not ask for DNA testing-related testimony.[208] He testified that analysts in his laboratory are asked to testify, on average, in 2% of the cases they analyze, even after the *Melendez-Diaz* decision.[209] Nonetheless, it is not known whether other private laboratories experiences with analysts being called to testify mirror those of Dr. Boschwitz's laboratory or whether trends will change in the future. Partnerships between public and private laboratories could also provide public laboratories with a temporary increase in capacity when there is a unexpected increase in requests for DNA analysis.[210] By outsourcing more work, public laboratories could increase their capacity to a level where they would be able to handle the regular demand for DNA analysis without having to develop a surplus capacity to handle occasional increases in demand. This could promote more efficient public laboratories since they would not need to hire additional staff or purchase additional equipment to prepare for instances where there is increased demand for DNA analysis.

Some advocates of reducing DNA backlogs by having private laboratories conduct more DNA analyses might propose allowing private laboratories to have direct access to CODIS. After all, private laboratories are held to the same standard as public crime laboratories. This proposal raises one of the key questions about any public-private partnership in criminal justice: are some functions of the system inherently governmental? As discussed above, in some states laboratories retain the DNA sample that is used to generate the databased profile. This is done so that samples can be retested to confirm matches made using CODIS and to allow laboratories to retest samples in case of any technological advancements. If private laboratories were allowed direct access to CODIS, it might raise a question about whether it would be prudent for private laboratories to retain DNA samples. Another consideration might be whether involving private laboratories in investigating criminal offenses could decrease the efficiency of the investigation. When a match is made between a forensic and offender profile, there are several additional steps that must take place before charging the suspect with the crime. These steps include, but are not limited to,

- obtaining the personal identification of the subject whose DNA profile matches the forensic profile,
- verifying whether the person is the correct sex and in the appropriate age range to be the perpetrator,
- confirming that the suspect was not in custody at the time of the offense and had access to the crime scene when the offense occurred,
- corroborating the match between the offender and forensic profiles after collecting a new sample from the suspect, and
- determining if other non-DNA evidence supports or refutes that the person could be a viable suspect.211

While private laboratories would be able to verify the match between forensic and sample profiles, it is likely that law enforcement would have to be involved with the other steps in moving from a cold hit to a criminal charge. The coordination between the private laboratory and law enforcement could be complicated if the private laboratory is in another city or even another state. Policymakers might consider whether cold hits could be investigated more efficiently if the analysis was done "in-house."

There could be some concern that private laboratories might cut corners as a way to increase profit margins, thereby resulting in flawed DNA profiles being included in the NDIS. While this is a valid concern—after all, flawed DNA profiles could mean that a crime goes unsolved or that the wrong person is convicted for the crime—thus far there has not been an evaluation of whether private laboratories are more likely than public laboratories to make mistakes in their analysis. Before deciding on whether to allow public laboratories to forego reviewing DNA profiles generated by private laboratories, or granting private laboratories access to CODIS, Congress could consider requiring an independent evaluation of whether private laboratories that are in compliance with the conditions of the QAS make errors at a different rate than that of public laboratories.

Annual Backlog Data

As discussed, there is a lack of comprehensive annual data on the size of the forensic casework and offender and arrestee sample backlogs. The data collected by NIJ only provide an estimate of the national forensic casework backlog. They do not include a state-by-state breakdown of the backlog, nor do they estimate the nature of the backlog. Having reliable data on the size and nature of the DNA backlog could provide policymakers with information about the efficacy of existing policies, and it could assist policymakers if they choose to consider other policies for reducing the backlog. Policymakers might consider creating a system to collect annual DNA backlog data. Currently, state and local governments that receive Debbie Smith grants are required to submit an annual report to DOJ that contains a summary of the activities carried out using grant funds and any other information DOJ might require;[212] but state and local governments are not required to submit data on backlogs, unless required by DOJ. If Congress chooses to consider legislation related to this issue, it might consider requiring states to report data on the size of their forensic casework and offender and arrestee (if applicable) sample backlogs as a condition of receiving Debbie Smith grants.[213] Congress could also consider reducing a state's allocation under the Edward Byrne Memorial Justice Assistance Grant (JAG) program (which provides assistance to state and local law enforcement for a variety of purposes) if a state did not submit annual backlog data to DOJ.[214] One advantage of these methods is that it provides a financial incentive for states to report backlog data. However, there might be a concern about making a state ineligible for Debbie Smith grants or reducing a state's allocation under the JAG program since funding under both programs can be used to help reduce DNA backlogs.[215] Congress could consider requiring BJS to conduct an annual survey of publicly funded crime laboratories to collect data on DNA backlogs. One advantage to this method is that it would not make a state ineligible for funding that could be used for DNA analysis. On the other hand, it would rely on states to voluntarily provide the data. There might be concern on the part of some policymakers that states will not submit backlog data unless there is a reason for them to do so. The concern is not unfounded because response rates to surveys can vary. However, surveys conducted by BJS tend to have high response rates. For example, 90% of laboratories contacted by BJS to participate in the 2005 Survey of Publicly Funded Forensic Crime Laboratories responded to the survey.[216] If policymakers choose to establish a system for collecting data on DNA backlogs, they might also consider the scope of the data collected. Would states only be required to submit a count of the number of forensic casework samples in their backlog, or would they also be required to provide a breakdown of the different types of forensic casework samples in their backlog, such as the number of sexual assault evidence collection kits? Should states be required to provide data on the amount of time it takes laboratories to complete the analysis of different types of samples? Should states provide data on backlogged DNA samples for the state in total or should states be required to submit data on backlogs at each local lab that provides DNA profiles to the SDIS?

Standardization of Sexual Assault Evidence Collection Kits

As discussed, the contents of a sexual assault evidence collection kit and the protocols governing whether the kit is submitted to a crime laboratory for analysis can vary from jurisdiction to jurisdiction, which could mean that the quantity and quality of evidence collected in a sexual assault case and the probability that the evidence will be analyzed could depend in part on where the sexual assault occurred. Policymakers might consider legislation that would provide for the standardization of the contents of sexual assault evidence kits and protocols for submitting the kits to crime laboratories for analysis.

One issue before Congress is whether it wants to make adopting a standardized sexual assault evidence collection kit a condition of receiving Debbie Smith grants. Another option involves reducing a state or local government's allocation under the JAG program if they choose not to use the standardized kit. If Congress chooses to consider such legislation, one question before policymakers might be whether to specify the exact contents of all sexual assault evidence collection kits used in the United States, or whether to establish a minimum standard for kits and allow jurisdictions to add additional elements if they choose. In their national protocol for sexual assault medical examinations, the Office on Violence Against Women (OVW) provided guidelines for the minimum content of sexual assault evidence collection kits. The guidelines state that, at a minimum, sexual assault evidence collection kits should include

- a kit container which has a label with blanks for identifying information and documenting the chain of custody;
- an instruction sheet or checklist to guide examiners in collecting evidence and maintaining the chain of custody;
- forms that facilitate evidence collection and analysis, including the patient's authorization for collection and release of evidence and information for law enforcement, the patient's medical history, and anatomical diagrams; and
- materials for collecting and preserving evidence, according to jurisdictional policy, including the patient's clothing and underwear and foreign material dislodged from clothing; foreign materials on the patient's body (e.g., blood, dried secretions, fibers, loose hairs, vegetation, soil/debris, fingernail scrapings and/or cuttings, matted hair cuttings, material dislodged from mouth using dental floss, and swabs of suspected semen, saliva, and/or areas highlighted by alternate light sources); hair evidence; vaginal/cervical swabs and smears; penile swabs and smears; anal/perianal swabs and smears; oral swabs and smears; body swabs; and a sample of the patient's DNA for comparison purposes.[217]

As a part of its national protocols, OVW recommended that sexual assault evidence collection kits be standardized within a jurisdiction, and within a state if possible.[218] OVW noted it could be beneficial to develop a standardized kit for use across the country, but that "[f]urther analysis is needed to assess the benefits and disadvantages of such a kit and the feasibility of development and implementation. Some challenges could include building consensus across communities regarding best practices and obtaining buy-in from involved agencies."[219] CRS did not find any evaluations of the benefits and disadvantages of a standardized sexual assault evidence collection kit and the feasibility of its development and implementation. Before requiring states and local governments to adopt a standardized sexual assault evidence collection kit as a condition of receiving funding under certain programs,

Congress could consider requiring OVW or another DOJ agency to evaluate the feasibility of a standardized sexual assault evidence collection kit.

Another issue that might come before Congress is whether to reduce or eliminate funding under certain DOJ grant programs for state and local governments that do not submit all sexual assault evidence collection kits booked into evidence to a crime laboratory for analysis. One question policymakers might consider while debating this issue is whether law enforcement agencies should be required to submit all sexual assault evidence collection kits they receive. As previously discussed, there are legitimate reasons why law enforcement agencies choose not to submit collected kits to crime laboratories for analysis. Since there is currently a backlog of forensic DNA evidence, it is likely a more efficient use of a crime laboratory's time to only conduct DNA analyses on evidence where a suspect cannot be identified through other methods or where the case will proceed after the analysis is conducted. Another question policymakers might consider is whether law enforcement would be required to work through their complete backlog of sexual assault evidence collection kits. In addition to potentially adding to the backlog of forensic evidence, there might be some concern that analyzing old kits might have a negative effect on victims who do not wish to revisit the crime.

Arrestee Collection Statutes

As outlined above, approximately one-half of states collect DNA samples from people arrested for certain crimes. Expanding the number of profiles in the NDIS is an important element of using the database to solve cold cases. As shown in *Table 1*, as the number of offender profiles in the NDIS increased, so did the number of offenders hits and investigations aided. However, the mandatory collection of DNA samples from arrestees is still contentious. In general, courts have found that collecting DNA samples from convicted offenders is constitutional because offenders have a lower expectation of privacy, and any intrusion into an offender's privacy is reasonable in order to maintain public safety.[220] However, people arrested for crimes have a legal presumption of innocence.[221] Opponents of arrestee collection statutes argue that collecting DNA samples from arrestees is an invasion of their privacy that cannot be justified because of the presumption of innocence, meaning that arrestees do not have a diminished expectation of privacy.[222] Proponents of collecting DNA sample from arrestees argue that the state has an interest in identifying criminals and deterring crime. More specifically, they argue that there is an administrative interest in creating an unequivocal record of who has been arrested.

They also suggest that people who are arrested based on probable cause have a diminished expectation of privacy because DNA can be used to identify individuals, which is not only relevant to solving the crime that the individual was arrested for, but it also provides a record of who was arrested that can be used to solve past and future crimes.[223] Arrestee collection statutes have been challenged in court, with some courts striking down the laws and other courts upholding them.[224]

Congress could consider whether it wants to promote statutes that require DNA samples to be collected from individuals arrested for certain offenses. The debate about whether to encourage states to collect DNA from arrestees is framed on one side by an interest in expanding DNA databases to include more profiles in the hope of generating more investigative leads, and on the other side by a concern about collecting more samples for analysis at a time when there is a persistent backlog of casework for analysis. If Congress chooses to consider legislation that would encourage states to adopt arrestee collection statutes, one policy option Congress could consider is providing additional funding to states to offset the cost of implementing such statutes.[225] This could be accomplished by either creating a new grant program that would provide grants to states that already have or pass arrestee collection statutes or amending the authorizing legislation for an existing grant program (the JAG program, for example) so that states receive additional funding if they already have or pass arrestee collection statutes. Congress could also consider making states ineligible to receive funding under an existing grant program, or reducing the state's funding under an existing program, unless the state has a law that requires the collection of DNA samples from arrestees. However, given that arrestee collection statutes might be challenged in court, if Congress chooses to consider this option for promoting arrestee collection statutes it might also consider exempting states from penalties if the state's arrestee collection statute is ruled unconstitutional.

Access to Post-conviction Testing

As discussed, nearly all states have laws that provide prisoners access to post-conviction DNA testing. However, the standards for gaining access to post-conviction DNA testing vary from state to state. A provision in the Justice for All Act of 2004 (P.L. 108-405) requires that applicants for funding under the DNA Research and Development Grants; DNA Training and Education for Law Enforcement, Correctional Personnel, and Court Officers; and the Kirk Bloodsworth Post-conviction DNA Testing Grant programs demonstrate that the state provides adequate procedures for providing post-conviction DNA testing for individuals under a sentence of imprisonment or death and preservation of biological evidence. However, the incentive for states to put such procedures in place might be diminished by the fact that Congress has not provided funding for either the DNA Research and Development Grants or the DNA Training and Education for Law Enforcement, Correctional Personnel, and Court Officers programs since FY2006.

Further, funding for the Kirk Bloodsworth Post-conviction DNA Testing Grant program has averaged $4.6 million since FY2006. Even if states do not have laws in place to provide adequate procedures for post-conviction DNA testing, they are still eligible to receive Debbie Smith grants, which accounts for most funding authorized under the act.

If Congress wants to promote greater access to post-conviction DNA testing for state prisoners, one policy option that could be considered is extending the prohibition on funding to all grant programs authorized by the Justice for All Act.

Congress could consider extending the prohibition in connection with increasing funding for the Kirk Bloodsworth Post-conviction DNA Testing Grant program to help offset the cost of extending DNA testing to a greater number of prisoners.

While extending the prohibition on funding to more grant programs might provide a greater incentive to states to expand access to post-conviction DNA testing, there might be some concern about prohibiting a state from receiving federal funding to reduce forensic and convicted offender backlogs and to increase laboratory capacity because the state chose a standard for access that is lower than the one set by Congress.

Also, providing additional funding to states through the Kirk Bloodsworth Post-conviction DNA Testing program to offset the cost of expanding post-conviction testing might mask the actual cost of running the state's post-conviction DNA testing program.

This could prevent the state from dedicating adequate resources to the post-conviction DNA testing program, which in turn could require Congress to keep funding grants in order to ensure that states are able to support post-conviction testing programs.

Federal Accreditation Standards

One issue policymakers might consider is whether there is a need for national accreditation standards for forensic laboratories.[226] Currently, nine states require forensic science laboratories in the state to be accredited.[227] However, even with the lack of state law requiring laboratories to be accredited, most publicly funded laboratories are still seeking and maintaining accreditation. BJS, in their 2005 census of publicly funded crime laboratories, found that 82% of crime laboratories were accredited, up from 71% of laboratories in 2002.[228] Furthermore, nearly all of the accredited laboratories were accredited by ASCLD/LAB. BJS reported that 78% of accredited laboratories were accredited by ACSLD/LAB.[229]

In addition, as discussed above, all CODIS laboratories must be accredited and audited annually and analysts at these laboratories are required to undergo semiannual proficiency testing; however, these standards only apply to DNA analysis functions and not to any other forensics analyses the crime laboratory might conduct (e.g., ballistics testing, fingerprint analysis, or toxicology). While national accreditation standards would ensure that all laboratories are held to the same standard and they might help prevent embarrassing incidents where faulty procedures or downright deception has resulted in erroneous DNA analyses, accreditation can only ensure that procedures and practices are in place that would reduce the likelihood of flawed results; it cannot eliminate them. Even if a laboratory is accredited, it is still possible that human error or malignant intentions would mean that the results of some DNA analyses are incorrect. It is also possible that a private accrediting organization like ACSLD/LAB would be able to amend its accreditation standards faster to reflect changes in technology or practice than a federal board could, especially one that is charged with overseeing accreditation standards for all forensic crime laboratories.

All laboratories that receive Debbie Smith grants are required to adhere to the FBI's QAS, which requires the laboratories to be audited annually to ensure that they adhere to the QAS and requires analysts to undergo semiannual proficiency testing. Also, in order to receive funding under the Paul Coverdell Forensic Sciences Improvement Grant program, any forensic science laboratory system, medical examiner's office, or coroner's office (including any laboratory operated by a unit of local government) receiving grant funding uses generally accepted laboratory practices and procedures established by accrediting organizations or appropriate certifying bodies. It appears that current law would prohibit unaccredited laboratories from receiving grants under most currently funded grant programs.

However, laboratories would not be prohibited from receiving funding under *all* grant programs (e.g., the Edward Byrne Memorial Justice Assistance Grant program) under which they could possibly receive support. Instead of creating a new office to promulgate and enforce national accreditation standards, Congress could consider amending current law to require state and local governments to certify that any funding they receive under any federal grant program for a forensic science laboratory has received and maintained accreditation from a nonprofit professional organization of persons actively involved in forensic science that is nationally recognized within the forensic science community.

APPENDIX. STATE DNA DATABASE LAWS

Table A-1. State DNA Database Laws, Qualifying Offenses (as of September 2011)

	Felony Convictions			Misdemeanor Convictions		Arrests			
State	All Convicted Felons	Juvenile Adjudications	Jail & Probation	Sex Crimes	Offenses Other Than Sex Crimes	Murder	Sex Crimes	Burglary	All Felony
Alabama	✓	✓	✓	✓		✓	✓	✓	✓
Alaska	✓	✓	✓	✓		✓	✓	✓	✓
Arizona	✓	✓	✓	✓		✓	✓	✓	
Arkansas	✓		✓	✓		✓	✓		
California	✓	✓	✓	✓		✓	✓	✓	✓
Colorado	✓	✓	✓	✓		✓	✓	✓	✓
Connecticut	✓		✓	✓					
Delaware	✓		✓	✓					
Florida	✓	✓	✓	✓		✓	✓	✓	✓
Georgia	✓	✓	✓	✓					
Hawaii	✓	✓	✓	✓					
Idaho	✓	✓	✓						
Illinois	✓	✓	✓	✓		✓	✓	✓	
Indiana	✓		✓						
Iowa	✓	✓	✓	✓					
Kansas	✓	✓	✓	✓		✓	✓	✓	✓
Kentucky	✓		✓						
Louisiana	✓	✓	✓	✓		✓	✓	✓	✓
Maine	✓	✓	✓						
Maryland	✓	✓	✓	✓		✓	✓	✓	
Massachusetts	✓	✓	✓						
Michigan	✓	✓	✓	✓		✓	✓		
Minnesota	✓	✓	✓	✓					
Mississippi	✓		✓						
Missouri	✓		✓	✓		✓	✓	✓	
Montana	✓	✓	✓						
Nebraska	✓		✓	✓					
Nevada	✓		✓	✓					
New Hampshire	✓	✓	✓						
New Jersey	✓	✓	✓	✓	✓				
New Mexico	✓	✓	✓			✓	✓	✓	✓
New York	✓		✓	✓	✓				
North Carolina	✓		✓	✓		✓	✓	✓	
North Dakota	✓		✓			✓	✓	✓	✓
Ohio	✓	✓	✓	✓		✓	✓	✓	✓
Oklahoma	✓		✓	✓	✓				
Oregon	✓	✓	✓	✓					
Pennsylvania	✓	✓	✓	✓					
Rhode Island	✓		✓						
South Carolina	✓	✓	✓	✓		✓	✓	✓	✓
South Dakota	✓	✓	✓	✓		✓	✓	✓	✓
Tennessee	✓	✓	✓	✓		✓	✓	✓	
Texas	✓	✓	✓	✓		✓	✓	✓	

Table A-1. (Continued)

	Felony Convictions			Misdemeanor Convictions		Arrests			
State	All Convicted Felons	Juvenile Adjudications	Jail & Probation	Sex Crimes	Offenses Other Than Sex Crimes	Murder	Sex Crimes	Burglary	All Felony
Utah	✓	✓	✓	✓	✓	✓	✓	✓	
Vermont	✓		✓	✓		✓	✓	✓	✓
Virginia	✓	✓	✓			✓	✓	✓	
Washington	✓	✓	✓	✓	✓				
West Virginia	✓		✓	✓					
Wisconsin	✓	✓	✓						
Wyoming	✓	✓	✓	✓					

Source: Gordon Thomas Honeywell Governmental Affairs, http://www.dnaresource. com/documents/ statequalifyingoffenses2011.pdf.

End Notes

[1] See CRS Report RL30717, DNA Identification: Applications and Issues, by Eric A. Fischer.

[2] U.S. Department of Justice, DNA Initiative: Possible Results from DNA Tests, http://www. dna.gov/basics/analysis/ types-of-results.

[3] U.S. Department of Justice, DNA Initiative: Forensic DNA Databases, http://www.dna.gov/ dna-databases/.

[4] Ibid.

[5] This report does not include a discussion of the use of DNA to identify missing persons and unidentified human remains, nor does it include an overview of grant programs to state and local governments for developing DNA profiles from samples from missing persons, close relatives of missing persons, or unidentified human remains. For more on this issue, see CRS Report RL34616, Missing Adults: Background, Federal Programs, and Issues for Congress, by Adrienne L. Fernandes-Alcantara.

[6] Statement of Dwight E. Adams, Deputy Assistant Director, Laboratory Division, Federal Bureau of Investigation, in U.S. Congress, House of Representatives, Government Reform Committee, Subcommittee on Government Efficiency, Financial Management and Intergovernmental Relations, How Effective are State and Federal Agencies Working Together to Implement the Use of New DNA Technologies?, hearing, 107th Cong., 1st sess., March 29, 2004, pp. 53-54, at http://www.fbi.gov/congress/congress01 /dwight061201.htm.

[7] John M. Butler, Fundamentals of Forensic DNA Typing (Burlington, MA: Academic Press, 2010), p. 265 (hereafter, Fundamentals of Forensic DNA Typing).

[8] Ibid.

[9] U.S. Department of Justice, Federal Bureau of Investigation, Fact Sheet—CODIS and the National DNA Index System, http://www.fbi.gov/about-us/lab/codis/codis-and-ndis-fact-sheet (hereafter "CODIS Fact Sheet").

[10] U.S. Department of Justice, Federal Bureau of Investigation, CODIS—NDIS Statistics, http://www.fbi.gov/about-us/ lab/codis/ndis-statistics.

[11] Ibid.

[12] U.S. Department of Justice, DNA Initiative: DNA Databases, http://www.dna.gov/dna-databases/levels.

[13] Ibid.

[14] Ibid. If an "offender hit" is obtained, that information typically is used as probable cause to obtain a new DNA sample from that suspect so the match can be confirmed by the crime laboratory before an arrest is made.

[15] Ibid.

[16] CODIS Fact Sheet.

[17] Ibid.

[18] U.S. Department of Justice, DNA Initiative: About Forensic DNA, http://www.dna.gov/ basics/analysis/str.

[19] Jules Epstein, "Genetic Surveillance—The Bogeyman Response to Familial DNA Investigations," University of Illinois Journal of Law, Technology and Policy, vol. 2009, no. 1, (2009), p. 143 (hereafter, "Epstein, 'Genetic Surveillance'").

[20] Henry T. Greely, Daniel P. Riordan, and Nanibaa' A. Garrison, et al., "Family Ties: The Use of DNA Offender Databases to Catch Offenders' Kin," Journal of Law, Medicine and Ethics, vol. 34, no. 2 (Summer 2006), p. 250 (hereafter, "Greely, Riordan, Garrison, et al., 'Family Ties'").

[21] Ibid.
[22] Fundamentals of Forensic DNA Typing, p. 270.
[23] Ibid., p. 271.
[24] U.S. Department of Justice, Federal Bureau of Investigation, Quality Assurance Standards for DNA Databasing Laboratories, Standard 15, http://www.fbi.gov/about-us/lab/codis/qas_ databaselabs. U.S. Department of Justice, Federal Bureau of Investigation, Quality Assurance Standards for Forensic DNA Testing Laboratories, Standard 15, http://www.fbi. gov/about-us/lab/codis/qas_testlabs (hereafter "QAS").
[25] Fundamentals of Forensic DNA Typing, p. 271.
[26] Ibid.
[27] Ibid.
[28] U.S. Congress, House Committee on the Judiciary, Subcommittee on Crime, Terrorism, and Homeland Security, Testimony of Jeffery S. Boschwitz, Ph.D., Hearing on "Rape Kit Backlogs: Failing the Test of Providing Justice to Sexual Assault Survivors", 111th Cong., 2nd sess., May 20, 2010, H.Hrg 111-115 (Washington: GPO, 2010), p. 81 (hereafter, "Testimony of Dr. Boschwitz")
[29] Matthew R. Durose, Census of Publicly Funded Forensic Crime Laboratories, 2005, U.S. Department of Justice, Office of Justice Programs, Bureau of Justice Statistics, NCJ 222181, Washington, DC, July 2008, p. 7, http://bjs.ojp.usdoj.gov/content/pub/pdf/ cpffcl05.pdf (hereafter, Census of Publicly Funded Forensic Crime Laboratories).
[30] QAS, Standard 17.
[31] CODIS Fact Sheet.
[32] QAS, Standard 17.
[33] Ibid.
[34] Fundamentals of Forensic DNA Typing, p. 270.
[35] Ibid., p. 262.
[36] Ibid.
[37] Tania Simoncelli, "Dangerous Excursions: The Case Against Expanding Forensic DNA Databases to Innocent Persons," Journal of Law, Medicine, and Ethics, vol. 34, no. 2 (Summer 2006), p. 392 (hereafter "Simoncelli, 'Dangerous Excursions'").
[38] Simoncelli, "Dangerous Excursions," p. 392.
[39] 42 U.S.C. §14135e(c).
[40] Frederick R. Bieber, "Turning Base Hits into Earned Runs: Improving the Effectiveness of Forensic DNA Data Bank Programs," Journal of Law, Medicine and Ethics, vol. 34, no. 2 (Summer 2006), p. 227.
[41] Ibid.
[42] Matthew Gabriel, Cherisse Boland, and Cydne Holt, "Beyond the Cold Hit: Measuring the Impact of the National DNA Data Bank on Public Safety at the City and County Level," Journal of Law, Medicine and Ethics, vol. 38, no. 2 (Summer 2010), pp. 396-411.
[43] Ibid., p. 397. "Cold cases" were defined as crimes where the investigation has not generated a named suspect(s) through traditional methods of police investigation (e.g., interviewing witnesses, identification through non-DNA physical evidence left at the crime scene, or tips from confidential informants).
[44] Ibid., p. 398.
[45] Ibid., p. 400.
[46] Ibid.
[47] Fundamentals of Forensic DNA Typing, p. 278.
[48] Ibid.
[49] Gordon Thomas Honeywell, Governmental Affairs, State DNA Database Laws, Qualifying Offenses, http://www.dnaresource.com/documents/statequalifyingoffenses2011.pdf.
[50] Ibid.
[51] Ibid.
[52] Ibid.
[53] Ibid.
[54] Edwin Zedlewski and Mary B. Murphy, "DNA Analysis for 'Minor' Crimes: A Major Benefit for Law Enforcement," NIJ Journal, vol. 253 (January 2006) (hereafter, "DNA Analysis for 'Minor' Crimes").
[55] Data from BJS show that approximately 1 in 5 property offenders released from prison in 1994 were rearrested for a violent crime with three years. Patrick A. Langan and David J. Levin, Recidivism of Prisoners Released in 1994, U.S. Department of Justice, Office of Justice Programs, Bureau of Justice Statistics, NCJ 193427, Washington, DC, June 2002, p. 9, http://bjs.ojp.usdoj.gov/content/pub/pdf/rpr94.pdf. In addition, a study of

DNA databases hits in Florida show that 52% of hits for homicide and sexual assault cases matched offenders who had prior convictions for burglary. DNA Analysis for "Minor" Crimes.

[56] NIJ defines a backlogged case as a case that has not been analyzed within 30 days of being submitted to the laboratory. Mark Nelson, Making Sense of DNA Backlogs—Myths vs. Reality, U.S. Department of Justice, Office of Justice Programs, National Institute of Justice, NCJ 323197, Washington, DC, February 2011, p. 3, http://www.ncjrs. gov/pdffiles1/nij/232197.pdf (hereafter, Making Sense of DNA Backlogs—Myths vs. Reality).

[57] Backlog data for 2005 were collected as a part of the Bureau of Justice Statistic's Census of Publicly Funded Forensic Crime Laboratories (see Matthew R. Durose, Census of Publicly Funded Forensic Crime Laboratories, 2005, NCJ 222181, July 2008, http://bjs.ojp.usdoj. gov/content/pub/pdf/cpffcl05.pdf). Data for 2007 were collected as a part of a NIJ funded study of DNA backlogs (see Lisa Hurst and Kevin Lothridge, "2007 DNA Evidence and Offender Measurement Analysis: DNA Backlogs, Capacity and Funding," NCJ 230328, January 2010, http://www.ncjrs.gov/ pdffiles1/nij/grants/230328.pdf). Data for 2008 were reported by applicants for NIJ's FY2009 DNA Backlog Reduction Program grant solicitation. Data for 2009 were reported by applicants for NIJ's FY2010 DNA Backlog Reduction Program. Ibid., p. 3.

[58] Ibid., p. 4.

[59] Ibid.

[60] Ibid., p. 1.

[61] Ibid., p. 7.

[62] Ibid., p. 8.

[63] Ibid., p. 2.

[64] Ibid.

[65] Kevin J. Strom, Jeri Ropero-Miller, and Shelton Jones, et al., The 2007 Survey of Law Enforcement Forensic Evidence Processing, U.S. Department of Justice, Office of Justice Programs, National Institute of Justice, Washington, DC, October 2009, pp. 3-2, http://www.ncjrs.gov/pdffiles1/nij/grants/228415.pdf.

[66] Ibid.

[67] Ibid., p. 3-7.

[68] Ibid., p. 3-6.

[69] Lisa Hurst and Kevin Lothridge, 2007 DNA Evidence and Offender Analysis Measurement: DNA Backlogs, Capacity and Funding, Final Report to National Institute of Justice Grant 2006-MU-BX-K002, Washington, DC, January 2010, p. 8, http://www.ncjrs.gov/ pdffiles1/nij/grants/230328.pdf (hereafter, 2007 DNA Evidence and Offender Analysis Measurement).

[70] Ibid.

[71] Ibid., p. 12.

[72] Ibid.

[73] In the 111th Congress, the House Subcommittee on Crime, Terrorism, and Homeland Security held a hearing on sexual assault evidence collection kit backlogs. See U.S. Congress, House Committee on the Judiciary, Subcommittee on Crime, Terrorism, and Homeland Security, Hearing on "Rape Kit Backlogs: Failing the Test of Providing Justice to Sexual Assault Survivors," 111th Cong., 2nd sess., May 20, 2010, H.Hrg 111-115 (Washington: GPO, 2010). Also, several pieces of legislation in the 111th Congress would have, among other things, provided funding for processing backlogged sexual assault evidence collection kits, made eligibility for funding under certain Department of Justice (DOJ) grant programs contingent on eliminating the backlog of sexual assault evidence collection kits, or provided for the collection of data on the backlog of kits. See H.R. 4114, S. 2736, H.R. 5640, H.R. 6085, S. 3842, and H.R. 6214 in the 111th Congress and S. 250 and S. 254 in the 112th Congress.

[74] CBS News conducted a five month investigation into the backlog of sexual assault evidence collection kits. The investigation collected data on the number of backlogged kits in 24 cities and states. CBS News reports that there are more than 20,000 sexual assault evidence collection kits that were never sent to crime laboratories and another 6,000 kits from cases that are under active investigation, but are still waiting to be tested. The results of CBS News' investigation are available online at http://www.cbsnews.com/stories/ 2009/11/ 10/cbsnews_investigates/ main5605770.shtml?tag=contentMain;contentBody.

[75] Nancy Ritter, "Solving the Problem of Untested Evidence in Sexual Assaults," NIJ Journal, no. 267 (Winter 2010), p. 18.

[76] Ibid., p. 19.

[77] Armen Keteyian, "Untested Rape Kits Lead to More Crimes," CBS News, November 10, 2009, http://www.cbsnews.com/stories/2009/11/10/cbsnews_investigates/ main5603492.shtml.

[78] Rape, Abuse and Incest National Network (RAINN), What is a Rape Kit, http://www.rainn. org/get-information/ sexual-assault-recovery/rape-kit (hereafter, What is a Rape Kit).

[79] U.S. Department of Justice, Office on Violence Against Women, A National Protocol for Sexual Assault Medical Forensic Examinations, NCJ 206554, September 2004, p. 65, http://www.ncjrs.gov/pdffiles1/ovw/206554.pdf (hereafter, A National Protocol for Sexua Assault Medical Forensic Examinations).

[80] What is a Rape Kit.

[81] Ibid.

[82] Ibid.

[83] Angela Wu, "Will Rape Kit Testing Laws Help Clear Cases?," Newsweek, July 27, 2010, http://www.newsweek. com/2010/07/27/will-rape-kit-testing-laws-help-clear-cases.html (hereafter, "Will Rape Kit Testing Laws Help Clear Cases?").

[84] Human Rights Watch, Testing Justice: The Rape Kit Backlog in Los Angeles City and County, 1-56432-461-3, New York, NY, March 2009, p. 22, http://www.hrw.org/node/ 81826.

[85] U.S. Department of Justice, Office of Justice Programs, National Institute of Justice, Untested Sexual Assault Evidence in Law Enforcement Custody, http://www.nij.gov/topics/forensics/ lab-operations/evidence-backlogs/lawenforcement-sexual-assault.htm.

[86] Department of Justice, Office on Violence Against Women, Eliminating the Rape Kit Backlog: A Roundtable to Explore a Victim-centered Approach, Washington, DC, May 10, 2010, p. 15, http://www.ovw.usdoj. gov/docs/rape-kitroundtable-summary-10262010.pdf.

[87] Dan Cantillon, Kathy Kopiec, and Heather Clawson, Evaluation of the Impact of the Forensic Casework DNA Backlog Reduction Program, ICF International, Fairfax, VA, February 2009, p. 10, http://www.ncjrs.gov/ pdffiles1/nij/ grants/225803.pdf (hereafter, Evaluation of the Impact of the Forensic Casework DNA Backlog Reduction Program").

[88] Ibid., p. 11.

[89] Ibid.

[90] Ibid., p. 10.

[91] Ibid., p. 11.

[92] Ibid., p. 14.

[93] Fundamentals of Forensic DNA Typing, p. 275.

[94] Ibid.

[95] Ibid.

[96] The FBI permits forensic profiles with 10 of the 13 CODIS loci to be uploaded into the NDIS for searching against the offender and arrestee indexes. CODIS Fact Sheet.

[97] David Lazer, Searching the Family Tree for Suspects: Ethical and Implementation Issues in the Familial Searching of DNA Databases, A. Alfred Taubman Center for State and Local Government, Cambridge, MA, March 2008, p. 1, http://www.hks.harvard.edu/var/ ezp_site/storage/fckeditor/file/pdfs/centers-programs/ centers/taubman/policybriefs/ lazer_final.pdf (hereafter, Lazer, Searching the Family Tree for Suspects).

[98] Natalie Ram, DNA Confidential: State Law Enforcement Policies for Genetic Databases Lack Transparency, Science Progress, October 2009, p. 2, http://www.scienceprogress.org/2009/ 11/dna-confidential/.

[99] Ibid., p. 1.

[100] Ibid., p. 3.

[101] Greely, Riordan, Garrison, et al., "Family Ties," p. 251.

[102] Ibid., p. 252.

[103] Ibid.

[104] Ibid., p. 253.

[105] Ibid., p. 252.

[106] Ellen Nakashima, "From DNA of Family a Tool to Make Arrests," The Washington Post, April 21, 2008.

[107] The FBI defines a "partial match" as a match between two single source profiles (i.e., offender profiles and forensic profiles that contain DNA from one perpetrator) having at least one allele in common at each locus. U.S. Department of Justice, Federal Bureau of Investigation, "Interim Plan for the Release of Information In the Event of a 'Partial Match' at NDIS," Bulletin #BT072006, July 20, 2006, http://www.bioforensics.com/ conference 08/Familial_Searches/ CODIS_Bulletin.pdf.

[108] Sonia M. Suter, "All in the Family: Privacy and DNA Familial Searching," Harvard Journal of Law and Technology, vol. 23, no. 2 (Spring 2010), p. 324 (hereafter, Suter, "All in the Family").

[109] The Innocence Project, http://www.innocenceproject.org/.

[110] Ala. Code §15-18-200; Alaska Stat. §§12.73.010 through 12.73.090; Ariz. Rev. Stat. Ann. §13-4240; Ark. Stat. Ann. §§16-112-201 through 16-112-208; Cal. Penal Code §1405; Colo. Rev. Stat §§18-1-411 through 18-1-

416; Conn. Gen. Stat. §54-102-kk; Del. Code Ann. Tit. 11 §4504; Fla. Stat. §§925.11 and 925.12; Ga. Code §5-5-41; Hawaii Rev. Stat. §§844D-121 through 844D-133; Idaho Code §§19-4901 through 19-4911; Ill. Rev. Stat. Ch. 725, §5/116-3; Ind. Code §§35-38-7-1 through 35-38-7-13 and 35-38-7-17 through 35-38-7-19; Iowa Code §81.10; Kan. Stat. Ann. §21- 2512; Ky. Rev. Stat. §§422.285 through 422.287; La. Code. Crim. Pro. §926.1; Me. Rev. Stat. Ann. Tit. 15 §§2136 through 2138; Md. Crim. Proc. Code Ann. §8-201; Mich. Comp. Laws §770.16; Minn. Stat. §§590.01 through 590.06; Miss. Code Ann. §99-39-5; Mo. Rev. Stat §547.035; Mont. Code Ann. §46-21-110; Neb. Rev. Stat §§29-2101 and 29- 4117 through 29-4124; Nev. Rev. Stat. §176.0918; N.H. Rev. Stat. Ann §§651-D:1 through 651-D:4; N.J. Rev. Stat. §2A:84A-32a; N.M. Stat. Ann §31-1A-2; N.Y. Crim. Pro. Law §440.30; N.C. Gen. Stat. §§15A-269 through 15A270.1; N.D. Cent. Code §29-32.1-15; Ohio Rev. Code Ann. §§2953.71 through 2953.83; Or. Rev. Stat. §§138.690 through 138.698; Pa. Cons. Stat. Tit. 42, §9543.1; R.I. Gen. Laws §§10-9.1-11 and 10-9.1-12; S.C. Code Ann. §§17- 28-10 through 17-28-120; S.D. Codified Laws Ann. §§23-5B-1 through 23-5B-17; Tenn. Code Ann. §§40-30-301 through 40-30-313; Tex. Crim. Proc. Code Ann. §§64.01 through 64.05; Utah Code Ann. §§78B-9-301 through 78B-9- 304; Vt. Stat. Ann. Tit. §§5561 through 5577; Va. Code §19.2-371.1; Wash. Rev. Code Ann. §10.73.170; W. Va. Code §15-2B-14; Wis. Stat. §974.07; and Wyo. Stat. §§7-12-302 through 7-12-315.

[111] Alaska Stat. §12.73.200; Ariz. Rev. Stat. Ann. §13-4221; Ark. Stat. Ann. §12-12-104; Cal. Penal Code §1417.9; Colo. Rev. Stat §§18-1-1101 through 18-1-1108; Conn. Gen. Stat. §54-102-jj; Fla. Stat. §925.11; Ga. Code §17-5-56; Hawaii Rev. Stat. §844D-126; Ill. Rev. Stat. Ch. 725, §5/116-4; Ind. Code §35-38-7-14; Iowa Code §81.10; Ky. Rev. Stat. §422.285; La. Code. Crim. Pro. §926.1; Md. Crim. Proc. Code Ann. §8-201; Mich. Comp. Laws §770.16; Minn. Stat. §590.10; Miss. Code Ann. §99-49-1; Mo. Rev. Stat §650.056; Mont. Code Ann. §46-21-111; Neb. Rev. Stat §29- 4125; Nev. Rev. Stat. §176.0912; N.H. Rev. Stat. Ann §651-D:3; N.M. Stat. Ann §31-1A-2; N.C. Gen. Stat. §15A-268; Ohio Rev. Code Ann. §2953.81; R.I. Gen. Laws §10-9.1-11; S.C. Code Ann. §17-28-70; S.D. Codified Laws Ann. §23- 5B-5; Tenn. Code Ann. §40-30-309; Tex. Gov. Code Ann. §411.052 and Tex. Crim. Proc. Code Ann. §38.43; Va. Code §19.2-270.4:1; Wash. Rev. Code Ann. §10.73.170; Wis. Stat. §968.205; and Wyo. Stat. §7-12-304.

[112] 42 U.S.C. §14131(a)(2).

[113] The most recent QAS took effect on July 1, 2009.

[114] 42 U.S.C. §14131(a)(3).

[115] Ibid.

[116] 42 U.S.C. §14133(a)(1)(A).

[117] CODIS Fact Sheet.

[118] Ibid.

[119] 42 U.S.C. §14132(a).

[120] Under the Violent Crime Control and Law Enforcement Act of 1994 (P.L. 103-322), the NDIS was only to include analyses of DNA samples collected from (1) individuals convicted of crimes, (2) crime scenes, and (3) unidentified human remains. The Justice for All Act of 2004 (P.L. 108-405) amended the authorizing legislation for the NDIS to allow analyses of DNA samples collected from persons who have been charged in an indictment or information with a crime and other persons whose DNA samples are collected under applicable legal authorities to be included in the NDIS, provided that profiles from arrestees who have not been charged with a crime and samples that are voluntarily submitted solely for elimination purposes are not included in the NDIS. The Violence Against Women and Department of Justice Reauthorization Act of 2005 (P.L. 109-162) amended the authorizing legislation for the NDIS to allow analyses of samples collected from arrestees to be included in the NDIS.

[121] According to the FBI, the American Society of Crime Laboratory Directors/Laboratory Accreditation Board (ASCLD/LAB) and Forensic Quality Services, Inc. (FQS) meet the definition specified at 42 U.S.C. §14132(b)(2)(A) for an accrediting organization. CODIS Fact Sheet.

[122] 42 U.S.C. §14132(b).

[123] 42 U.S.C. §14132(d)(1)(A)(i).

[124] 42 U.S.C. §14132(d)(1)(A)(ii).

[125] 42 U.S.C. §14132(d)(2)(A)(i).

[126] 10 U.S.C. §1565(e).

[127] 42 U.S.C. §14135a(a)(1)(A).

[128] The DNA Analysis Backlog Elimination Act of 2000 (P.L. 106-546) required BOP and U.S. probation offices to collect DNA samples from anyone in their custody who was convicted of qualifying federal offenses. The act defined a "qualifying federal offense" as murder, voluntary manslaughter, or other offenses relating to homicide; an offense relating to sexual abuse, sexual exploitation or other abuse of children, or transportation for illegal sexual activity; an offense relating to peonage or slavery; kidnapping; an offense relating to robbery

or burglary; any offense committed in Indian country relating to murder, manslaughter, kidnapping, maiming, a felony sexual abuse offense, incest, arson, robbery, or burglary; or any attempt or conspiracy to commit any of these crimes. The Uniting and Strengthening America by Providing Appropriate Tools Required to Intercept and Obstruct Terrorism (USA PATRIOT) Act of 2001 (P.L. 107-56) expanded the definition of "qualifying federal offense" to include crimes of terrorism, crimes of violence, or any attempt or conspiracy to commit either crime. The Justice for All Act of 2004 (P.L. 108-405) amended the definition of "qualifying federal offense" to include any felony, sexual abuse offense, crime of violence, or attempt or conspiracy to commit any of these crimes. The Violence Against Women and Department of Justice Reauthorization Act of 2005 (P.L. 109-162) authorized DOJ to collect DNA samples from arrestees and non-citizens who are detained under the authority of the United States. The Adam Walsh Child Protection and Safety Act of 2006 (P.L. 109-248) authorized DOJ to also collect DNA samples from individuals facing charges in addition to those who have been arrested or convicted.

[129] As defined at 18 U.S.C. §16.

[130] 42 U.S.C. §14135a(a)(1)(B).

[131] 42 U.S.C. §14135a(a)(2).

[132] 42 U.S.C. §14135a(b).

[133] 42 U.S.C. §14135b(a)(1).

[134] 42 U.S.C. §14135b(a)(2).

[135] 42 U.S.C. §14135b(d).

[136] 42 U.S.C. §14135b(b). The following are considered qualifying offenses under the D.C. Code: (1) any felony; (2) any offense for which the penalty is greater than one year imprisonment; (3) lewd, indecent, or obscene acts knowingly committed in the presence of a child under 16 years of age (D.C. Code §22-1312(b)); (4) certain obscene activities involving minors (D.C. Code §22-2201); (5) sexual performances using a minor (D.C. Code §22-3102); (6) misdemeanor sexual abuse (D.C. Code §22-3006); (7) misdemeanor sexual abuse of child or a minor (D.C. Code §22- 3010.01); or (8) any attempt or conspiracy to commit any of theses crimes. D.C. Code §22-4151.

[137] 10 U.S.C. §1565(a)(1).

[138] The requirement to collect DNA samples for people convicted of certain offenses under the Uniform Code of Military Justice is separate from the DNA samples the Department of Defense collects to aid in the identification of human remains.

[139] 10 U.S.C. §1565(b).

[140] 18 U.S.C. §3600(a).

[141] There is a rebuttable presumption of timeliness if the motion is made within 60 months of the enactment of the Justice for All Act of 2004 (October 30, 2004) or within 36 months of conviction, whichever comes later. The presumption of timeliness may be rebutted upon a showing that the applicant's motion for DNA testing is based solely upon information used in a previously denied motion or of clear and convincing evidence that the applicant's filing is done solely to cause delay or harass. For any motion that is not made within 60 months of the enactment of the Justice for All Act of 2004 or within 36 months of conviction, there is a rebuttable presumption against timeliness. The presumption against timeliness can be rebutted upon the court's finding (1) that the applicant was or is incompetent and such incompetence substantially contributed to the delay in the applicant's motion for a DNA test; (2) the evidence to be tested is newly discovered DNA evidence; (3) that the applicant's motion is not based solely upon the applicant's own assertion of innocence and, after considering all relevant facts and circumstances surrounding the motion, a denial would result in a manifest injustice; or (4) upon good cause shown. 18 U.S.C. §3600(a)(10)(B).

[142] 18 U.S.C. §3600(c)(1).

[143] 18 U.S.C. §3600(c)(2).

[144] 18 U.S.C. §3600(c)(3).

[145] 18 U.S.C. §3600(e)(2).

[146] 18 U.S.C. §3600(e)(3)(A).

[147] 18 U.S.C. §3600(e)(3)(B).

[148] 18 U.S.C. §3600(e)(3)(C).

[149] 18 U.S.C. §3600(f)(1).

[150] Each prisoner serving a term of imprisonment of more than one year, but not prisoners serving a life sentence, can receive a good time credit of up to 54 days per year to count toward serving the sentence. The amount of the credit is subject to the determination of BOP. 18 U.S.C. §3624(b).

[151] 18 U.S.C. §3600(f)(2).

[152] 18 U.S.C. §3600(f)(3).

[153] 18 U.S.C. §3600(g)(1).
[154] 18 U.S.C. §3600(g)(2).
[155] 18 U.S.C. §3600A(a).
[156] "Biological evidence" is defined as a sexual assault forensic examination kit, or semen, blood, saliva, hair, skin tissue, or other identified biological material. 18 U.S.C. §3600A(b).
[157] 18 U.S.C. §3600A(c).
[158] The DNA Backlog Elimination Act of 2000 (P.L. 106-546) authorized grants to increase the capacity of state and local government laboratories to conduct DNA analysis of biological samples from crime scenes.
[159] On March 11, 2003, President George W. Bush announced his DNA Initiative, "Advancing Justice Through DNA Technology," which provided "funds, training, and assistance to ensure that DNA technology reaches its full potential to solve crimes, protect the innocent, and identify missing persons." From FY2004 to FY2007, Congress appropriated funding for the President's DNA initiative, although the initiative was not authorized in statute.
[160] U.S. Congress, House Committee on Appropriations, Subcommittee on Science, The Departments of State, Justice, and Commerce, and Related Agencies, Making Appropriations for Science, the Departments of State, Justice, Commerce and Related Agencies for the Fiscal Year Ending September 30, 2006, and for Other Purposes, Report to Accompany H.R. 2862, 109th Cong., 1st sess., November 7, 2005, H.Rept. 109-272 (Washington: GPO, 2005), p. 121.
[161] See P.L. 110-161, P.L. 111-8, and P.L. 111-117.
[162] A breakdown of awards under each of these programs is available online at http://www.dna.gov/funding/.
[163] Frederick R. Bieber, Charles H. Brenner, and David Lazer, "Finding Criminals Through DNA of Their Relatives," Science, vol. 312, no. 5778 (June 2, 2006), p. 1316 (hereafter, Bieber, Brenner, and Lazer, "Finding Criminals Through the DNA of Their Relatives," p. 1316).
[164] Ibid.
[165] Ibid.
[166] Erin Murphy, "Relative Doubt: Familial Searches of DNA Databases," Michigan Law Review, vol. 109, no. 3 (December 2010), p. 306 (hereafter, Murphy, "Relative Doubt").
[167] Ibid., p. 307.
[168] Suter, "All in the Family," p. 361.
[169] Greely, Riordan, Garrison, et al., "Family Ties," p. 257.
[170] Epstein, "Genetic Surveillance," p. 171.
[171] Murphy, "Relative Doubt," p. 308.
[172] Ibid.
[173] Bieber, Brenner, and Lazer, "Finding Criminals Through the DNA of Their Relatives," p. 1316.
[174] Kimberly A. Wah, "A New Investigative Lead: Familial Searching as an Effective Crime-Fighting Tool," Whittier Law Review, vol. 29, no. 4 (Summer 2008), p. 976 (hereafter, Wah, "A New Investigative Lead").
[175] Y-chromosomes and mitochondrial DNA (mtDNA) are considered "lineage markers" because they are passed down from one generation to the next without changing (except for cases where they mutate). The Y-chromosome is only found in males. Paternal lineages can be traced using Y-chromosome markers because a father passes his Y-chromosome on to his son. Therefore, fathers and sons and brothers will share the same Y-chromosome (except for cases where the chromosome mutates). In contrast, mtDNA is only passed from a mother to her children, therefore all siblings with the same biological mother will share the same mtDNA (except for cases where the mtDNA mutates), as opposed to the Y-chromosome, which is only shared by male siblings. In addition, children will share mtDNA with maternal relatives. Therefore, mtDNA can be used to trace maternal lineage. John M. Butler, Forensic DNA Typing: Biology, Technology, and Genetics of STR Markers (2nd ed.) (Burlington, MA: Elsevier Academic Press, 2005), pp. 201-204 and 247-249.
[176] Murphy, "Relative Doubt," p. 300.
[177] Suter, "All in the Family," pp. 368-370.
[178] Ibid., p. 370.
[179] Greely, Riordan, Garrison, et al., "Family Ties," p. 259.
[180] Suter, "All in the Family," p. 370.
[181] Lazer, Searching the Family Tree for Suspects, p. 6.
[182] Ibid., pp. 6-7.
[183] Epstein, "Genetic Surveillance," p. 163.
[184] Ibid., p. 165.

[185] Jessica D. Gabel, "Probable Cause from Probable Bonds: A Genetic Tattle Tail Based on Familial DNA," Hastings Women's Law Journal, vol. 21, no. 1 (Winter 2010), p. 41 (hereafter, Gabel, "Probable Cause from Probable Bonds").
[186] Epstein, "Genetic Surveillance," p. 165.
[187] Gabel, "Probable Cause from Probable Bonds," p. 42.
[188] Suter, "All in the Family," p. 328.
[189] Lazer, Searching the Family Tree for Suspects, p. 6.
[190] H.R. 6011, the Utilizing DNA Technology to Solve Cold Cases Act of 2010 (111th Congress) would have required the Attorney General to adopt policies and procedures to allow the NDIS to be used for familial searching. The bill would have allowed FBI familial searches to be conducted only if no identical match for a DNA sample collected from a crime scene can be identified in the offender index, and the investigation for which DNA samples are collected involves murder, manslaughter, a sex offense against a minor, sexual assault, or an offense that involves a sexual act or sexual contact with another and is punishable by imprisonment for more than one year.
[191] U.S. Congress, House Committee on Appropriations, Agriculture, Rural Development, Food and Drug Administration, and Related Agencies Programs for the Fiscal Year Ending September 30, 2012, and for Other Purposes, Conference Report to Accompany H.R. 2112, 112th Cong., 1st sess., November 14, 2011, H.Rept. 112-284 (Washington: GPO, 2011), p. 238.
[192] Ibid.
[193] See, for example, in the 112th Congress, H.R. 1523, the Sexual Assault Forensic Evidence Registry (SAFER) Act of 2010; S. 250, the Justice for All Reauthorization Act of 2011; and S. 254, the Justice for Survivors of Sexual Assault Act of 2011. In the 111th Congress, see H.R. 2462, the Convicted Child Sex Offender DNA Index System Support Act; H.R. 4114 and S. 2736, the Justice for Survivors of Sexual Assault Act of 2009; H.R. 6085, the Sexual Assault Forensic Evidence Registry (SAFER) Act of 2010; and H.R. 6214, the Justice for Rape Victims and Improving Use of DNA Evidence Act of 2010.
[194] 2007 DNA Evidence and Offender Analysis Measurement, p. 9.
[195] Ibid.
[196] Ibid., p. 13.
[197] Census of Publicly Funded Forensic Crime Laboratories, p. 7.
[198] S. 250, the Justice for All Reauthorization Act of 2011, would, among other things, extend the currently authorized funding for the Debbie Smith DNA Backlog Grant program ($151 million per fiscal year) to FY2016.
[199] Evaluation of the Impact of the Forensic Casework DNA Backlog Reduction Program, p. 50.
[200] Ibid.
[201] Making Sense of DNA Backlogs, 2010—Myths vs. Reality, p. 8.
[202] Ibid., p. 49.
[203] Dr. Boschwitz is a vice president and executive officer of Orchid Cellmark, Inc., "one of the largest worldwide providers of human DNA testing." Testimony of Dr. Boschwitz, p. 81.
[204] Ibid.
[205] Ibid., pp. 82-83.
[206] Ibid., p. 83.
[207] Melendez-Diaz v. Massachusetts, 129 S.Ct. 2527 (2009).
[208] Testimony of Dr. Boschwitz, p. 83.
[209] Ibid.
[210] Ibid., p. 84.
[211] Ranajit Chakroborty and Jianye Ge, "Statistical Weight of a DNA Match in Cold-hit Cases," Forensic Science Communications, vol. 11, no. 3 (July 2009).
[212] 42 U.S.C. §14135(f).
[213] S. 250, the Justice for All Reauthorization Act of 2011, would, among other things, require state and local governments receiving grants to submit annual data on the current backlog for DNA casework within the jurisdiction in which the funds are used. The bill would require DOJ to publish the data submitted by state and local governments. S. 254, the Justice for Survivors of Sexual Assault Act of 2011, would, among other things, amend the authorizing legislation for the Debbie Smith DNA Backlog Grant program to require state and local laboratories that receive funding under the program to submit data to DOJ on the number of sexual assault cases that are in a backlog for DNA case work.
[214] For more information on the JAG program, see CRS Report RS22416, Edward Byrne Memorial Justice Assistance Grant (JAG) Program, by Nathan James.

[215] The National Criminal Justice Association (NCJA) surveyed states to collect data on how states used their JAG funding in 2009. States reported that they used nearly $21.9 million of the JAG funding available to them in 2009 on "forensic sciences, crime labs, DNA analysis, and cold cases." For more information on the NCJA's survey, see http://www.ncja.org/ NCJA/Navigation/PoliciesPractices/Byrne_JAG_Data/Byrne_JAG_ 4_ Pager.aspx and http:// www.ncja.org/NCJA/Navigation/PoliciesPractices/Byrne_JAG_ Data/Byrne_JAG_ Spending_by_Purpose_Area_and_Project_Type.aspx.

[216] Census of Publicly Funded Forensic Crime Laboratories, p. 8.

[217] A National Protocol for Sexual Assault Medical Forensic Examinations, pp. 65-66.

[218] Ibid., p. 66.

[219] Ibid.

[220] Gabel, "Probable Cause from Probable Bonds," p. 28.

[221] Suter, "All in the Family," p. 311.

[222] Ibid., p. 339.

[223] Ibid.

[224] Ibid., pp. 340-341.

[225] Two bills, H.R. 988 and S. 517, the Katie Sepich Enhanced DNA Collection Act of 2011, would, among other things, provide funding to states that collected DNA samples from people arrested for certain crimes.

[226] S. 132, the Criminal Justice and Forensic Science Reform Act of 2011, would establish an Office of Forensic Science (Office) and a Forensic Science Board (Board) within DOJ. One of the responsibilities of the Board would be to recommend standards and procedures for the accreditation of forensic science laboratories and certification of relevant personnel in forensic science disciplines. After the Director of the Office received the recommendations of the Board, the Director would establish standards and procedures for the accreditation of forensic science laboratories and certification of relevant forensic science personnel. Under the bill, both the Board and the Office could consider whether any relevant national accreditation standards that are currently in effect would be sufficient, with or without supplemental standards, for the accreditation of forensic science laboratories. The standards would also include educational and training requirements for relevant laboratory personnel, proficiency and competency testing requirements for relevant personnel, and maintenance and auditing requirements for accredited forensic science laboratories. The bill would prohibit a laboratory from receiving any federal funding unless it is accredited and its relevant personnel are certified under the standards that would be established by the Office. The bill would require the Director of the Office, at least once every five years, to review whether a laboratory accredited under the standards promulgated by the Office is still eligible to receive federal funding.

[227] National Conference of State Legislatures, Table 6. Accreditation of Forensic Laboratories, http://www.ncsl.org/ portals/1/Documents/cj/Table6AccredLaboratories.pdf.

[228] Census of Publicly Funded Forensic Crime Laboratories, p. 3.

[229] Ibid.

INDEX

C

D

E

F

G

H

M

N

O

P

Q

R

S

Y